Studies in Inorganic Chemistry 4

Gas Hydrates

Studies in Inorganic Chemistry

1 Phosphine Arsine and Stibine Complexes of the Transition Elements
by C. A. McAuliffe and W. Levason

2 Phosphorus: An Outline of its Chemistry, Biochemistry and
Technology
(Second Edition) by D. E. C. Corbridge

3 Solid State Chemistry 1982
edited by R. Metselaar, H. J. M. Heijligers and J. Schoonman

4 Gas Hydrates
by E. Berecz and M. Balla-Achs

Studies in Inorganic Chemistry 4

Gas Hydrates

by

E. BERECZ and M. BALLA-ACHS

Department of General and Physical Chemistry
Technical University for Heavy Industry, Miskolc, Hungary

ELSEVIER
AMSTERDAM—OXFORD—NEW YORK 1983

Joint edition published by
Elsevier Science Publishers B.V., Amsterdam, The Netherlands and
Akadémiai Kiadó, The Publishing House of the Hungarian Academy of Sciences,
Budapest, Hungary

Translated by
L. PAKSY

English manuscript revised by
D. DURHAM

The distribution of this book is being handled by the following publishers:
for the U.S.A. and Canada
Elsevier Science Publishing Company, Inc.
52 Vanderbilt Avenue
New York, N.Y. 10017, U.S.A.
for the East European countries,
Democratic People's Republic of Korea, Republic of Çuba,
Socialist Republic of Vietnam and People's Republic of Mongolia
KULTURA Hungarian Foreign Trading Company P.O.B. 149 H–1389 Budapest 62, Hungary
for all remaining areas
Elsevier Science Publishers
1 Molenwerf
P.O. Box 211, 1000 AE Amsterdam, The Netherlands

Library of Congress Cataloging in Publication Data
Berecz, Endre.
  Gas hydrates.

  (Studies in inorganic chemistry; 4)
  Translation of: Gázhidrátok.
  Bibliography: p.
  Includes index.
  1. Hydrates. 2. Gases. I. Balla-Achs, Márta.
II. Title. III. Series.
QD541.B4713 1983   541.3'422   83-5557
ISBN 0-444-99657-5

ISBN 0-444-99657-5 (Vol. 4)

ISBN 0-444-41750-8 (Series)

© *Akadémiai Kiadó, Budapest 1983*

Printed in Hungary

CONTENTS

Preface   7

Chapter 1. Historical survey of the literature on gas hydrate studies   9

Chapter 2. Fundamentals of the structures of gas hydrates   30

2.1. Gas hydrates and clathrate structures   30
2.2. The structures of water and ice   33
2.3. Effects of solutes on the structure of water   39
2.3.1. Effects of molecularly dissolved materials on the structure of water   40
2.3.2. Effects of ionically dissolved materials on the structure of water. Funda-
        mentals of the structure of aqueous electrolyte solutions   42
2.4. Structures, compositions and systematization of crystalline gas hydrates   48
2.4.1. Simple gas hydrates (H_I)   56
2.4.2. Liquid hydrates (H_{II})   57
2.4.3. Double hydrates   58
2.4.4. Mixed hydrates   59

Chapter 3. Stability relations of gas hydrates   60

3.1. Conditions of formation and decomposition of gas hydrates   60
3.2. Effect of third component on stability of gas hydrates   79
3.2.1. Effects of gases on stability   83
3.2.2. Effects of electrolytes on stability   87
3.2.3. Effects of alcohols on stability   95
3.2.4. Effects of a condensed apolar phase and of other compounds on stability.
        Hydrates of liquid hydrocarbons. Stabilities of mixed hydrates   109

Chapter 4. Thermodynamics of gas hydrates   114

4.1. Chemical thermodynamic investigations of gas hydrates   114
4.2. Statistical thermodynamic investigations of gas hydrates   140

Chapter 5. Properties and characteristic data of individual gas hydrate systems 171

5.1. Systems with a single gas component   171
5.1.1. Nitrogen hydrate and oxygen hydrate   171
5.1.2. Hydrates of the noble gases   174

5.1.3.   Hydrates of the halogen elements   183
5.1.4.   Carbon dioxide hydrate   184
5.1.5.   Sulphur dioxide hydrate   188
5.1.6.   Hydrogen sulphide hydrate   191
5.1.7.   Hydrates of pure hydrocarbons   194
5.1.7.1. Methane hydrate   194
5.1.7.2. Ethane hydrate   204
5.1.7.3. Ethylene hydrate   206
5.1.7.4. Propane hydrate   209
5.1.7.5. Propylene hydrate   211
5.1.7.6. Butane hydrates   213
5.1.8.   Freon hydrates   215
5.2.   Hydrate systems containing more than one gas component   219
5.2.1.   Methane–carbon dioxide hydrate   219
5.2.2.   Methane–hydrogen sulphide hydrate   228
5.2.3.   Methane–propane hydrate   230
5.2.4.   Methane–ethylene hydrate   232
5.2.5.   Propane–hydrogen sulphide hydrate   234
5.2.6.   Propane–nitrogen hydrate   235
5.2.7.   Propane–carbon dioxide hydrate   237
5.2.8.   Methane–propylene hydrate   239
5.2.9.   Propane–propylene hydrate   243
5.2.10. Chloroform–hydrogen sulphide hydrate   245
5.2.11. Carbon tetrachloride–hydrogen sulphide hydrate   247
5.2.12. Difluoroethane–hydrogen sulphide hydrate   247
5.3.   Hydrates of natural gases   249

Chapter 6.   Some technological aspects of the role and application of gas hydrates   267

6.1.   Role and application of gas hydrates in the natural gas industry   272
6.1.1.   Gas hydrates and natural gas production   273
6.1.1.1. The role of porous medium in gas hydrate formation   273
6.1.1.2. Conditions of gas hydrate plug formation in the tubing, and the methods
         employed to prevent this   277
6.1.2.   Gas hydrates and the transportation of natural gas   283
6.1.3.   Application of gas hydrates for gas storage purposes   288
6.1.4.   Methods for the separation of components in the gaseous and liquid states by
         means of gas hydrates   289
6.2.   Desalination of water via gas hydrate formation   292
6.3.   Other technological possibilities for the application of gas hydrates 294

**Chapter 7.   Fundamental methods for the experimental determination of the conditions of
         formation and decomposition of gas hydrates**   298

References   311

Acknowledgements   327

Author index   329

Subject index   337

PREFACE

In a review published in 1927, Schroeder [1] stated that one of the important tasks of modern chemistry was the investigation of the specific affinities occurring in the interaction of different chemical materials with water. At that time, a vast amount of practical work had already proved that not only water-soluble solid materials, but also many extremely poorly-soluble gases and liquids, interact with water as a solvent, and that under appropriate circumstances solid-state hydrates, known as gas hydrates and liquid hydrates, separate out. Schroeder concluded that these gas hydrates are products of the "secondary valence" (Nebenvalenz) effects, and display characteristic features; in comparison with their real importance they have given very little attention in the systems that have been investigated.

The discovery of crystalline gas hydrates dates from the experiments of Davy [2] in 1810. Up to the end of the 19th century, some 40 papers were published in which information was given about the conditions of crystallization of gas hydrates, their suspected or actual compositions and properties, and their thermodynamic evaluation. After the turn of the century, a large proportion of the earlier investigations were reassessed, and new theories were elaborated for the interpretation of the conditions of existence and the structures of gas hydrates. At the same time, however, this area of research was forced into the background: it did not provide sufficient results of practical interest to draw attention to itself, while in addition the analysis of gas hydrates involved very tedious labour. Nevertheless, from a theoretical point of view these strange material systems did appear interesting enough to continue to attract research workers, with procedures becoming more and more systematic than the earlier ones. Between 1940 and 1950, impulse was given to the investigation of gas hydrates by the practical phenomenon that natural-gas pipelines, then rapidly finding wider application, became clogged in cold weather; this was due to the deposition of crystalline gas hydrates under these operating conditions.

The research needed in connection with this phenomenon prepared the way for the knowledge of the phase characteristics of non-stoichiometric inclusion compounds, their statistical thermodynamic description and the determination of their structures by means of X-ray diffraction. In the past decade, wide-ranging research has revealed an isomorphic series connection in the various clathrate compounds with substance types similar to, or related to, the gas hydrates; moreover, an isostructural relationship similar to those detected in the gas hydrates has been observed in certain salt hydrates too.

More systematic research has revealed many areas where the practical importance of gas hydrates is already increasingly more evident. Although research aimed at the elimination of gas hydrate clogs in natural-gas pipelines remains important, economical sea-water desalination processes are also nowadays based on the formation of gas hydrates, and the investigation of this type of behaviour in the freon coolants is by no means unimportant. Further, it has been found that the hydrates of certain pure gases can be detected in interplanetary space, in the heads of comets, so that hydrates play a role in the chemistry of the solar system. Indeed, as can be proved by the comparison of some physical properties of protein molecules, it is now evident that the potential characteristics of hydrate structures can even be recognized in living organisms, and experimental data prove the importance of gas hydrates in the process of anaesthesia.

In addition to these selected examples, it can justifiably be assumed that in the near future improved methods will bring this group of materials more into the limelight in many areas of chemistry. In this book, our aim has been to survey the literature currently available on the topic of gas hydrates. Attention is drawn both to the practical and to the theoretical importance of this class of compounds. Further, our own investigations are drawn upon to discuss the conditions of hydrate formation, the properties of these hydrates, and the conclusions suggested by the results in the case of gaseous hydrocarbon systems containing CO_2, which have so far been comparatively little studied elsewhere. This is an important field, because many gas wells contain appreciable amounts (10–80%) of CO_2, which exerts a powerful influence on the hydrate-forming properties, and hence on the transportability and applicability of these hydrocarbon systems.

The special literature has been surveyed up to the end of 1980, thereby providing an overall picture of the present situation of investigations into gas hydrates, and the results obtained in this field.

We sincerely hope that the aim of our book will be achieved, and that it will be of benefit not only to those whose researches are devoted to gas hydrates, but also to scientists and tehnologists who are confronted by gas-hydrate problems in the course of their work.

The Authors

Chapter 1

HISTORICAL SURVEY OF THE LITERATURE
ON GAS HYDRATE STUDIES

As already mentioned in the Preface, the first information on a gas hydrate, that of chlorine, was published by Davy [2] in 1811. His findings regarding the nature of the hydrate were corroborated in 1832 by Faraday [3], whose first experiments suggested the composition $Cl_2.10H_2O$. The hydrate of bromine was first detected by Löwig [4], and 48 years later Alexeyeff [5] determined the composition as $Br_2.10H_2O$. Simultaneously with the discovery of liquid SO_2, de la Rive [6] produced its hydrate and found its composition to be $SO_2.7H_2O$; however, nearly 20 years later, Pierre [7] modified this to $SO_2.11H_2O$, while Schoenfeld [8] gave the composition of the hydrate as $SO_2.14H_2O$. In the years 1882–1883, the scientific dispute between Ditte [9] and Maumené [10] was directed to the accurate definition of the composition of chlorine hydrate.

The hydrate-forming properties of CS_2 were recognized by Berthelot [11], Millon [12] and Duclaux [13], and the early determinations gave rise to much debate, in which the research activities of Wartha [14], Balló [15], Mijers [16], Decharme [17] and Tanret [18] played an important role. However, the true discoverer of the hydrates of CS_2 and diethyl ether can be considered to be Stokris [19], who in 1870 unambiguously demonstrated the properties of both compounds. The gas hydrate of chloroform was produced by Chancel and Parmentier [20] in 1885, in the course of solubility investigations.

With the application of pressure, Wroblewski [21] was the first to produce CO_2 hydrate, and in fact his investigations led to the later deeper understanding of gas hydrates. Wroblewski established that, for CO_2 hydrate at $0\,°C$, 1.28×10^6 Pa is a critical pressure, equal to the dissociation pressure of the hydrate. The value of critical pressure increases with increasing temperature, and is independent of the relative quantities of the components present. His indirect analytical method yielded a gas hydrate composition corresponding to the formula $CO_2.8H_2O$.

Cailletet [22] and Cailletet and Bordet [23] observed hydrate formation on the liquefaction of acetylene and phosphine in the presence of water. They produced the first mixed hydrate from equal proportions of carbon dioxide and phosphine, and found that hydrogen sulphide and phosphine form a double hydrate.

During the same period, de Forcrand and Villard, perhaps the most successful early researchers into gas hydrates, turned their attention to this very disputed system. Supplementing the previous investigations by Wöhler [24], de Forcrand [25] carried out detailed studies to determine the tendency of hydrogen sulphide to form a hydrate. His experiments showed the composition of the hydrate to vary between $H_2S.16H_2O$ and $H_2S.12H_2O$, and from considerations of the mother liquor inclusions he held the latter to be the more probable.

As a result of other investigations too, enough material was soon available to verify the ability of H_2S hydrate to react with many other compounds, the properties of the products being very similar to those of simple hydrates. The first such reference can be found in the work of Loir [26] in 1852; he performed experiments on the production of the double hydrate of chloroform and H_2S, but also gave information on the existence of the hydrate of chloroform and hydrogen selenide. The use of methyl chloride, ethyl chloride, ethyl bromide or ethylene chloride in place of the chloroform results in analogous crystalline products. With halogenated hydrocarbons, de Forcrand prepared some 30 double hydrates and found that their stabilities are higher than those of the individual hydrates, so that they can be produced more easily; however, his investigations indicated that double hydrates with hydrogen sulphide are formed only by those halogen-substituted derivatives with boiling-points lower than 110 °C. Besides those of alkyl halides, de Forcrand also produced double hydrates of nitromethane and nitroethane, and found the compositions of all double hydrates to be $X.2H_2S.23H_2O$.

From 1884 on, Roozeboom [27] attempted to clarify many of the disputed questions connected with the gas hydrates, in the hope that the general exploration of the dissociation phenomena would explain the problems of the formation and decomposition of gas hydrates. By that time, the Debray rule had already been formulated: "At a given temperature, all solid materials which decompose into a solid and a gas do so at a definite dissociation pressure, which varies with the temperature". The validity of this rule had previously been recognized by Wroblewski, Cailletet and de Forcrand with regard to the hydrates of CO_2, PH_3 and H_2S, and Roozeboom later found it true for SO_2 too. He experimentally proved that, for each of the gas hydrates he investigated, a definite dissociation pressure existed at a given temperature, and the dissociation equilibrium was established if the gas pressure was equal to the dissociation pressure. On increase of the gas volume, the dissociation will be higher, but as a result of a volume decrease the gas will be absorbed, and new hydrate is formed until the pressure reaches the equilibrium value corresponding to the given conditions. Variation of the temperature leads to decomposition or reformation of the hydrate until the equilibrium dissociation pressure corresponding to the new temperature is attained. In the case of aqueous solutions, the equilibrium is also influenced by the fact that the solubility of the gas (i.e. the gas content of the solution) changes with the temperature. On increase of the temperature, part of the hydrate must undergo decomposition, not only to such an extent that the free gas pressure reaches the dissociation pressure, but also so that the saturation of the liquid assumes the equilibrium value.

According to Roozeboom's findings, the maximum dissociation pressure of a hydrate that can be reached in a closed space is that pressure value at which the gas liquefies, for all the free gas will condense at a pressure higher than this. In the course of his experiments, Roozeboom determined the critical decomposition parameters of a number of hydrate-forming components, making use of the solubility or vapour pressure data of other authors too. For all hydrates, the hydrate vapour pressure curve exhibited a sharp break at the solidification-point of water.

After Roozeboom had thrown light on the phase theory of gas hydrates, Villard and de Forcrand made further theoretical investigations, in which they determined the temperature-dependence of the dissociation pressure of the hydrates of methyl chloride and H_2S [28, 29]. A short time later, the discovery of newer gas hydrates was reported by Villard [30], who demonstrated that the hydrates of CH_4, C_2H_6, C_2H_4, C_2H_2 and N_2O can be produced under pressure. At the same time, he claimed that, of the gases that dissolve weakly in water and can easily be condensed, cyanogen, ammonia, propylene, butylene and allene are not able to form solid crystalline hydrates.

After 1890, Villard investigated the hydrate-forming properties of halogen-substituted compounds. With ethyl bromide and ethyl iodide, his experiments were not successful, but in the cases of ethyl chloride, methyl iodide and methyl fluoride he succeeded in producing well-defined hydrates.

In the same year, he also prepared the hydrates of methane, ethane and propane [31]. The latter compound formed readily when gaseous propane was abruptly cooled under 0 °C in the presence of water. He further observed that, in the presence of a little air, the pressure-dependence of the vapour tension of propane hydrate increases, though the critical parameters of the hydrate are not influenced by the presence of foreign gas molecules. Villard also made the interesting discovery that the critical temperatures of the hydrates of hydrocarbons are decreased by the increase of the relative molecular mass of the component forming the hydrate.

After the discovery of the new hydrates, Villard surveyed the results obtained, and this subsequently furnished a good basis for answers to the questions of the structures of the gas hydrates.

A review of the experimental results of the initial period reveals that the more accurate gas hydrate compositions obtained by the more sophisticated analytical methods or by indirect analyses show an increasingly higher water content as compared to the first data. Moreover, the newer formulae approximated more and more closely to the hexahydrate structure assumed by Villard. This called attention to the presumption that all gas hydrates display a symmetrical crystalline form. This was corroborated by the fact that polarized light did not influence any of the crystalline gas hydrates, while analogous phenomena could be observed in investigations of the heats of formation.

In 1896 it was considered an extraordinary discovery that the chemically indifferent argon was classified by Villard [32] among the hydrate-forming gases, since it formed a solid hydrate with a dissociation pressure of 2.126×10^7 Pa at 8 °C. Relying on this result, Villard boldly postulated that at higher pressure nitrogen and oxygen were also presumably capable of forming hydrates.

Villard's novel findings aroused lively discussion among research workers, and everyone strived to provide evidence regarding his point of view with newer experimental results. This was one of the reasons for the significant progress made at that time concerning both simple and mixed hydrates.

Particular importance was attached to the work of de Forcrand and Thomas [33], who attempted to elucidate which organic materials form compounds of double hydrate type. In this work they produced mixed hydrates of CCl_4 with acetylene, carbon dioxide or ethylene. The CCl_4 could be replaced by numerous halohydrocarbons (e.g. $CHCl_3$, $C_2H_4Cl_2$, CH_3I, CH_3Br, CH_2Cl_2, $C_2H_3Cl_3$, CH_2I_2, etc.), similarly as for the double hydrates formed with hydrogen sulphide.

The research trends of the early 1900's were characterized by the determination of the physical properties of the gas hydrates. Le Chatelier devised thermochemical relationships on the analogy of the Trouton rule. These were developed further by de Forcrand, who used the hydrate formation heats to draw conclusions on the number of water molecules bound in the compounds and to check the compositions of the known gas hydrates.

In 1919, Scheffer and Meijer [34] elaborated a new thermodynamic method for the indirect analysis of gas hydrates. By constructing the three-phase equilibrium curves, with the application of the Clausius–Clapeyron equation, they were able to postulate the compositions of the hydrates.

An important result from the 1920's was the discovery of the hydrates of krypton and xenon. In this respect, de Forcrand [35, 36] assumed that increase of the relative molecular mass of the noble gas increased not only the stability of the hydrate, but also the number of water molecules bound.

This was likewise the period of Villard's efforts to produce iodine hydrate [37], which led to the finding that this hydrate can only form at an elevated pressure in the presence of inert gas at a temperature above 0 °C. This study was of special interest because it suggested the possibility of the production in inert gas of hydrates of other substances less volatile than iodine.

The phase equilibrium investigations of Bouzat [38] at this time again involved the heats of formation, with conclusions on this basis as to the water contents of the hydrates. In joint work with Aziniers [39], a new experimental method based on direct analysis was developed, by means of which they succeeded in proving the theoretical composition $X.6H_2O$ proposed by Villard, with an accuracy superior to those of the previous methods. The mean number of water molecules taking part in the hydrate formation was calculated from very many experimental data to be 6.3. In opposition to the conception of Villard, de Forcrand defended the higher water contents he had determined in the cases of the hydrates of ethylene, sulphur dioxide and chlorine, proving that the crystals formed from the mother liquor in the liquid gases are able to include the hydrate-forming component. The results of Villard were later reinforced by Tammann and Kriege [40], who dealt in particular with the eutectic compositions of the hydrates.

From the beginning of the 1930's, attention turned to the formation of mixed hydrates in the pipelines carrying natural gases, because the clogging of these pipelines as a consequence of gas hydrate formation caused many problems in the developing gas industry. As a pioneer of these researches, Hammerschmidt reported [41] that, in addition to the natural formation conditions, the separation of the hydrates as solids was also influenced by secondary factors. Of these, primarily the gas flow rate and pressure fluctuations had to be taken into account. Deaton and Frost [42, 43] too published the results of their systematic investigations in connection with the natural gases, and reported experiments relating to the methods (e.g., gas dehydration or addition of inhibitors) and extents of reduction of hydrate formation for given operating parameters. They also investigated the conditions of hydrate formation in mixtures of natural gases under operating circumstances.

The phase characteristics of formation of the hydrates of the natural gas constituents methane and ethane were analyzed in detail by Roberts and his co-workers [44]. After carrying out experiments for many years and amassing a wealth of empirical data, Parent [45] was able to point out the most likely directions for the further advance of investigations into natural gas hydrates.

Carson and Katz [46] studied the four-phase equilibria of gas mixtures in the presence of gas hydrates and of liquids rich in hydrocarbons. This was the first publication stressing that the gas hydrates of the natural gas constituents behave as solid solutions, and that pentane and higher homologues do not enter the solid phase.

However, despite the fact that the hydrate-forming properties of hydrocarbons were already being investigated in detail by this time, and accordingly a great many data were available, it did not prove possible to develop a method of predicting the pressure limit at which natural gases saturated with water vapour at a given temperature form the hydrate. Employing gas analysis data, Carson and Katz [46] attempted to find a possibility of calculating the conditions of hydrate formation via the equilibrium constant. In 1944, Katz [47] reported the hydration behaviour of natural gases of various densities, giving at the same time graphical and tabulated methods of predicting the pressure limits to which the different natural gases can be expanded without hydrate separation. The entropy and enthalpy data necessary for the construction of the expansion diagrams were previously determined by Brown [48], from temperature, pressure and density data.

The conditions for water condensation at the temperatures and pressures of natural gas reservoirs were investigated by Sage and Lacey [49], as were the possibilities of the natural gas constituents undergoing hydrate formation in the field, and the associated problems of pipeline transportation.

Miller and Strong [50] described their ideas for the economic storage of natural gases in hydrate form, and for their possible transportation in this state, and they reported experiments relating primarily to propane hydrate. Their research work was stimulated by the fact that a given quantity of gas hydrate occupies a much smaller volume than that of the gas necessary for its formation. In their work they dealt with the influence of non-electrolytes (sucrose, methanol, ethanol, acetone) dissolved in the water on the critical decomposition parameters of the hydrates formed. They compared their experimental results with the data of Frost and Deaton [51], who determined the compositions and equilibrium data of hydrates of various pure gases at temperatures below 0 °C. They made a critical study of several methods for the laboratory production of hydrates, and developed an analytical procedure for the determination of the compositions of hydrates at temperatures below 0 °C. By indirect analysis they also proved that the composition of the hydrate is not identical with that of the compressed gas mixture.

Besides investigations into the properties of the natural gas hydrates, studies of the hydrates of the noble gases and the inert gases came into the foreground at this time. In this field, the works of Bradley [52], Godchot and his co-workers [53], and Nikitin [54–58] are outstanding.

The crystalline structures of clathrates of the noble gases with water or hydroquinone were investigated by Palin and Powell, whose results appeared in a series of publications [59–62].

The conditions under which freons F11, F12 and F22 (with halon numbers 113, 122 and 121), methyl chloride (with halon number 101) and sulphur dioxide form hydrates were studied by Chinworth and Katz [63] in order to shed light on these features of the gases used as coolants.

A considerable proportion of the natural gases contain CO_2 (on the average 0.1–0.2%), which is itself a hydrate-forming component too. It was assumed or established, on the basis of the previous natural gas investigations, that this gas enters the solid solution in the process of hydrate formation. Direct determination of the CO_2 content of the solid hydrate involves a very difficult separation procedure, with an appreciable accompanying error. Unruh and Katz [64] therefore tried an indirect approach to find experimental and calculation possibilities relating to the conditions of hydrate formation for CO_2-containing natural gas constituents and their mixtures, and from the mole fraction ratio $x_{CO_2}^{vapour}/x_{CO_2}^{hydrate}$ they determined the equilibrium constant for the formation of CO_2 hydrate.

In his initial experiments, Villard [30–32] considered that methane hydrate cannot exist above 21.5 °C at any pressure. In contrast with this, in 1949 Kobayashi and Katz [65] published experimental data referring to the methane hydrate formation equilibrium up to a pressure of 7.88×10^7 Pa, and they verified by means of theoretical considerations the formation of methane hydrate at a pressure of 2.80×10^8 Pa at about 37.4 °C.

From 1949, several important reviews appeared on the production, structures and physical chemistry of gas hydrates. These works summarized the results of the research by von Stackelberg and his co-workers [66–77] and the investigations of Claussen [78, 79] and Pauling and Marsh [80] dealing with the inert gases and the structure of the gas hydrate of chlorine. These latter authors determined the structure of the almost symmetrical, cubic clathrate hydrate, with lattice constants of 1.2 nm and 1.7 nm. From the point of view of most hydrates, this was a very important discovery, and it could be confirmed by X-ray powder diffraction.

Simultaneously with the elucidation of the structure, Records and Seely [81] made proposals for the practical use of hydrate formation in the dehydration of natural gases at low temperature. They proved experimentally that the natural gases undergo effective dehydration in the course of expansion. In their experiments, they used natural gas at a pressure of $6–10 \times 10^6$ Pa for dehydration in a suitably dimensioned hydrate separator provided with a heat exchanger, and determined the optimum heat exchanger temperatures and gas flow speeds to be used with natural gases of various origins.

At about this time, the necessity arose for systematic investigations as to how the formation conditions for a number of hydrates vary in the presence of various molecular or ionic materials. Bond and Russell [82] examined the effects of NaCl, $CaCl_2$, CH_3OH, C_2H_5OH, ethylene glycol, sucrose and dextrose (as freezing-point depressants) on the formation of the hydrate of hydrogen sulphide. In the course of their investigations, they revealed certain errors and deficiencies in the previous knowledge about this hydrate.

Since water and hydrogen sulphide occur in almost every gas field, increasingly more detailed investigations of these systems were a continuous feature of the research in the late 1930's and the 1940's. For example, the volumetric properties and vapour pressure were determined by Murphy [83], while the thermodynamic properties were reviewed by West [84]. Reamer and his co-workers later published new data referring to the volumetric properties and vapour pressure [85].

Selleck and his co-workers [86] gave an account of their investigations of the three-phase equilibrium. With a measurement method based on a dynamic principle, they found that the three-phase equilibrium state can be determined unequivocally from the discontinuity in the first derivative of the isochore pressure vs. temperature relationship. They set themselves the task of determining the equilibrium pressures of homogeneous and heterogeneous systems as a function of the temperature, the pressure-dependence of the specific volume measured at a given temperature, and the compositions of the individual phases.

Lippert and his co-workers [87], at Oklahoma University, carried out studies over several years on the effect of a third component on the hydrates of liquid hydrocarbons with low melting-points.

Making use of the experimental data of Hammerschmidt [41] and of Powell [88], Reamer and his co-workers [89] investigated the effects of olefinic hydrocarbons on the hydrate formation of paraffins, particularly in the systems propane–propylene–water and ethane–ethylene–water. They found that, under the conditions of hydrate formation in the propane–propylene–water system, the hydrate phase forms a solid solution in which the distribution of propane and propylene is approximately equal to that in the coexisting liquid hydrocarbon phase.

Smirnov [90] and Czaplinski [91] analyzed the possibilities of hydrate formation under normal operating parameters in the Kuybishev natural gas pipeline. Using Smirnov's data, Czaplinski made a critical evaluation of the effects of inhibitors applicable for the prevention of pipeline clogging. He too found methanol to be the most suitable inhibitor and, by taking into account the previous literature data [92–94], he elaborated a simple graphical method for the determination of the necessary quantity of methanol.

An approximate correlation between the heats of formation of gas hydrates, their compositions, and the equilibrium temperature decreases induced by inhibitors was established by Pieroen [95]. In his calculations, he assumed the presence of a third component (e.g. ethanol) which does not form a gas hydrate or a mixed hydrate and does not dissociate in water, and determined the conditions of the three-phase equilibrium as a function of the concentration of the third component at constant pressure. In a further part of his work, he applied his calculation method to the experimental data obtained by Lippert and his co-workers [87] for the chloroform–ethanol–water system, and found it suitable for this.

Between 1957 and 1965, important theoretical works were published on the statistical thermodynamic properties of clathrate compounds in general, including the properties of gas hydrates; these form a good basis for the physico-chemical description of the non-stoichiometric inclusion compounds too.

The characteristics of the inclusion compounds of water and phenol were surveyed by von Stackelberg [96], but as regards the statistical thermodynamic description of phenol and water clathrates, the works of van der Waals [97] and van der Waals and Platteeuw [98] are perhaps the most important. Barrer and Stuart [99] and Barrer and Ruzicka [100] dealt successfully with the theory of non-stoichiometric clathrate compounds of water.

The statistical theory of van der Waals was based on the three-dimensional generalization of ideal localized adsorption. The relationships he derived for the establishment of the equilibrium conditions stemmed from the generalization of the Langmuir isotherm or of Raoult's law: in his model, the determining factors in the stability of gas hydrates are the interactions of the characteristics of the solvent and the localized adsorption. The theory of Lennard-Jones and Devonshire [101] referring to liquids gave a possibility for the quantitative determination of the distribution functions of the molecules situated in the cavities, and hence for the expression of the equilibrium vapour pressure, the hydration number and the chemical potential. From the experimental data of Evans and Richards [102], the heat of formation and bond energy were determined, concrete heterogeneous equilibrium calculations were performed for several binary and ternary systems, and an attempt was made to give a thermodynamic interpretation of the stabilizing effects of "auxiliary gases" too. Though a significant proportion of their statistical mechanical deductions refers to the hydroquinone clathrates, numerous general relationships can be formulated for the gas hydrates as well.

The theory of dissociation pressure was elaborated for some gas hydrates by McKoy and Sinanoglu [103] on the basis of the cell model of Lennard-Jones and Devonshire, with the aid of the Lennard-Jones 12–6 and 28–7 and the Kihara [104] potentials.

A series of papers on the crystalline structures of polyhedral clathrate hydrates was published by McMullan and Jeffrey [105] in 1965, and a short time later the results were summarized in a review [106].

Child [107] investigated and interpreted the molecular interactions in clathrates. His main purpose was to interpret the possibility of movement of the enclosed molecules on the basis of the interaction energy of the clathrate-former and the enclosed molecules (calculated from thermodynamic data) and from dielectric and infrared spectral data. He used his calculations to determine the dissociation enthalpies and entropies of hydroquinone clathrates and gas hydrates.

Both Wilson and Davidson [108, 109] and Bertie and Othen [110] carried out dielectric and infrared spectrophotometric studies on hydrates, with resulting advances in the knowledge of their structures, especially in the case of ethylene oxide hydrate.

Similarly on a statistical mechanical basis, the hydrates of CH_4, Ar and N_2 were studied by Saito and his co-workers [111]. They made use of the experimental data to apply the theory of solid solutions in the construction of the three-phase equilibrium data in the region above the quadruple-point of the four-phase system ice–hydrate–liquid rich in water–gas. They determined the temperature-dependence of the fugacity of the hydrate, the chemical potential, the hydration number and the heat of formation values. The Lennard-Jones–Devonshire force constants they determined were in good agreement with those of the earlier authors.

This process of the ever deeper understanding of the properties of the gas hydrates drew attention once more to practice. For instance, the properties of natural gases were investigated by McLeod and Campbell [112] in the case of high-pressure systems. With a modified form of the Clausius–Clapeyron equation, they attempted to predict the temperature of hydrate formation. Carson and Katz [46] gave the complete hydrate curve for the high-pressure region by means of the vapour–solid equilibrium constant. They found that at high pressures the crystalline structure of the hydrate becomes sensitive to pressure.

By investigating the CH_4–H_2S–H_2O system, Noaker and Katz [113] promoted the more detailed recognition of the hydrate-forming features of natural gases containing hydrogen sulphide, and also the possibilities of application of the equilibrium constants.

From the thermodynamic functions of the dissolution of methane in water, Glew [114] constructed the equilibrium relationships of this gas hydrate. He elaborated a new type of equation for the temperature-dependence of the dissolution of methane in water and for the calculation of the molar heat capacity. With this solution model, he determined the molar volume of methane dissolved in water, and hence drew conclusions on the structure.

Meanwhile, in the investigation of hydrocarbon systems, study of the effect of olefinic constituents again came into prominence. Otto and Robinson [115] examined the system methane–propylene–water, analyzing the phase diagrams of the system in a very detailed way for propylene contents of 0.5–24%. Snell and his co-workers [116] extended this knowledge with investigations of the ethylene–water, methane–ethylene–water, and methane–ethylene–propylene–water systems. Clarke and his co-workers [117] published the results of detailed examinations of the propylene gas hydrates, the aim of their experiments being primarily to determine the conditions of hydrate formation more accurately than previously.

In 1969, King [118] briefly surveyed the knowledge that had accumulated about gas hydrates, and a further review was published by Bhatnagar [119] in 1970. In 1972, Byk and Fomina [120] gave an overall picture of the gas hydrates, mainly from a thermodynamic aspect, including reference to their own earlier work [121, 122].

In the 1960's, many publications appeared in connection with investigations of the conditions of formation of hydrates of natural gases and their constituents, and the possibilities of preventing hydrate formation in pipelines.

Sherwood [123] investigated the possibility of the addition of glycol or methanol to prevent hydrate formation, and also studied the economics of various types of gas drying procedures.

An indirect method for the equilibrium investigation of methane–water and argon–water systems at very high pressures (up to 7×10^8 Pa) was developed by Marshall [124].

The conditions for the prevention of hydrate precipitation in pipelines with a capacity of 10^6 m³/day, and the optimum choice for the site of addition of methanol or the site of separation, were studied by Trebin and Makogon [125]. In another work [126], they reported a simple relationship for determining the rate of formation of gas hydrates.

Marshall and his co-workers [127] carried out high-pressure investigations, and attempted to use measured data to determine the compositions of hydrates, but they did not succeed in attaining sufficient accuracy.

Musayev and Chernikhin [128] investigated the various compositions in the propane–ethylene–butane–water systems, with the aim of establishing the equilibrium parameters of the mixed hydrates. They determined the effect of the cooling rate and the modifying effect of nitrogen (as an auxiliary gas) on the temperature and rate of hydrate formation and on the stability of the hydrate formed. Musayev [129] similarly studied the conditions of hydrate formation from liquefied hydrocarbon gases.

Trebin and his co-workers [130] discussed the kinetics of hydrate formation from the natural gases. They found that, after the first hydrate precipitation, the free cavities in the crystalline lattice of the hydrate are slowly further saturated, and until this process is completed the change in time of the phase transition is of a parabolic character. For a natural gas hydrate, the degree of gas saturation no longer undergoing change was reached after 9 hours. On the basis of their experiments, it can be stated that the duration of contact of the hydrate and the gas phase cannot be disregarded as a parameter in pipeline investigations.

Korotayev and his co-workers [131] investigated the conditions of hydrate formation in a propane–butane mixture in the presence of nitrogen. Their experimental data demonstrated the significant elevation of the temperature of hydrate formation from the liquid hydrocarbon mixture in the presence of nitrogen and on the increase of its pressure. In a later work [132], they proved that the application of anti-freezers is not necessary in the pipeline transportation of liquid technical butane, because of the non-occurrence of hydrate precipitation under the transportation conditions.

Hydrate formation in condensed hydrocarbon–methanol–water systems in the neighbourhood of −40 °C was investigated by Musayev and Korotayev [133]; they determined how the temperatures of formation of ice and of the hydrate crystals on the interface of the two liquid phases depend on the alcohol concentration.

Similarly, Fomina [134] studied the hydrate–gas-phase equilibrium of paraffin–olefin mixtures in systems containing isobutylene. She reported the pressure-dependence of the equilibrium constants of the reactions of the individual components, and also the calculated heat of evaporation of isobutylene hydrate.

As a characteristic of the stability of gas hydrates, the data on the heat of formation of certain hydrates were systematized by Korotayev [135], who also gave the dependence of the stability of the hydrates of condensed hydrocarbons on the pressure of the auxiliary gas.

The hydrate–gas-phase equilibrium in the entire concentration range in the system propane–propylene–water was further investigated by Fomina [136] under isothermal conditions at 0.3 °C. She published the equilibrium pressure vs. composition diagram, the relative volatility values, and the diagram demonstrating the pressure-dependence of the vapour–hydrate equilibrium constant. With the aid of the latter, a method was given for calculating the equilibrium constant to be expected in the mixing of different hydrocarbon components.

Susummi [137] published informative diagrams, primarily relating to halohydrocarbons; these were constructed from the boiling-points of the hydrate-forming compounds and the ratio $T_{\mathrm{cr.,H}}^{\mathrm{decomp.}}/T_{\mathrm{cr.}}$ (where $T_{\mathrm{cr.,H}}^{\mathrm{decomp.}}$ is the critical decomposition temperature of the hydrate, and $T_{\mathrm{cr.}}$ is the critical temperature of the hydrate-forming component).

In his attempts to use petroleum as a hydrate inhibitor in the transportation of natural gases, Dzhavadov [138] discovered an interesting phenomenon. In the presence of petroleum, the hydrate formed was of far looser structure than that obtained in the presence of pure water. The cause of the structural change was explained as the formation of a thin film on the surface of the water by both the condensed hydrocarbons and the petroleum. This film inhibits the contact of the water and the gas, and is an obstacle to the growth of the nuclei after the precipitation of a crystal. Dzhavadov established that this is a suitable method of inhibiting or diminishing hydrate formation, particularly in systems containing much condensed hydrocarbon.

Yorizane and Nishimoto [139] utilized Hildebrand's rule [140] and the data of Frost and Deaton [51] to construct a general equation for the corresponding pressure and temperature values of hydrocarbon hydrates.

The pressure vs. temperature diagram for the system propane–isobutane–water in various mixing proportions between 0 and 7 °C was constructed by Ichiro [141]. He concluded from his experiments that, for all hydrocarbons and their hydrates, an empirical relationship can be given whereby the pressure vs. temperature diagrams describing the decomposition and formation processes can be preconstructed.

Bukhgalter [142] stated that a semi-empirical relationship can be derived that is suitable for the calculation of the temperatures of hydrate formation from hydrocarbon mixtures.

From a quantitative examination of the gas separation accompanying a hydrate formation process in a gas pipeline, Kedzierski and Chowaniec [143] found that the C_3 and C_4 components are considerably concentrated in the hydrate, whereas the quantities of the C_1 and C_2 constituents and nitrogen in the solid hydrate are far less than in the gas phase.

Aliyev and his co-workers [144] examined the mechanism of action of inhibitors of hydrate formation, and came to the conclusion that there is a connection between the freezing-point and concentration of the inhibitor and the temperature of hydrate formation. They derived empirical correlations for this in the cases of calcium chloride, methanol, ethylene glycol and diethylene glycol.

In place of methanol and the glycols, which are widely used in practice, Andryushchenko and Vasilchenko [145] dealt with the search for cheaper and less dangerous inhibitors of hydrate formation. They aimed at achieving a modification of the structure of the water or of the crystalline gas hydrates as a consequence of the presence of the third component. From this point of view, it appeared easier to modify the structure of the water, primarily by adding various electrolytes. This involved determination of the bonding forces between the ions and the water molecules by comparing them with the bonding forces in the crystalline hydrates, about which information is given by the entropies of formation of the hydrates and the dissolution entropies of the ions. Their work was performed with NaCl, $CaCl_2$, $MgCl_2$, KNO_3 and LiCl. They determined the decrease in the temperature of hydrate formation and the dew-point of natural gas from Shebelinsk on the action of electrolyte solutions, as a function of the electrolyte concentration. They found that for all types of electrolytes there was a "critical concentration", at which hydrate formation did not occur. They proved experimentally, therefore, that the structure developing in the water as a result of the electrolyte plays a determining role in the process of hydrate formation.

Musayev [146] and Bukhgalter [147] published studies almost simultaneously on the separation of hydrocarbon gases and liquids by means of the hydrate formation process. Musayev was looking mainly for an effective, cheap method of satisfying the propane–butane gas requirements. Taking into account the

data of Nikitin [54] and Hammerschmidt [41], his first aim was to find a solution
to the calculation of the composition of the hydrate phase in the knowledge of
the thermodynamic parameters of hydrate formation. He proved experimentally
that the composition of the hydrate is affected much more by the temperature than
by the pressure, but it is also strongly influenced by the concentrations of the components in the gas phase. Another important finding was that the greater the number of components in the gas phase, the higher the quantity of methane incorporated in the hydrate phase formed from the natural gas. In the hydrate of the
natural gas used in his experiments, the amount of propane–butane mixture
was concentrated by a factor of about twelve, while higher hydrocarbons were
not detectable in the hydrate.

The first patents in connection with the separation of gases and liquids via
hydrate formation processes appeared in the middle of the 1940's [148, 149].
The separation of H_2, N_2 and He from natural gas and the recovery of organic
liquids (especially halohydrocarbons) were planned.

Bukhgalter [147] designed an apparatus suitable for the separation of methane
and ethane. With a slight modification, this could also be used to separate ethane–
hydrogen mixtures. A periodically operating device was built for the separation
of propane–propylene mixtures too, the propane content being converted practically completely into the hydrate phase.

The recovery of helium from natural gas in the two-step hydrate-formation
process of Kinney and Kahre [150] was of particular industrial importance.

In 1969, a huge field of natural gas in the solid hydrate state was found in West
Siberia and Yakutia. The discovery and the possibilities of exploiting the enormous
hydrocarbon reserves were reported by Chersky and Makogon [151] and Belov
[152].

Katz [153] described the conditions for the existence of gas hydrates in the
depths of the Polar regions, giving a method for the determination of the depth
and thickness of the hydrate deposits.

At almost the same time, Evrenos and his co-workers [154] carried out laboratory
experiments to impermeate water-saturated porous rock with an appropriate
hydrate-forming gas so that, by means of convective cooling and mixing, gas
hydrate should be formed in the pore space in a locally limited part of the water
(where the water is in contact with the gas). This method is interesting because it
is of potentially great importance for the local sealing of holed gas reservoirs.

One possibility for hydrate occurrence in a gas well is the formation of a hydrate
plug in the tubing under the temperature and pressure conditions prevailing after
the well has been shut down. Banks [155] elaborated a method for calculating
the time period during which a hydrate plug will not be formed in such
a well.

The deeper exploration of the hydrate-forming properties of natural gases was continued in the 1970's. One publication resulting from this work was that of Robinson and Mehta [156], who investigated the equilibrium and phase relations of the system CO_2–C_3H_8–H_2O in a broad concentration range.

The research activity at the Gubkin Technical University for Petrochemistry and the Gas Industry (in Moscow) from the middle of the 1950's to the present day is of great importance. In addition to some previously-mentioned references, this work has given rise to reviews and to papers dealing with special part-problems. The research on this topic by Makogon and his co-workers Musayev, Teodorovich, Tesner, Bogayevski, Trebin, Khoroshilov, Ginzburg, Dyegtyarev, Lutoshkin, Bukhgalter, Efremov, Khalikov, Lisichkin, Trofimuk, Chersky and Sarkisyants is particularly noteworthy. The books by Makogon and Sarkisyants [157] and Makogon [158] present a good account of this activity.

Van Cleeff and Diepen [159] dealt with the hydrates of oxygen and nitrogen at relatively high pressures (5–10×10^7 Pa).

From a technological point of view, the role of the gas hydrates in fresh-water production can no longer be disregarded. Many experimental methods have been elaborated for desalination with n-butane and isobutane. Such procedures have been described, for example, by Gilliland [160], Wiegandt [161] and Hendrickson and Moulton [162].

Barduhn and his co-workers [163] established that the economics of the desalination process depends to a large extent on the properties of the hydrate-forming component, and they therefore discussed the relevant properties of several hydrate-forming compounds in detail. They mainly examined methyl bromide, freon 21 ($CHCl_2F$; with halon number 112) and freon 31 (CH_2ClF; with halon number 111).

Propane was investigated as a hydrate-forming agent by Knox and his co-workers [164], but they also described the conditions of layer formation and the equilibrium conditions. They likewise made concrete calculations regarding the extent to which the energy requirements of the freezing and evaporation processes decrease in the case of a hydrate solidifying at a temperature above 0 °C, and also regarding the economic parameters connected with a heat-exchanger surface area, which can be diminished because of this.

Requirements to be met by hydrate-forming materials of use for the desalination of sea-water were summarized by Barduhn and Towlson [165]. The degree of solubility of the hydrate-forming material in water, and the reaction-kinetic effect of this, were given as uncertainty factors. In connection with this, it was automatically assumed that, since hydrate formation is essentially an interfacial phenomenon, the degree of solubility must influence the rate of hydrate formation. At the same time, the solubility of the hydrate-forming substances previously used in practice was only between some hundredths of a g/t and 1%, and in the 1960's the advantages assumed to follow on the basis of higher solubility were not yet known.

Wittstruck [166] investigated the hydrate-forming properties of freon 11 ($CFCl_3$; halon number 113), freon 12 (CF_2Cl_2; halon number 122), freon 13 B1 (CF_3Br; halon number 134) and freon 22 (CHF_2Cl; halon number 121). These gases may be saturated with water vapour, and dissolve to an appreciable extent in liquid water. With consideration to this, correction factors were applied in the calculation of the heats of formation and decomposition of the hydrates. The hydrates of these freons exist at pressures below atmospheric, and characteristically contain 12.6–15.6 water molecules per freon molecule. Their critical decomposition-points lie in the interval 10–16 °C, which corresponds to the mean temperature of sea-water. $CBrF_3$ hydrate decomposes most easily of them. During his experimental work, Wittstruck determined a factor characteristic of the relative stability of the freon hydrates and correlated with the molecular size of the freon. In X-ray investigations, he proved that the hydrates of all these halomethanes are of type H_I; they are not stoichiometric compounds, because the molecules of freon 22, for instance, are capable of filling only the larger cavities in the water lattice, to an extent depending on the circumstances.

The natural occurrence of gas hydrates on the Earth is known only in the case of the hydrates of natural gases and their components, as the pressure and temperature combinations and the gas composition of the Earth's atmosphere do not ensure the conditions necessary for the spontaneous formation of gas hydrates.

Elsewhere in the Solar System, however, there are different conditions, and it was therefore worthwhile to deal intensively with the possibility of spontaneous gas hydrate formation. In this respect, a comprehensive analysis was published by Miller [167], who dealt with the planets and comets. In his calculations, he used the data of Brown [168] (e.g. in the cases of Uranus, Neptune and Pluto), accepting the assumption that these planets are comprised of an inner metal core, surrounded by a silicate layer, enclosed in turn by a closely-fitting water–ammonia–methane layer. Their atmospheres consist of hydrogen, helium and neon, together with very small quantities of water vapour, ammonia and methane. Miller considers that, if water vapour condenses on these planets, it is conceivable that hydrate precipitation begins.

The findings of De Marcus [169] indicated that Jupiter and Saturn consist chiefly of hydrogen and helium, with approximately 1 wt.% of water, but the situation of the latter was not proved. It appeared likely that the largest proportion of the water was in the interior of the planets, but it was also possible that a certain amount was forced out under the cloud layer, which was considered in all probability to consist of solid ammonia. It was further assumed that ethane, ethylene or acetylene may also be found in the atmosphere of Jupiter, because electric discharges could be induced in a model of its atmosphere. More recent information, transmitted from the spacecraft Voyagers 1 and 2, demonstrates that Jupiter has no solid surface, and consists mainly of liquid hydrogen and helium. The cloud layer is believed to be made up of an inner zone of water ice crystals, separated by hydrogen and helium gas from a zone of ammonium hydrosulphide crystals, separated in turn by hydrogen and helium gas, together with traces of ammonia, from a third zone of ammonia crystals, the whole several hundred kilometres thick [170]. Individual hydrates of hydrogen and helium have not yet been prepared, but it is a proved fact that these gases (containing small molecules and atoms) play an important role in the stabilization of hydrocarbon hydrates.

The rings of Saturn are in general considered to be white frost deposited on silicate powder grains, or finely-distributed ice.

The moons of Jupiter, Io and Europa, presumably consist of silicates and elements of the iron–nickel group. Their high reflectivity probably indicates the presence of some ice layer on their surfaces. The surface of Io is likely to be rich in solid sulphur and sulphur dioxide, possibly covered by sulphur dioxide ice and gas clouds [170]. The lower densities of the moons Ganymede and Callisto suggest a higher water content. Each of the moons of Saturn is of low density, and they presumably consist mainly of water and ammonia. Since methane can be detected beyond doubt in the atmosphere of Titan, it is feasible that all of the methane did not escape before the formation of the satellites, and that hence the hydrate of methane is present on other satellites of Jupiter and Saturn too.

The investigations of Sagan and Strong [171] indicate that the clouds of Venus consist of ice crystals. At the top of the cloud layer a temperature of 230 K was measured, and the partial pressure of CO_2 in the layer was estimated as about 3×10^3 Pa. Pure CO_2 hydrate cannot form under such conditions, even if the effects of nitrogen, nitrogen oxide and argon auxiliary gas are manifested.

Hydrate formation can be helped here only by sulphur dioxide, the dissociation pressure of the hydrate of which at a temperature of 230 K is 2×10^3 Pa. However, this gas cannot be present on Venus, and hence mixed hydrates are presumably not present either, unless formation of metastable ice with coordination number 7 occurs, instead of stable hexagonal ice. In this case, the dissociation pressure of the hydrate would be significantly decreased, and thus mixed hydrates could be formed with CO_2 and other gases present. This can be concluded from the reflectivity of Venus.

At a height of 15 km above the surface of the Earth, the temperature is about 209 K and the pressure is 1.32×10^4 Pa. Though a mixed hydrate of nitrogen, oxygen and argon can be formed only at a pressure of at least 2.13×10^6 Pa at such a temperature, if metastable ice is present there is a possibility for hydrate formation here too. A similar possibility would be given if the water content of the air were present in a supercooled liquid state instead of as ice, for in this case the calculations show that the hydrate of nitrogen, for example, could exist at a pressure of 8.1×10^3 Pa at a temperature of 209 K. The resulting air hydrate is very unstable, however, in comparison to ice and air, and readily decomposes. If such a hydrate formation does occur, it remains in the upper atmosphere for a time and may even reach the cold regions of the Earth.

Scholander [172] demonstrated that the composition of gas extracted from glaciers differs from that of the air, its CO_2 and Ar contents being higher. He attempted to justify this in part by the condensation of the hydrate formed in the atmosphere of the Earth, and also by the composition of the gas mixture absorbed in glaciers differing from that of the mixture of the air and the gases released from the hydrate.

Urey [173] estimated the atmospheric pressure on the surface of Mars to be about 9.1×10^3 Pa, the atmosphere consisting of nitrogen, argon and carbon dioxide. The temperature is around 180 K. Under such conditions, here too a mixture of gas hydrates can be formed only in the presence of unstable ice formations, just as on Venus or in the upper atmosphere of the Earth. (It is interesting that it proved possible experimentally to supercool very pure water in the liquid state to approximately 230 K, which approximates fairly well to the surface temperatures of several planets.)

In their study of the dissociation pressure of CO_2 hydrate, Miller and Smythe [174] proved that the clathrate $CO.6H_2O$ is more stable than solid CO_2 and even ice if the temperature is lower than 121 K. The hydrates they produced form ice and gaseous CO_2 during various periods at 150 K had similar structures to the solid formation to be found in the polar icecaps of Mars. On the basis of their experiments, they assumed that the icecaps of Mars consist of ice $+CO_2$ hydrate or solid $CO_2 + CO_2$ hydrate, but definitely not of a mixture of ice and solid CO_2.

In his comet model, Whipple [175] postulated that the comet cores contain solid H_2O, NH_3, CH_4. CO_2, possibly CO, C_2 and C_3 hydrocarbons, other compounds volatile at room temperature, e.g. hydrogen cyanide, dicyanogen and nitrogen, and other meteoric materials. The comet core initially has a temperature below 50 K, but on proceeding towards the Sun it warms up, its ice-like components evaporating. In all probability, the possibility for gas hydrate formation is given here too, so long as ice is present as such. Primarily nitrogen, carbon monoxide and methane can be reckoned with as hydrate-forming components, for in the neighbourhood of 50 K their vapour pressures are high enough for evaporation from the comet core and for their stabilization as hydrates in the presence of free ice. Though CO hydrate has not yet been prepared under terrestrial conditions, its dissociation pressure is presumably close to that of nitrogen hydrate.

Delsemme and Swings [176] claimed that the comets do not contain mixtures of the pure hydrates, because they considered that the pure hydrates are not stable in the presence of one another: instead, a mixed hydrate is formed. Their claim was based on their spectral investigations of comets, which indicated that the constituents can always be detected in constant proportions. This permitted the conclusion that the gases involved in the individual hydrate crystals are released simultaneously.

If the comets originate from space materials, then these hydrates must contain noble gases, hydrogen and helium too. Because of their small sizes, hydrogen, helium, and even neon, can diffuse out from the cavities, and thus evaporate from the comet. However, argon, krypton and xenon remain in the hydrate, and can be observed spectroscopically.

From indirect evidence, Delsemme and Miller [177] later assumed that a significant amount of water is present in comet cores. Accordingly, they interpreted the possibility of hydrate formation as a special absorption phenomenon in the water-snow, with its very large surface area of approximately 500 m^2 g^{-1}. They proposed a model system to explain the processes occurring in the cores of comets.

The thermodynamic properties of CH$_4$ hydrate at 100 K were studied by Delsemme and Wenger [178]; this substance was prepared not from ordinary water or ice, but from "snow-powder", with a morphologically characteristic appearance. It had a planned and definite size distribution, such that the grains were at most 0.1–1 mm in diameter, while their apparent density lay in the range 0.33–0.54 g cm^{-3}. When the conditions in comets were modelled in vacuum, the original structure of the grains progressively decomposed and CH$_4$ was released, similarly as in the dissociation of ordinary gas hydrates. Their experiments led to the conclusions that the phenomenon can ensue if ice grains penetrate to the interior of the comet cores, and that the formation and dissociation of CH$_4$ hydrates play important roles in the production of halos.

The investigations of Mendis [179] suggested that, in all comets containing excess water, there is a possibility for hydrate formation. The properties of the Kohutek comet were studied on a nuclear model, and the presence of water in a quantity less than six times that of the more volatile components was confirmed. Hence, the hydrate structure could not form, and the more volatile components therefore concentrated around the outer wall of the comet, as a consequence of outward diffusion. This layer would evaporate in the vicinity of the Sun, and an empty water lattice would remain on the surface. The cross-section of the Kohutek comet was determined on the basis of these experiments.

Delsemme [180] subsequently dealt in detail with the regularities of hydrate structures produced in comets.

A very interesting investigation was reported by Pena and de Pena [181], who described the clathrate-forming effects of CO_2, Ar, CH_4 and SO_2 at the freezing-temperature of atmospheric water drops, and hence evaluated the possible hydrate properties of clouds. They studied the effects of dissolved gases on the increase of supercooling, and the correlation of the resulting depression of the freezing-point of the water. It was found that, so long as the atmospheric concentrations of the above gases do not increase significantly, the clathrate-like structural arrangement of the water drops forming the clouds cannot develop.

There is a high probability that some of the interstellar dusts are gas hydrates, as the conditions of formation of methane hydrate, for instance, are satisfied; the only question is that of how the stabilities of such hydrates are affected by cosmic radiation.

Even from this brief survey of the most recent cosmic chemical research, it emerges that gas hydrates may occur in many parts of the Solar System. They are distinguished from ice by their light dispersion, their polarization properties and even their infrared spectra. Certain research workers attribute an important role to gas hydrates in the chemistry of the development of the planets.

There are numerous indications that gas hydrates are of importance in biological systems too.

In the course of his researches, Miller [182] observed that some of the hydrate-forming gases, such as ethylene, nitrogen oxide, chloroform and xenon, possess anaesthetic action. He found that there is a correlation between the dissociation pressures of the gas hydrates at $0\,°C$ and the pressure necessary to sustain the anaesthesia. For 25 investigated gases, the ratio of the anaesthetic pressure and the dissociation pressure of the hydrate has a value between 0.1 and 0.5. In the concrete investigations, it was necessary to determine the dissociation pressures at $37\,°C$ of the hydrates of the gases exhibiting anaesthetic action. However, these are such high pressures that they automatically exclude hydrate formation in the organism at body temperature.

Nevertheless, a number of researchers [183–185] have pointed out that the small solubilities of non-polar gases in water, the large entropies of dissolution and the high partial molar heat capacities of the dissolved gases all verify the more ordered state of the water molecules surrounding the dissolved gas molecules than that of the other water molecules in the solution. Some theories relating to the structure of water accept that liquid water contains ice-like formations ("icebergs") which constantly undergo stationary decomposition and rearrangement (these will be treated in more detail in Section 2.2). It has also been reported that some of the water in proteins is present in such an ice-like state, and plays an important role in the properties of these large molecules [186–188].

Miller [182] calculated what fraction of a given liquid surface is covered by ordered water molecules. He found that, at a given temperature, the ratio of the ordered and the free water molecules in the various states of anaesthesia is proportional to the pressure of the gas applied, and that this is proportional to the dissociation pressure of the gas hydrate. From this, he concluded that the cavities in the "icebergs" are identical with those necessary for gas hydrate formation, and thus his model is suitable for the interpretation of the phenomenon.

However, there are many gaseous anaesthetics, hydrates of which cannot be produced. The best known of these is diethyl ether, but hydrocarbons higher than propane, aromatic hydrocarbons, certain halohydrocarbons and various other ethers can also be included here. Their molecules are too large to fit into any cavities in a water structure. In their aqueous solutions, however, these large molecules are surrounded by "icebergs", just as for hydrate-forming gases. In the mechanism of anaesthesia, therefore, it may be assumed that these surface "ice-sheets" lower the conductivity of the cytoplasm, block the pores of the lipoid membrans, and hence stiffen the cell walls (membranes). It is also conceivable that anaesthetics increase the conductivity threshold value along the nerve tracts, with simultaneous increases in the capacity of the cell membranes and the association of the double water layer. It can be proved experimentally that the hydrate-forming anaesthetic gases change the permeability of the cells too.

To summarize the results of Miller, it appears that no gas hydrate is produced in the living organism under the conditions of anaesthesia, but in the presence of these gases the quantity of "icebergs" in water or in the cytoplasm is increased. The decrease of temperature further increases the "ice-coverage" of the cell surfaces. The effect of this on hypothermia has similarly been proved experimentally.

Similar ideas were published by Pauling [189], but he assumed the formation of hydrate microcrystals in the course of anaesthesia, these being stabilized by the protein side-chains. The microcrystals postulated by Pauling seem to correspond to the "icebergs" enclosing the gas molecules.

In later work, Pauling and Hayward [190] investigated the structural properties of the xenon clathrates in particular, and stated that xenon is an excellent anaesthetic.

Dorsch and his co-workers [191, 192] re-evaluated the hydrate theory in connection with gaseous anaesthesia. By introducing suitably chosen force constants, they determined the possible anaesthetic pressure and the free energy value relating to the body temperature for various species. Their experiments led to the elaboration of a generalized hydrate mechanism referring to anaesthesia, and they confirmed the possibility of formation of a certain gas hydrate-like structure.

Chapter 2

FUNDAMENTALS OF THE STRUCTURES OF GAS HYDRATES

2.1. Gas hydrates and clathrate structures

Gas hydrates are two- or multicomponent crystalline materials. The molecules of one component (always water) form a framework containing relatively large cavities, which are occupied by molecules of other components, these being individual gases or gas mixtures.

The structures of a considerable proportion of gas hydrates involving gases of importance in industrial and laboratory practice can be characterized by the formula $X.nH_2O$, where X denotes the hydrate-forming molecule, while the number of water molecules in the compound is $n \geqslant 5.67$. Such hydrates can be generally formed only in the presence of condensed water, i.e. liquid water or ice (and comparatively rarely in supersaturated water vapour). The hydrogen-bonded water molecules comprise a "host" lattice around one or more species of the "guest" molecules. A physical enclosing process, accompanied only by weak interactions, take place between the host and guest constituents when the latter enter the cavities in the host lattice, and they are released from the cavities only under appropriate circumstances, when the host lattice breaks down.

Thus, the gas components filling the cavities are not directly bonded to the water molecules of the framework. It is for geometrical reasons that they cannot leave the hydrogen-bonded water molecule lattice until this collapses.

Accordingly, in their stable state, the gas hydrates are always two- or multicomponent clathrate compounds. Nevertheless, the various authors use different classifications and terminologies, which are not always unambiguous. The reasons for this are the special structures, as well as the fact that the components of the gas hydrate molecules are not linked by normal chemical bonds. For instance, in some publications the gas hydrates are mentioned as inclusion compounds and are classed in the large group of molecular compounds.

Molecular compounds were defined by Hertl and Römer [193] in the following manner: "A molecular compound is a material arising from two different components, both of which have individual crystalline structures, and in solution (or in the vapour phase) it decomposes into its components in accordance with the law of mass action. The forces holding together molecular compounds are secondary valence bonds, or residual affinity." The theory of Hertl and Römer is no longer completely acceptable. It contains many contradictions and unjustifiable views concerning in particular the bonding forces producing various molecular compounds.

Robertson [194], Clapp [195] and Ketelaar [196] demonstrated the roles of van der Waals forces, coordinative covalent chemical bonds, characteristic hydrogenbonds and donor–acceptor interactions in the formation of molecular associations. Schlenk [197] mentioned gas hydrate clathrates as inclusion compounds, or in some cases occlusion compounds. Baron [198] developed this theory, and classified these compounds more concretely in the group of polymolecular inclusion compounds with "cage-like" structures.

In the view of Powell [199], the general features of the clathrate structure can indeed be applied to the gas hydrates, where the components enclose each other via a complex mechanism. However, the cohesive forces between the host and guest molecules are not sufficient for the formation of a clathrate. Independently of the cohesive forces, there are two essential criteria for clathrate formation: the tendency of water molecules to form a lattice must be satisfied, and the guest species must be of suitable size and shape to enter the cavities in the hydrogenbonded water framework. A further requirement for the formation of the structure is that there should be no chemical reaction between the guest species and the water molecules, i.e. during the crystallization of the material hydrolysis should not occur, nor hydration leading to a structure the total energy of which is less than that of the clathrate.

Jeffrey and McMullan [106] considered that clathrate hydrates can be systematized on the basis of the chemical nature and stereochemistry of the guest species, and also on the basis of the properties of the water host lattice. The large variety of interactions to be expected between the dissolved species and the water molecules in their environment cannot be the basis of such a systematization.

Depending on the chemical properties of the guest species, the clathrate hydrates can be divided into four groups:

(a) *Hydrophobic compounds:* gases or liquids only very slightly soluble in water. On the formation of the clathrate, an apparently increased "solubility" can be observed. In this case, primarily van der Waals interactions occur between the host and guest molecules.

(b) *Water-solube, acidogenic gases:* Their clathration can be regarded as a hydration process under appropriate circumstances. The solubility in water of a number of such gases is low, as a consequence of their small dipole moments. In this case, there is a possibility for the formation of gas hydrates.

(c) *Water-soluble, polar compounds.*

(d) *Water-soluble ternary or quaternary alkylammonium salts:* In this case, the cation is the guest species, while the anion is incorporated into the host lattice.

Both hydrophobic gases and liquids and water-soluble, acidogenic gases form gas hydrates if they mix well with water at a given temperature and a pressure higher than the decomposition pressure. Precipitation of the crystals occurs at the phase boundary. If the active surfaces are renewed as a result of mixing, the process is rapid, but the product grows slowly and it has a microcrystalline structure.

The water-soluble compounds generally form hydrate crystals with simple structures. However, a study of the phase behaviour is made difficult by the fact that most salts can form various hydrates, depending upon the circumstances.

Table 2.1. Gas hydrate-forming materials occurring most commonly in practice and the corresponding gas hydrates [66–70, 106, 159, 201–203]

Structure and ideal composition	Hydrate-forming material, X				
	hydrophobic guest gases or liquids at 0 °C				
1.2 nm cubic lattice $8X.46H_2O$	Ar H_2S CH_4	Kr PH_3	Xe CH_3F	C_2H_4 O_2	N_2 H_2Se
1.2 nm cubic lattice $6X.46H_2O$	AsH_3 C_2H_4 C_2H_6	N_2O CH_3Cl Cl_2	CH_3Br COS CH_3SH	BrCl Br_2 CH_2CHF	C_2H_5F CHF_3 CH_3CHF_2
1.7 nm cubic lattice $8X.136H_2O$	CH_3I CH_3CHCl_2 CF_2Cl_2	C_3H_8 CH_3CF_2Cl CF_2Br_2	C_2H_5Br CH_3NO_2	$CFCl_3$ C_2H_5Cl	CH_2Cl_2 $(CH_3)_3CH$
1.7 nm cubic lattice $8X.16H_2S.136H_2O$	CH_3Br C_2H_5Cl SF_6 CH_2Cl_2	CS_2 $CFCl_3$ C_6H_6 CCl_4	$(CH_3)_2S$ CH_3CF_2Cl $CHFCF_2$ CCl_3Br	C_2H_5Br CH_3I $CHCl_3$	COS CH_3CHF_2 CCl_3NO_2
	water-soluble acidogenic gases				
1.2 nm cubic lattice $6X.46H_2O$	CO_2	SO_2	ClO_2		

The clathrate compounds with a gas hydrate character belong to groups (a) and (b). A list of the gas hydrates occurring most frequently in practice is given in Table 2.1.

These data were compiled from the publication of Jeffrey and McMullan [106] using the experimental results of de Forcrand [200], Villard [201], von Stackelberg and his co-workers [66–70], van Cleeff and Diepen [159], Glew [202] and Lippert and his co-workers [203].

It was considered by von Stackelberg and Müller [66, 71] that materials able to form gas hydrates must satisfy the following requirements:

(a) the molecules should be of suitable size and shape to enter the cavities in the water host lattice;

(b) their solubility in water should be low;

(c) they should be sufficiently volatile;

(d) they should be of homopolar character;

(e) the van der Waals forces of the molecule should not be too large, its heat of evaporation should be less than 31 400 J mol^{-1}, and the boiling-point should not exceed 60 °C (or for double hydrates 115 °C);

(f) the component forming the hydrate should not contain hydrogen atoms capable of yielding further hydrogen-bonds (this latter condition also includes the limitation that the component forming the hydrate must not be a gas which has a good solubility in water, e.g. NH_3 or HCl, or a liquid readily miscible with water, e.g. CH_3OH).

The structural properties of the gas hydrates are influenced decisively by the arrangement and binding mode of the water molecules forming the host lattice framework, and we therefore have to deal with the structures of ice and water.

2.2. The structures of water and ice

Very wide-ranging studies have been performed in an attempt to clarify the structure of water, but in spite of this, a uniform theory capable of interpreting all the properties of water unambiguously and completely has not yet been developed. The results of these investigations are contradictory or are suitable only for the interpretation of certain phenomena.

The structure of free, monomeric water is well known. The three atoms form an isosceles triangle, with the valence angle scarcely deviating from the regular tetrahedral vertex–centre–vertex angle of 109.5°. The investigations of Mulliken [205] and Lennard-Jones and Pople [206] revealed that an individual water molecule has an electronic structure in which the eight outermost electrons are situated in four elongated ellipsoidal orbitals (Fig. 2.1 (a)). The axes of two of these orbitals coincide with the O—H bonds, which enclose an angle of 104.5°, corresponding closely to the vertex–centre–vertex angle of a regular tetrahedron. The axes of the other two orbitals are perpendicular to the H—O—H plane, and lie in a plane passing through the nucleus of the central oxygen atom of the tetrahedron. The directions of the axes of these two orbitals enclose an angle of approximately 109°, and thus the axes of the four ellipsoidal orbitals are directed towards the vertices of a tetrahedron (Fig. 2.1 (b)), the centre of which does not coincide with

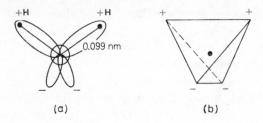

(a)   (b)

Fig. 2.1. The electronic structure of an individual water molecule [204]

the position of the oxygen atom of the water molecule. The equilibrium position of the oxygen atom is shifted from the geometric centre of the symmetrical tetrahedron, in the plane containing the two non-bonding electron lone pairs, towards the vertices of the tetrahedron corresponding to the non-bonding electron pairs. The centre of gravity of the two positive charge centres situated on the periphery of the water molecule, i.e. on the corresponding edge of the tetrahedron, lies towards the protons involved in the two $s–p$ bonded H—O orbitals. At the same time, the lone pairs correspond to the other orbitals; they ensure the linkage of the individual water molecules to one another by means of hydrogen-bonds, and form the centres of gravity of negative charges on this side.

Hydrogen-bonds are formed in bulk water when a hydrogen atom of a water molecule is linked with a lone pair of a neighbouring water molecule. Thus, the oxygen atoms take part as donors to two protons in the hydrogen-bonds between the water molecules, while the two protons act as acceptors. The donor and acceptor effects mutually strengthen each other, so that the individual hydrogen-bonds can also deform one another to a certain extent.

From the relatively early X-ray structural investigations of Dennison [207], Bragg [208] and Barnes [209], but mainly from the fundamental work of Bernal and Fowler [210] and the subsequent investigations (Bernal [211], Bjerrum [204], Pauling [212] and Fox and Martin [213]), the structure of ice is known with an even higher degree of certainty.

Under normal conditions, ice has a hexagonal structure, the ice I structure (Fig. 2.2 (a)), where every water molecule has four immediate neighbours (Fig. 2.2 (b)). This is a quartz-like layer structure, a tridymite lattice with a relatively loose packing, in which three-quarters of the bonds are in a central-symmetrical arrangement (Fig. 2.2 (c)) and one-quarter is in a mirror-symmetrical arrangement

(Fig. 2.2 (d)), and symmetrical cavities are formed between them. Each cavity is surrounded by six water molecules at a distance of 0.294 nm from the centre. Around each water molecule there are six cavities at a distance of 0.347 nm, these forming uninterrupted channels (Fig. 2.2 (e)).

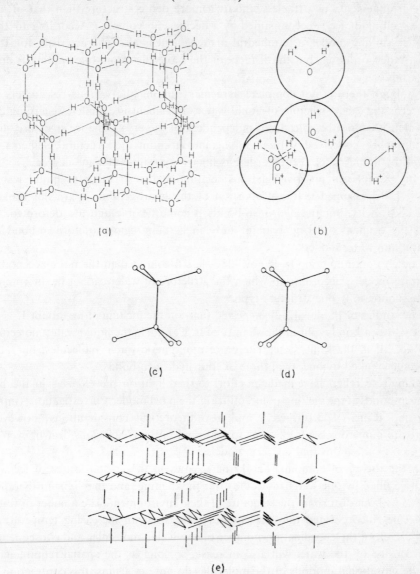

Fig. 2.2. The structure of ice I (a, b, c and d [214]; e [215]). Reproduced by permission of the author, M. V. Tracey, The structure of water. *Austral. J. of Sci.,* Vol. 31. p. 418, Fig. 2

In the investigations of Bernal and Fowler [210], it was found that liquid water also has a modified quartz-like structure. On melting of the ice lattice with long-range ordering, most of the hydrogen-bonds split, but after the collapse of the ice lattice, the water molecules in the liquid state do not assume the most compact arrangement; nevertheless, this is a more dense structure than that of ice. (The melting of ice is accompanied by a decrease in volume.) According to this quasicrystalline model, a tetrahedral arrangement is also characteristic for this short-range ordering, but the distance of the molecules from one another is only 0.276 nm.

Although there is not complete agreement among the various researchers as concerns the exact structure of liquid water, agreement is complete regarding the fact that it cannot be considered independent of the crystalline ice structure, and it cannot be conceived without taking into account the structural elements of ice in the liquid state, also regarded as micro-crystalline to some degree.

For example, in his investigations relating to the structure of liquid water, Pople [216] assumed that a proportion of the hydrogen-bonds are not broken when ice melts, but the hydrogen-bonds become flexible and are deformed, so that the original ordering remains only in certain regions (deformed-bond or continuum water model).

Pople's continuum model is nowadays less favoured than the more acceptable mixture models, according to which the structure of water cannot be interpreted by assuming only one structure type.

The studies of Pauling [212] suggested that, on the melting of ice, about 15% of the hydrogen-bonds split in the vicinity of 0 °C, and so the liquid water molecules are not all in the same state. There are monomeric water molecules, but also molecules linked by one, two, three or four hydrogen-bonds.

Many researchers have made an effort to shed light on the changes in the hydrogen-bonds at the melting-point, utilizing the most modern investigations (infrared and Raman spectra, etc.), and also theoretical considerations; however, there are numerous contradictions to be found in the conclusions drawn, even in the case of the individual mixture models.

In the theory of Samoilov [214], the cavities in the crystal lattice of ice will each be filled by one water molecule after the melting, and they remain there for a time without disturbing the structure of their environment. The number of water molecules incorporated into the cavities increases with increasing temperature, and therefore the coordination number of the water molecules and consequently the density of the water will also increase, so long as the spatial requirements of the vibrational motions of the molecules do not act against this contraction as a result of the temperature rise. The increase of pressure promotes the filling of the cavities in the structure.

This interstitial interpretation of Samoilov is very interesting; for example, Pauling [217] states that the structures of gas hydrates are approximately the same as the water hydrate structure he proposed, with the difference that the cavities in the water are filled not by hydrate-forming foreign atoms or molecules, but by

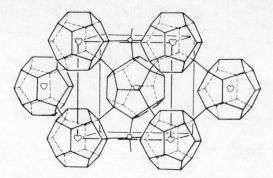

Fig. 2.3. The structure of gas hydrates, according to Pauling [212]

monomeric water molecules (Fig. 2.3). Moreover, the investigations of von Stackelberg and his co-workers [66–70] and Pauling and Marsh [218] prove that, in the crystalline hydrates of non-electrolytes, the bond angles and the mutual distances between the molecules hardly differ from the values corresponding to ice.

In the opinion of Marchi and Eyring [219], in water the molecules are similarly in two states, which undergo interchange with each other continuously. Molecules of the one type have a tetrahedral structure and an ice-like arrangement where the hydrogen-bond predominates, whereas the other type is comprised of monomeric water molecules capable of free rotation, which find room in the cavities in the hydrogen-bonded framework.

Forslind [220] considered that the cavities in the ice-like structure are lattice defects of Frenkel and Schottky type, and at 0 °C and 20 °C about 9% and 19%, respectively, of the lattice sites are vacant.

Several authors have attempted to interpret the special properties of liquid water in terms of the association of monomeric water molecules in the form $(H_2O)_n$. For instance, Eucken [221] claimed that H_2O and molecular aggregates $(H_2O)_2$, $(H_2O)_4$ and $(H_2O)_8$ can be found in water. This opinion has encountered many opponents, because the spatial cavity structure cannot exist side by side with the formation of associations.

In the theory of Frank and Wen [222], in addition to the monomeric water molecules, flickering clusters are present in water, the constant formation and decomposition of which occur in the sequence of formation and splitting of the hydrogen-bonds.

In liquid water, therefore, continuous changes, vibrations, dynamic interactions and transformations take place between the water molecules in clusters and those in the monomeric state. In the view of Frank and Wen [222], the clusters originate when and where "cold" regions with low thermal energy are formed locally as a consequence of the energy fluctuation. These clusters decompose at the sites of local increases in the thermal energy. The molecules are kept together within the

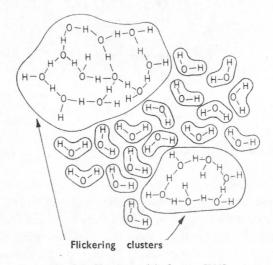

Flickering   clusters

Fig. 2.4. Cluster model of water [222]

flickering clusters by means of hydrogen-bonds, and therefore a lattice structure can develop. At low temperature, the tridymite-like molecular arrangement predominates in all probability, because the number of hydrogen-bonds per unit volume is the highest for this structure.

Cluster models for water with the double structure were proposed by Davis and Litovitz [223], Hall [224] and Némethy and Scheraga [225]. Of these, the investigations of Némethy and Scheraga [225] aroused the greatest interest. They suggested that about 70% of all the water molecules are linked in clusters (23% involving quadruply, 20% triply, 4% doubly and 23% singly hydrogen-bonded forms), and about 46% of the clusters have an ice-like structure, as calculated on the basis of the non-broken hydrogen-bonds (Fig. 2.4).

On the basis of his X-ray diffraction studies, one of the most modern water theories was elaborated by Hajdu [226], as a development of the method of Morgan and Warren [227] and the water structure model proposed following the investigations by Danford and Levy [228]. In the view of Hajdu, the structure of water can be interpreted and described as a closely-packed aggregation of clusters which have a framework with tetragonal symmetry. His model combines the rational elements of the "interstitial" and the "deformed-bond" water models; it provides perhaps the best possibility to date for interpretation of the phenomenon $\varrho_{water} >$ $> \varrho_{ice}$, the experimental results of infrared and Raman spectroscopy, the high relative permittivity of water, its specific heat capacity, etc., and for calculation of the average cluster size, the molecular distances and the coordination number.

Otherwise, the interpretation of water as a non-uniform quasi-crystalline structure has been brought into direct connection with the gas hydrates, in that Malenkov [229] proved that the gas hydrates crystallize in two kinds of structure. He reported that the water molecules are situated at the apices of a polyhedron bounded by hexagonal and pentagonal faces. In the lattice structure of the hydrogen-bonded water molecules, there are cavities of approximately spherical form with radii of 0.39 and 0.43 nm, respectively, in which gas molecules can be incorporated in accordance with the interpretations of Samoilov and Pauling, without seriously disturbing the lattice structure.

This review of the results of researchers who have dealt with the structure of water can obviously not be complete here, but this was not our aim. In this respect we refer to some good monographs and reviews published in recent years [230–232], which treat this question exhaustively. However, the most important results alluded to definitely establish the fact that hydrogen-bonds remain present in the liquid state, and that water molecules mainly linked in this way form a certain quasi-crystalline structure in the liquid state too, which can be described or at least approximated by means of this conception. In essence, we have to set out from this structure and from this conception if we wish to interpret the effects of dissolved materials on water.

2.3. Effects of solutes on the structure of water

Materials coming into contact with water and dissolving in it disturb or even totally change the original water structure brought about by the solvent–solvent interactions. The extent of the change in the structure on the action of the solute, and the resulting changes in the properties, are determined fundamentally by two parameters. One of these main factors involves the nature and the magnitude of the new interactions between the solute and solvent (in the case of electrolytes, for example, ion–solvent interactions and the resulting solvation and hydration), and how these compare to the original solvent–solvent interactions. The other main factor is how the dissolved species (gas atoms, gas and liquid molecules, simple or complex ions) can fit into the original structure of the water, into the cavities in the clusters, or into the place of water molecules. In the case of certain solutes (chiefly ions), especially at high solute concentrations, where the determining role of the water structure is strongly decreased, the importance of the interactions between the individual solute species (in electrolyte solutions the ion–ion interactions) increases to an ever greater extent.

The structure and properties of the solution formed, which clearly depend on the temperature, the pressure and the compositions of the components, are determined essentially by these three types of interaction, the initial structures of the water, and the charge, polarity and size of the individual solute species.

A uniform and fully satisfactory interpretation of the structural changes due to the solute has not yet been achieved. For the time being, it cannot be expected perhaps: depending on the solute and the composition, many variations can occur as regards the ion–ion and ion–solvent interactions, and it is extremely difficult to fit these into a uniform theory.

For a more detailed discussion and interpretation of the interactions of water and solutes, and the related phenomena, the reader is referred to some previously-mentioned [230–232] and more specific monographs and reviews [233–244].

2.3.1. Effects of molecularly dissolved materials
on the structure of water

An important role is generally played in the formation of gas hydrates by molecularly dissolved materials. Since the "iceberg" theory of the interaction between solute and solvent is particularly important from the point of view of gas hydrates, this must be discussed in some detail. From investigations of the heat of dissolution and entropy of dissolution of polar molecules, Eley [245] found that, on the dissolution of gas molecules in water, the water structure contains sufficient cavities in which the gas molecules can fit; in the case of large gas molecules, there is a possibility for the increase in volume of the cavities too. The energy of the hydrogen-bonds split during such cavity enlargement is compensated by that of the new hydrogen-bonds formed around the cavities, or by that of the essentially new structure developing. In the theory of Eley, the incorporation of large, non-polar solute molecules makes the water structure more closely fitting and diminishes the number of the original hydrogen-bonds.

In contrast, Frank and Evans [183] considered that the dissolution of gas atoms or non-polar gas molecules enhances the crystalline character of the water structure by increasing the number of hydrogen-bonds. The investigations by Powell [199] and Powell and Latimer [246] suggested that this increase in the crystalline structure results from the vibration of the solute molecules being diminished by the neighbouring water molecules.

Even though there are certain contradictions between the details of the individual theories, all these investigations demonstrate that an "iceberg" of submicroscopic dimensions is formed around the apolar solute molecules; this is supported by data relating to the changes in enthalpy and entropy. This phenomenon is also known as 'hydrophobic hydration'.

The formation of "icebergs" could be interpreted by Némethy and Scheraga [225] on a statistical thermodynamic basis. For instance, their calculations proved that the proportion of intact hydrogen-bonds around hydrocarbon molecules dissolved at 20 °C is 12% greater than in pure water, and primarily the number of quadruply and doubly hydrogen-bonded structures increases. From the aspect of investigations of the various gas hydrates and their behaviour, special interest is attached to those cases where the solute is not a completely apolar organic substance, e.g. mono-, di- or trihydric alcohols, consisting of molecules similarly linked only by hydrogen-bonds in the liquid phase.

It is evident that, since these compounds contain polar groups, they actively influence the structure of the water, and exert their actions in another manner than do the apolar gases with their quasi-inert molecules; this effect is strongly dependent on the composition and the temperature, but less so on the pressure. The validity of this is illustrated most strikingly by the anomalies (strong local deviations from linear change, or extreme values) observed in the dependences of the transport and other physicochemical properties (mainly the viscosity) of these water–organic liquid mixtures on the temperature and the composition. Particularly interesting from this point of view are the theoretical studies of Erdey-Grúz [233] dealing with transport properties, as well as those of Frank and Ives [247] and Némethy [232] relating to other properties too. Although their work reflects the lack of a fully comprehensive and satisfactory structure theory for even the water–alcohol systems, these researchers agree that, in water–alcohol mixtures of various concentrations, the water and alcohol molecules are interlinked by hydrogen-bonds, these and the structure formed in the solution being very strongly dependent on the concentration.

The water molecule is able to form two independent hydrogen-bonds; even if one of the hydrogen atoms is bonded to the end of a polymeric water molecule chain, or (in terms of the currently most generally accepted cluster ⇌ monomer mixture theory) to the edge of an "ice-like" cluster, the other hydrogen atom can form a hydrogen-bond, with a monoalcohol, for example. Thus, the alcohol molecules more or less seal in and stabilize the structure, decreasing the molecular exchange and creating a system with higher viscosity. With elevation of the temperature, the probability of formation of such water—alcohol bonds diminishes.

Bonding between the water and the alcohol similarly occurs in the case of di- and trihydric alcohols (e.g. glycol and glycerol). However, since these alcohols have additional possibilities for hydrogen-bond formation, they do not seal off the structure as monohydric alcohols do.

The alcohol–water interaction is not limited to the sealing-off effect of the water–alcohol hydrogen-bonds. If it is not too large and its concentration is not too high, there is a possibility for the dissolved alcohol to penetrate into the cavities in the water structure, where it is in a stable situation. In this case, especially because of its size, it nevertheless deforms the original structure of the water. Indeed, some data suggest that the alcohol inside the cavities forms hydrogen-bonds with the neighbouring water molecules.

If some quasi-inert gas is dissolved in a mixture of water and such an organic liquid, the formation of gas hydrate occurs to a lower extent than in the case of pure water, since the cavities in the water structure are partially or totally occupied by alcohol molecules, which generally interact more strongly than the gases with water.

This is even more the case if the solvent is not pure water, but, for example, an electrolyte solution containing "structure-maker" or "positively-hydrating" ions (for more details, see the following section). One such electrolyte solution is the brine in natural hydrocarbon reservoirs.

In this situation, the hydration of the dissolved salt, occurring to an extent depending on its concentration, causes the quantities of both the structured and the monomeric fractions of the water to decrease. The cavities in the remaining structured water are thus filled by less alcohol, and accordingly the possibility of hydrate formation by gas dissolved in the solution is even lower than in the water–alcohol mixtures.

The physicochemical processes of inhibition of hydrate formation by means of alcohol can essentially be interpreted on the basis of the above facts, as well as by taking into consideration the relative magnitudes of the interactions between the individual components.

2.3.2. Effects of ionically dissolved materials on the structure of water. Fundamentals of the structure of aqueous electrolyte solutions

During their dissolution in water, electrolytes dissociate fully or partially into their ions, and an electrolyte solution results. For the interpretation and investigation of the structure and properties of aqueous electrolyte solutions, let us start from two fundamental conceptions, referring to the water, as solvent, and to the dissolved ions. For this purpose, the modern models can be summarized most simply, but with good regard to the essence, if the structure of water is interpreted in the limiting case as a mixture of two fractions with differing structural states. One involves molecules bound by hydrogen-bonds; this is a "structured" water fraction (ice-like, existing in clusters), which in a statistical sense is in a quasi-crystalline state. The other is a monomeric water fraction, consisting of individual water molecules practically without hydrogen-bonds [230–235]. The water molecules in these two fractions are in dynamic equilibrium with each other [248, 249] this being characterized by the equilibrium constant.

$$K = [H_2O]_{structured}/[H_2O]_{monomeric}$$

On elevation of the temperature, the hydrogen-bonds undergo progressive splitting, so that the quantity of the structured water fraction and the number of its individual species decrease, leading to a lower value of K. As a consequence of the hydration occurring in the presence of ions, since the probability of hydration by the monomeric fraction is greater than that by the structured fraction, the number of free monomeric water molecules decreases more strongly in the latter case. In order that the value of K should remain constant at a given temperature, this is followed by a corresponding decrease in the structured water fraction. Thus, the quantity of the structured fraction decreases both when the temperature is raised, and when ions are dissolved in the water.

The other fundamental conception is that the ions can be divided into two groups on the basis of their effect on the structure of water: the "structure-maker" [250] or "positively-hydrating" [241] ions in one group, and the "structure-breaker" [250] or "negatively-hydrating" [214] ions in the other.

In general, the ions with large z/r^2 values (where z is the charge number, and r is the ionic radius), e.g. Li^+, Na^+, Mg^{2+}, Ca^{2+}, F^-, etc., belong in the group of structure-maker or positively-hydrating ions, while the ions with low z/r^2 values (mainly larger ions of low charge), e.g. K^+, NH_4^+, Cs^+, Cl^-, etc., are structure-breaker or negatively-hydrating ions (see Table 2.2).

These two kinds of terminology follow from the different approaches in the two currently most widely accepted possibilities for the interpretation of the phenomenon of hydration: those of Frank and Wen [250] and Samoilov [214].

In the view of Frank and Wen [250], ions dissolved in water form a new structure, consisting in the first hydrate layer (and possibly in the second layer too in the case of small ions of high charge) or radially strongly oriented, quasi-motionless hydrated water molecules, bound by the ion (A region). This is followed by a second (B) region, where the water molecule dipoles are oriented by the field of the ion, but to a much lower extent than in the A region or in the bulk water itself. In the third (C) region around the ion, the field of the ion is scarcely felt; at best it polarizes the water molecules a little better. It is evident that there is not a dis-

Table 2.2. Data relating to various ions and salting-out activities [158, 298]

		Al^{3+}	Be^{2+}	Mg^{2+}	Ca^{2+}	Li^+	Na^+	K^+	NH_4^+	Cs^+	F^-	Cl^-	NO_3^-	SO_4^{2-}
1	z	3	2	2	2	1	1	1	1	1	1	1	1	2
2	r(nm)	0.050	0.031	0.065	0.099	0.060	0.095	0.133	0.143	0.169	0.136	0.181	0.189	0.230
3	$\dfrac{z}{r^2}$ (nm^{-2})	1200	2081	473	204	278	111	57	49	35	54	31	28	38
4	Molar mass of cation	27	9	24	40	7	23	39						
5	A_m	5.26	5.88	2.56	1.89	1.28	1.02	0.75						
6	A_p	0.20	0.65	0.11	0.05	0.18	0.04	0.02						
7	Molar mass of chloride salt	133.5	80	95	111	42.5	58.5	74.5						
8	$A_{M_a Cl_b}$	0.041	0.073	0.028	0.018	0.030	0.016	0.010						
9	$A'_{M_a Cl_b}$	322	700	176	86	155	70	43						
10	$A''_{M_a Cl_b}$	2.4	8.7	1.8	0.8	3.6	1.2	0.6						

A_m, A_p : salting-out activity referred to 1 mole and 1 gram cation, respectively.
For definitions of $A_{M_a Cl_b}$, $A'_{M_a Cl_b}$ and $A''_{M_a Cl_b}$, see Equations (3.10), (3.11) and (3.12).

tinct border between the different regions, and hence their extents cannot be given exactly. However, for structure-maker ions the formation of region A is characteristic, as a consequence of the strong orientation effect, while for structure-breaker ions region B predominates, the radially orienting interaction then being much lower; in fact, the interaction is the weaker, the larger the ion.

In the classification introduced by Samoilov [214], the ions are divided into two groups on the basis of the change in mobility resulting when they replace one water molecule in the water structure; the mobility of the water molecules in the environment of the given ion will be lower (positive hydration) or higher (negative hydration) than it would be if the original water molecule had remained (i.e., if a water molecule were present at the position of the selected ion, taken as centre). The approach by Samoilov has the advantage that it better takes into consideration the actually fully dynamic character of liquid water, in spite of its quasi-crystalline state.

In the case of aqueous electrolyte solutions, the main factors determining the structure and properties are the extent and nature of the ion–solvent and ion–ion interactions. In addition, the solvent–solvent interactions can also play a role (which decreases as the concentration of the solute is increased), but only until the composition corresponding to "the limit of full hydration" is reached in the course of the addition of solute. This concept, introduced by Mishchenko [251], is the limiting concentration at which a complete hydration sphere is just able to form around all of the dissociated ions of the added electrolyte, and the "free" water with the original structure has just been totally consumed from the system. If further electrolyte is added, a new situation arises, for the dissociation of the electrolyte then occurs at the expense not of the free water, but of the water already bound in the hydrate sheath. From this point on, the further increase of the solute concentration leads to a decrease in the hydration number (this is essentially the number of water molecules within the range of action of the ion). As regards a cation and an anion of the same size, the cation is generally the more strongly hydrated (because of the structure of the water dipoles), and hence the hydration numbers of anions are normally smaller than those of cations.

The hydration number, which is to some extent characteristic of the properties of the electrolyte, is thus itself concentration-dependent. Even though they relate to the same concentration value, the hydration numbers determined by various authors by theoretical means, or from the dependence of various physical and chemical properties on the concentration and the temperature, often differ from one another.

One cause of this is that the hydrate sheath takes part in different ways in determining the various properties (e.g. the electric conductivity, viscosity, molar volume, relative permittivity, thermodynamic functions, etc.). In addition, with regard to the value of the hydration number, and hence the determination of the limiting concentration of full hydration, a decisive role is played by what is considered to be water really bound in the hydrate sheath. Should it be taken only as that which, radially oriented in a geometrically arranged manner, is linked directly to the immediate surface of the ion (i.e. region A), or should it perhaps include the much more weakly bound and oriented water molecules in region B too?

Accordingly, although the hydration of ions through ion–solvent interactions is an unquestioned fact, the clear-cut determination of the extent of hydration is problematical; in many cases it can be characterized only by the primary effects produced by hydration (e.g. the thermal effect and the molar volume change).

No matter which hydration model is considered, after the limit of full hydration is reached, subsequent addition of electrolyte leads to the progressive break-down of the maximum hydrate sheath. During this, the ions come nearer and nearer to one another, the ion–ion interactions increase, and the new structure, determined virtually only by the ion–ion interactions, and at most merely modified by the ion–solvent interactions, becomes more and more predominant. With further increase of the concentration, a structural preordering develops, tending towards the structure of solid phase in equilibrium with the saturated solution [252].

The investigations of Berecz [253] indicated that the linearly-increasing addition of electrolyte to water, or even to some aqueous electrolyte solution, does not cause the primary properties of the solution, nor the secondary ones calculated from them, to change in a linear way. In the concentration range between $m_{salt} = 0$ and $m_{salt} = m_{sat.}$, the curves of the concentration-dependence of the individual properties (mainly the calculated secondary ones) can be divided into two or more, practically linear sections with different slopes. This shows that the change in structure of the aqueous electrolyte solution in the case of a uniform increase in the salt concentration does not occur uniformly. Each of the linear sections observed in the curves relates to a certain solution structure, which predominates in the given section of the concentration range, but passes over into another predominating solution structure at a certain limiting concentration.

For chlorides of cations with large z/r^2 values (e.g. Li^+ or Mg^{2+}), four such predominating sections can generally be found [252], whereas for chlorides of cations with small z/r^2 values (e.g. K^+, NH_4^+ or Cs^+), only two, or at most three such predominating structure intervals occur. In the case of gas hydrates, this clearly results in the fact that a salt added to water to inhibit hydrate formation does not exert a uniformly increasing inhibitory effect, even if its concentration is increased uniformly. Below the limit of full hydration, when the structural effect of the water predominates, the effect of the salt will evidently be different from that at a concentration above this limit, when only the water bound in the hydrate sheath is present.

The effect of a salt on non-electrolyte solutions (e.g. aqueous solutions of gases or alcohols) is that the less polar material is salted out, as it were, its quantity in the solution decreasing. The cause of this is that the salt interacts much more with the water than with the apolar, or less polar solute. Thus, at a given temperature, the constancy of the activity of the saturated solution results in an increase of the activity coefficient of the less polar material, or in a decrease of its concentration. This effect is one of the factors in the process of inhibition of gas hydrate formation by means of an electrolyte. The end-conclusion is that the stronger the interaction of the electrolyte with the solvent as compared to that of the given gas hydrate-forming component, the greater the effectivity of inhibition at a given salt concentration.

The real structure of an electrolyte solution, and the nature of the possible species in it, are actually even more complicated than outlined above. Depending on the concentration, not only simple ionic species, but also more complex charged and uncharges species may be present in the solution.

As soon as the predominating role of the ion–solvent interactions becomes sufficiently weakened and the ion–ion interactions become strong enough, ion associations can occur and ion pairs can be formed. With the further decrease of the ion–solvent interactions, the ions can interact with the ion pairs too, leading to the formation of triple ions. If the ion–solvent interactions are very weak because the dielectric constant of the solvent is small, the ion pairs can interact with each other, and quadrupole or even multiply-charged complexes or neutral polyionic species can arise.

Consequently, if a pure solid electrolyte is added to a solvent of high dipole moment and not too great particle size, ion association does not generally occur because of the considerable ion–solvent interaction. If strong electrolyte is added to a solvent of low relative permittivity, ion association does occur to an increasing extent with decreasing electrolyte concentration. The lower the relative permittivity of the solvent, the greater the degree of association for a given electrolyte addition.

To summarize: the properties of electrolyte solutions are mainly determined by the ion–ion and ion–solvent interactions. These interactions depend on the size, charge, structure and polarizability of the ion, and the size, structure, dipole moment and polarizability of the solvent molecule. Besides the questions of economy, corrosion and other technologically interesting problems, these latter physical factors govern the selection of the materials necessary for inhibition of gas hydrate formation, and their optimum quantity.

2.4. Structures, compositions and systematization of crystalline gas hydrates

The study of the forms and dimensions of the cavities formed on the linkage of tetrahedral water molecules is closely connected with the investigation of the structures of gas hydrates. Claussen [78, 79] demonstrated that, if water molecules are linked in such a way that every molecule is situated approximately at the centre of the tetrahedron formed by the four neighbouring water molecules, cavities of various sizes can be formed. The simplest of these is the regular pentagonal dodecahedron. This has 12 regular pentagonal faces (F), 20 vertices (V) and 30 edges (E). By Euler's theory relating to convex polyhedra, therefore:

$$12F + 20V = 30E + 2$$

This water structure includes a unit with composition $H_{40}O_{20}$, where oxygen atoms can be found at the vertices, and hydrogen-bond linkages O—H ... O are formed on the edges. In this way, 30 of the 40 hydrogen atoms are located on the edges.

As a result of minor deformations, other structures containing cavities can be formed besides the regular pentagonal-dodecahedral ones. The cavity types occurring in the water structure were systematized by Jeffrey and McMullan [106] in the following way (Fig. 2.5):

1. The regular pentagonal dodecahedron (D).
2. The tetrakaidecahedron (T), a body with 2 hexagonal and 12 pentagonal (in all 14) faces, involving 24 water molecules:

$$14F + 24V = 36E + 2$$

3. The pentakaidecahedron (P), with 3 hexagonal and 12 pentagonal faces, involving 26 water molecules:

$$15F + 26V = 39E + 2$$

4. The hexakaidecahedron (H), with 4 hexagonal and 12 pentagonal faces, involving 28 water molecules:

$$16F + 28V = 42E + 2$$

The enumerated polyhedra can be combined so as to have common faces in four different ways:

5. Four tetrakaidecahedra $(4T)$, with 40 pentagonal and 4 hexagonal faces involving 70 water molecules:

$$44F + 70V = 112E + 2$$

No.	Symbol.	Denomination	Shape	No.	Symbol.	Denomination	Shape
1	D	Pentagonal dodecahedron $12F + 20V = 30E + 2$	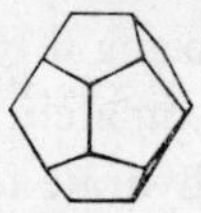	5	4 T	4 Tetrakaidecahedron $44F + 70V = 112E + 2$	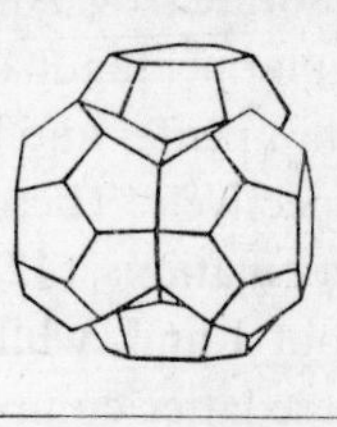
2	T	Tetrakaidecahedron $14F + 24V = 36E + 2$	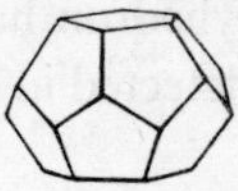				
3	P	Pentakaidecahedron $15F + 26V = 39E + 2$	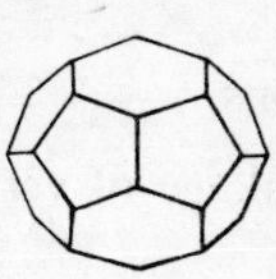	6	3 T 1 P	3 Tetrakaidecahedron + 1 Pentakaidecahedron $45F + 72V = 115E + 2$	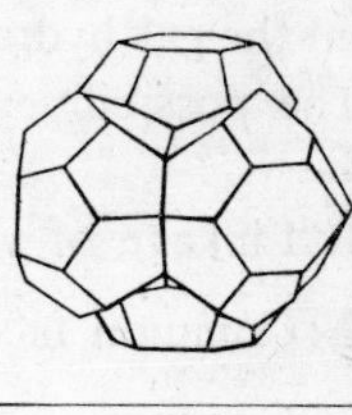
4	H	Hexakaidecahedron $16F + 28V = 42E + 2$	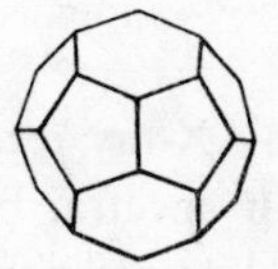	7	2 T 2 P	2 Tetrakaidecahedron + 2 Pentakaidecahedron $46F + 74V = 118E + 2$	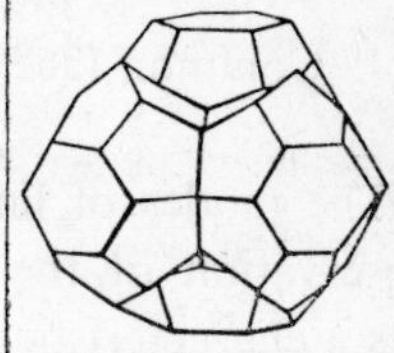

Fig. 2.5. Cavity types occurring in the water structure [106]

6. Three tetrakaidecahedra + one pentakaidecahedron ($3T.1P$), with 40 pentagonal and 5 hexagonal faces, involving 72 water molecules:

$$45F + 72V = 115E + 2$$

7. Two tetrakaidecahedra + two pentakaidecahedra ($2T.2P$), with 40 pentagonal and 6 hexagonal faces, involving 74 water molecules:

$$46F + 74V = 118E + 2$$

8. An even more irregular polyhedron, formed by the linkage of 96 water molecules, with 48 pentagonal, 10 hexagonal and 2 rhomboidal faces:

$$60F + 96V = 154E + 2$$

These regular or less regular cavities are capable of holding guest molecules by clathration.

For an approximation to the regularity of polyhedra, a calculation method was elaborated by Allen [254]. According to this, the smallest cavity is that of the regular pentagonal dodecahedron, its volume being 0.168 nm^3. The volumes of the tetra-, penta- and hexakaidecahedra are approximately 0.230, 0.260 and 0.290 nm^3, respectively. These cavities play a role in gas hydrate formation. The larger cavity combinations ($4T$, $3T.1P$ and $2T.2P$), as complex polyhedra, have volumes of about 1 nm^3, while the largest, irregular polyhedron has a volume of 1.60 nm^3. These latter complex polyhedra can only be detected in clathrate systems differing from the gas hydrates.

The possibilities for the linkage of pentagonal dodecahedra are:

1. Linkage of vertices in three dimensions.

2. Common faces in two dimensions; linkage of vertices in third dimension.

3. Common faces in three dimensions.

4. Common face and vertex linkages.

The modes of linkage can be detected by X-ray examinations, by the direct observation of the single-crystal diffraction spectra, because each distribution has a characteristic diffraction symmetry and periodicity.

The first group includes the cubic structures produced when 2 pentagonal dodecahedra and 6 tetrakaidecahedra join together with formation of a host structure consisting of 46 water molecules; these cubic structures have a lattice constant of 1.2 nm, which is very characteristic of the simple gas hydrates.

The deformation-free pentagonal dodecahedra are not able to combine through face-face linkages, but if the distortion of the angle between the faces reaches 3°, linkage results in a hexagonal system belonging to the second group. An example is the (iso-C$_5$H$_{11}$)$_4$NF.38H$_2$O system.

Among the hexagonal faces of a polyhedron consisting of common faces in three dimensions, the water molecules are arranged in such a manner that the polyhedron enclosed by the faces is mainly a pentagonal dodecahedron or hexakaidecahedron. This includes the large group of gas hydrates with cubic crystalline structure and a lattice constant of 1.7 nm, where the unit cell contains 136 water molecules and consists of 16 pentagonal dodecahedra and 8 hexakaidecahedra. Apart from the gas hydrates, other clathrate compounds do not normally belong in this linkage group, unless the structure is strongly deformed.

Other forms of cubic face and vertex linkages lead to host structures not characteristic of the gas hydrates.

Studies of the structures of the clathrate and semiclathrate hydrates formed from three-dimensional convex polyhedra were also reported by King [255].

The X-ray investigations by von Stackelberg and his co-workers [66–70] revealed two basic types of gas hydrates. Both of them exhibit a cubic elementary lattice structure.

The first hydrate crystal lattice type (H_1): this type characteristic of hydrates of gases with relatively small molecules. The lattice constant of the unit cell, formed by the linkage of 46 water molecules, is about 1.2 nm. The two pentagonal-dodeca-

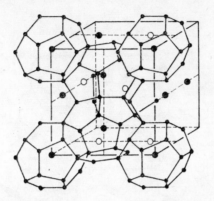

Fig. 2.6. Linkage of cavity structures of types
$1D$ and $2T$ in hydrate crystal of type H_1 [66–70]

hedral cavities are spherical in form, and are surrounded by 20 water molecules at a distance of 0.42 nm from the centre of the cavity (see Fig. 2.5(1)). If the size of the water molecule ($r=0.13$ nm) is taken into consideration, a free cavity radius of about 0.29 nm is available for guest molecules here. Consequently, these cavities can accommodate only those gas molecules with molecular diameters $\leqslant 0.58$ nm. The coordination number of the six tetrakaidecahedral cavities (see Fig. 2.5(2)) is 24. Their structure can be considered rather ellipsoidal, because the distance of 8 water molecules from the centre is 0.37 nm, that of another 8 is 0.48 nm, and that of the final 8 is 0.52 nm; the mean distance from the centre is therefore 0.46 nm. By subtracting from this the molecular radius of water, 0.33 nm is found for the average free cavity radius. Consequently, in this cavity type, molecules with a diameter of 0.66 nm can be incorporated. The eight cavities in the hydrate are thus not equivalent geometrically. The interlinkage of the two sorts of cavity structure is shown in Fig. 2.6.

In the event of total occupation of the cavities, the empirical formula of the resulting gas hydrate is $8X.46H_2O$ or $X.5.75H_2O$ (where X denotes the hydrate-forming molecule). In this case, two gas molecules occupy the small cavities of the dodecahedra, while six occupy the larger (T) cavities. However, if the diameter of the gas molecules to be incorporated is larger than the capacity of the small cavities (0.58 nm), only the larger cavities are occupied, and the composition will then be $6X.46H_2O$ or $X.7.67H_2O$. For gas mixtures where the system contains molecules with diameters larger and smaller than the critical molecular diameter, double or mixed hydrates can be formed, with composition $6X.2Y.46H_2O$.

The second hydrate crystal lattice type (H_{II}): in this type the unit cell consists of 136 water molecules, enclosing 16 pentagonal-dodecahedral and 8 hexakaide-

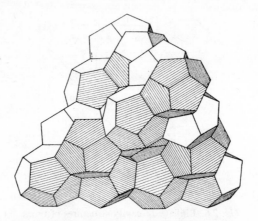

Fig. 2.7. Schematic representation of the structure
of gas hydrates of type H_{II} [66–70]

cahedral cavities. The coordination numbers of these cavities are 20 and 28, respectively. The lattice constant is about 1.73 nm, and the free cavity diameters are 0.48 and 0.69 nm, respectively.

In this hydrate type, which is more complicated than the previous one, the composition of the resulting hydrate in the case of small guest molecules is $24X.$ $.136H_2O$ or $X.5.66H_2O$ if all the cavities are occupied. If the gas molecules can be accommodated only in the larger cavities, the composition is $8.X.136H_2O$ or $X.17H_2O$. From gas mixtures, a composition $8X.16Y.136H_2O$ or $X.2Y.$ $.17H_2O$ can be formed.

The schematic structure and unit cell of gas hydrates of type H_{II} are shown in Fig. 2.7.

Experience indicates that a proportion (about 10–20%) of the cavities that are theoretically available remain unoccupied during hydrate formation; accordingly, the above hydration number values of 5.75, 7.67 and 17, respectively, can be considered only as lower limits.

On this basis, the gas hydrates can be classified in fact among the non-stoichiometric clathrate compounds, since it is not absolutely necessary that all the cavities should be occupied for a stable crystalline compound to be formed. However, the investigations of Barrer and Ruzicka [100] indicate that consideration must be paid to the assumption that, from the point of view of the formation of cohesion forces, the development of a stable clathrate requires that at least the majority of the larger cavities should be occupied.

Glew [256] found that a possible cause of the non-stoichiometry of gas hydrates is either that the bulk of the larger guest species impedes their occupation of the smaller cavities in any case, or the crystalline structure is built up too quickly, before the gas atoms or molecules can all occupy suitable positions for enclosure.

Of the compounds with ideal stoichiometry, double hydrates with a 1.2 nm cubic structure and composition $6X.2Y.46H_2O$ have not yet been detected.

In the 1.7 nm cubic lattice for the composition $24X.136H_2O$, the structure is a very special one. Guest molecules are necessary here which are small enough to fit between the dodecahedra too, but the molecules of such type rather form a cubic structure with a lattice constant of 1.2 nm.

Under equilibrium conditions, the composition $X.5.75H_2O$ is best approximated by the hydrates of Ar, Kr, Xe, CH_4, H_2S, H_2Se, PH_3 and AsH_3, when every cavity in the 1.2 nm cubic structure is occupied. In the case of Cl_2, C_2H_6 and ethylene oxide, mostly only the larger cavities of the H_I hydrate type are filled, and thus the water content of the gas hydrate will be between 5.75 and 7.67 molecules per gas molecule.

In the 1.2 nm cubic structures formed with the larger gas molecules (e.g. CO_2 or COS), only the large cavities can be filled; with molecules the size of which approximates to the cavity size of the hexakaidecahedron (e.g. CH_3Br, CH_3CHF or Br_2), the value of the hydration number will vary, depending on the degree of occupation of the cavities, but it will always be greater than 7.67.

The molecular size of compounds forming a 1.7 nm cubic structure is too large for the pentagonal dodecahedra too to be filled, and therefore varying hydration numbers are mostly found here; thus, the nomenclature non-stoichiometric compounds best fits this type of gas hydrate.

The structures and compositions of the polyhedral clathrate hydrates were studied by Klotz [257], who classified these substances in three main groups, with general formulae $X_2X_6.46H_2O$, $X_8X_{16}.136H_2O$ and $X.(5-40)H_2O$. This systematization differs only formally from those described previously.

A very important task in investigations relating to gas hydrates is the determination of the stoichiometric limits of existence of a stable structure in the case of guest molecules of various sizes and geometrical shapes. However, very many of the relevant experimental data are contradictory. One instance of this is the case of ethylene oxide hydrate. Examinations of the electron density of the guest molecule on a single-crystal gave occupancy factors of 1.0 for the tetrakaidecahedron, and 0.2 for the dodecahedron, which corresponds to a coordination number of

$7.2H_2O$. At the same time, the calculations of Glew and Rath [258], based on accurate density measurement, indicated that the composition of the hydrate system under conditions corresponding to atmospheric pressure is C_2H_4O. $.6.75H_2O$, or at the eutectic composition $C_2H_4O.(7.21\pm0.07)H_2O$. This latter composition is equivalent to the lower occupancy limit.

For gases or liquids soluble in water, certain anomalies occur, because the interactions of the water with the solute influence the formation of the host lattice. Glew and Rath [258] assumed that the hydrate of ethylene oxide can form a 1.2 nm cubic structure with the ideal composition $6X.40H_2O$ ($X.6.67H_2O$). The crystal structure analysis of ethylene oxide hydrate by McMullan and Jeffrey [105], however, demonstrated that no hydrogen-bonding can be detected in the guest–host linkage of the compound, and that the clathrate properties of the hydrate coincide with those of the hydrates of various cyclopropane derivative.

The accurate determination of the compositions of gas hydrates is otherwise a very difficult task: direct analysis can lead to uncertain results, partly because of the decomposition of the hydrate under the conditions of sampling, and partly because inclusions of mother liquor can be incorporated in the structure. Hence, even the least inaccuracy can point to the non-stoichiometric character of the compound. This can also be caused by the conditions of preparation, sampling or analysis. This is presumably why the results of the various authors differ as regards the concrete compositions of the individual gas hydrates.

The basic parameters of the systematization of von Stackelberg and his co-workers [66–75], based on X-ray diffraction investigations of gas hydrates, are listed in Table 2.3.

In this way, therefore, von Stackelberg and his co-workers divided the gas hydrates into five sub-groups:

1. *Simple gas hydrates* and mixed crystals of these. This group contains hydrates of gases with an atomic or molecular size less than 0.58 nm, with an ideal composition $X.5.75 H_2O$.

2. *Liquid hydrates*. This includes the hydrates of readily liquefying gases with molecular diameters of 0.50–0.69 nm. They contain 7.67, 5.67 or 17.0 water molecules per gas molecule, depending on the hydrate structure formed and on the cavity types undergoing occupation.

3. *Double hydrates*. In the event of the formation of structure type H_{II}, the larger cavities are occupied by larger molecules, and the smaller ones by H_2S or H_2Se molecules. The resulting compound has the composition $X.2H_2S.17H_2O$ or $X.2H_2Se.17H_2O$, and forms a single phase.

4. *Mixed hydrates*. In these gas hydrates, with structure H_{II}, the small cavities can be occupied by other gas molecules (e.g. H_2, N_2, O_2, CO_2, etc.), but the final composition of the hydrate is determined here by the gas pressure, and

Table 2.3. Basic parameters of systematization of hydrates [66–75]

Type of hydrate	Sign of type	Lattice constant (nm)	Small cavities		Large cavities		Ideal composition
			number	diameter (nm)	number	diameter (nm)	
Gas hydrate	H_I	1.19–1.23	2	0.58	6	0.66	$8X . 46H_2O$ or $X . 5.75H_2O$
Liquid hydrate	H_I		2	0.58	6	0.66	$6X . 46H_2O$ or $X . 7.67H_2O$
Mixed or double hydrate	H_I		2	0.58	6	0.66	$6X . 2Y . 46H_2O$
Liquid hydrate	H_{II}	1.73–1.75	16	0.48	8	0.69	$24X . 136H_2O$ or $X . 5.67H_2O$; or $8X . 6Y . 136H_2O$ or $X . 17H_2O$
Mixed or double hydrate, or hydrate mixture	H_{II}		16	0.48	8	0.69	$8X . 6Y . 136H_2O$ or $X . 2Y . 17H_2O$

stoichiometric formulae such as can be written for the double hydrates are not necessarily observed.

5. *Mixtures of mixed hydrates.* Here, solid solutions are formed through the mixing of hydrates with structures H_I, and H_{II}.

This systematization of von Stackelberg and his co-workers, made on the basis of their X-ray powder diffraction analysis and single-crystal investigations, has proved to be so fundamental and durable that, even today, it is referred to in virtually all publications relating to this topic. These results are supported by the more recent hydrate structure researches, and essentially only minor correction and enlargement of these basic systems has followed, or is possibly to be expected in subsequent gas hydrate investigations.

2.4.1. Simple gas hydrates (H_I)

Simple gas hydrates can be formed most explicitly at a guest molecule size of less than 0.50 nm. Their less regular variations occur at a maximum molecular size of 0.58 nm. In reality, larger gas molecules than this cannot be incorporated in this hydrate structure. The cause of this is presumably that, in the course of the development of the tetrakaidecahedra forming the larger cavities, they are to some extent deformed, and therefore their free cavity diameter can lie in the range 0.52–0.58 nm; thus, even gas particles only a little larger than 0.50 nm are not always accommodated in the smaller cavities.

If the sizes and numbers of the cavities are taken into consideration for this hydrate system, in the event of full occupancy of the cavities the ideal composition per hydrate-forming component is X.5.75 H_2O, which corresponds to 8X.46 H_2O.

This composition approximates to that found experimentally by Villard [259], but it does not attain the mean water quantity content of 6 H_2O that he observed. The difference can be understood, however, if it is assumed that, in the course of the experiments, it is not sure that the simultaneous, full occupancy of all the cavities does occur; consequently, the number of water molecules per hydrate-forming component will be higher.

The investigations by Scheffer [260], de Forcrand [35] and Braun [261] showed that the very small krypton (0.39 nm) and hydrogen sulphide (0.41 nm) always form hydrates containing less than six water molecules per gas molecule.

Studies on the hydrates of the larger chlorine (0.517 nm) and sulphur dioxide (0.50 nm) began in the early research period. In 1823, Faraday [262] produced a hydrate of composition $Cl_2.10H_2O$, with a chlorine content of 27.7%. In a large chlorine excess, Maumené [10] obtained chlorine hydrates containing 4 and 6.67 water molecules per chlorine molecule. In 1902, on the basis of thermodynamic considerations, de Forcrand [263] proved that the water content of chlorine hydrate can be $6.91–7.12H_2O$. These data led Villard [259] to the conclusion that a large family of gas hydrates have the same structure, and can be characterized by the general formula $X.6H_2O$. He included sulphur dioxide hydrate in this family, despite the fact that the results of de la Rive [6], Pierre [7], Döpping [264], Schoenfeld [8], Roozeboom [265], Geuther [266] and de Forcrand [263] between 1829 and 1902 had indicated a composition of $SO_2 \cdot (7–10)H_2O$. In support of the investigations by Tammann and Kriege [40], it is now evident that the Villard formula holds well for these two hydrates, and that the water content of $6H_2O$ per hydrate-forming component corresponds to an occupancy of 96% of the cavities.

A number of researchers have dealt with the determination of the composition of the hydrate of bromine, which has a molecular length of 0.568 nm. The hydrate composition $Br_2.10H_2O$ early postulated by Löwig [267], Alexeyeff [5] and Roozeboom [27] was modified by Giraud [268] to $Br.8.0H_2O$, by Mulders [269] to $Br.8.4H_2O$, and later by Zernike [270] to $Br.7.0H_2O$, presumably as a consequence of the development of the technical experimental possibilities. From this composition, it can be concluded that only the six larger cavities are occupied here, the two smaller ones remaining vacant. The X-ray examinations of bromine hydrate single-crystals by Allen and Jeffrey [271] indicated that the lattice structure is saturated at a composition $Br_2.(8.47\pm0.05)H_2O$, a formula confirmed by density measurements. A similar phenomenon is experienced in the case of CH_3Br (0.533 nm), as verified by van der Waals and Platteeuw [272]. If certain of their properties (primarily the critical molecular size and the stability) are taken into consideration, these materials can also be classified as liquid hydrates, but on the basis of the lattice structure they are hydrates of type H_I.

2.4.2. Liquid hydrates (H_{II})

This structural type, involving crystals of composition $X.17H_2O$, is produced by the hydrates of the larger hydrate-forming molecules, such as C_3H_8, iso-C_4H_{10}, CH_3Cl and other relatively volatile liquids, if the hydrate-forming component occupies only the large cavities. Detailed investigations of hydrates of this type have been carried out mainly by Tammann and Kriege [40], Giraud [268] and Harris [273], but the results of Claussen [79] are particularly important as regards the elucidation of the structure.

The X-ray studies by von Stackelberg and Jahns [70] demonstrated that there is a decrease in the molar volume of the lattice-forming water in the liquid hydrates, in spite of the fact that the cavity size is larger in the H_{II} system, with a lattice constant of 1.7 nm, than in the H_I system; here the H_2O —H_2O bond distances decrease from 0.283 nm to 0.275 nm. This results in the decrease in size of the smaller cavities and at the same time a contraction tension arises in the large cavities.

2.4.3. Double hydrates

The group of double hydrates displays characteristic features. Here, the framework is mainly of type H_{II}, corresponding to the liquid hydrates (though hydrate type H_I may occur rarely). If the larger cavities are occupied by large gas molecules, and the small cavities by H_2S or H_2Se molecules, the resulting composition is $X.2H_2S.17H_2O$ or $X.2H_2Se.17H_2O$; this stoichiometry is a constant one [66]. In place of the H_2S or H_2Se, other molecules (CO_2, N_2 or O_2) may be situated in the small cavities, but the above constant composition can then no longer be observed; instead, a solid solution is formed with varying composition, depending on the conditions. These latter hydrates are not double hydrates, but are known as mixed hydrates.

In the formation of double hydrates, it can be assumed that the H_2S or H_2Se molecules are bound in the cavities of the water molecule lattice with higher energy than other hydrate-forming molecules.

The existence of double hydrates was postulated by Loir [26] in 1852. Then, in 1883, de Forcrand [200] produced about 30 double hydrates containing hydrogen sulphide, mainly with some alkyl halide as the other component. His results are given in Table 2.4, together with the findings of von Stackelberg and Müller [66] and the lattice constant data of Müller [274].

Table 2.4. Main data on some double hydrates containing H_2S [66. 200. 274]

Double hydrate-forming component	Boiling point (°C)	Dissociation temperature of double hydrates at 10^5 Pa (°C)	Lattice constant (nm)
COS	−50	0	1.730
C_3H_8	−45	+8	1.740
CH_3Br	+5	—	1.731
C_2H_5Cl	13	7.2	1.726
$(CH_3)_2S$	38	14.0	1.739
C_2H_5Br	38	13.0	1.726
CH_2Cl_2	42	14.0	1.728
CH_3I	43	14.2	1.737
CS_2	46	8.0	1.730
$CHCl_3$	61	16.3	1.729
$n\text{-}C_3H_7Br$	71	—	1.742
CCl_4	77	17.3	1.746
C_6H_6	80	3.5	1.748
$CH_2Cl.CH_2Cl$	82	—	1.751
CCl_3Br	104	11.3	1.757
CCl_3NO_2	112	13.0	1.760

2.4.4. Mixed hydrates

If the empty small cavities of liquid hydrates formed by large components are filled with small gas molecules (with the exceptions of H_2S and H_2Se), mixed hydrates are produced. Under certain conditions, mixtures of mixed hydrates can be formed too; this involves the mixing of different (H_I and H_{II}) hydrate

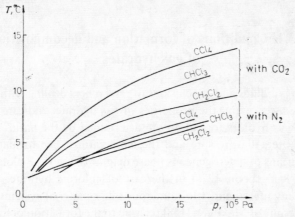

Fig. 2.8. Stabilizing effects of the auxiliary gases N_2 and CO_2 [37, 201]

structures. These systems do not have stoichiometric compositions. The quantity of small gas molecules incorporated depends strongly on the pressure.

It is an interesting phenomenon that the stability of a hydrate is generally enhanced by gas incorporation of this type, i.e. these complex hydrate systems can exist at higher temperatures under isobaric conditions, and it can even occur that species originally incapable of hydrate formation yield stable hydrates in the presence of such gases with small molecules. As an example, in the presence of oxygen the hydrate of iodine is stable up to a temperature of 8 °C at a pressure of 3.34×10^7 Pa. In the nomenclature of Villard [37], this phenomenon is known as the stabilizing effect of auxiliary gases (though Villard stated that the auxiliary gases do not participate in the hydrate formation). The degree of stabilization depends on the nature and the pressure of the auxiliary gas. The stabilizing effects of nitrogen and carbon dioxide are shown in Fig. 2.8, again on the basis of measurements by Villard [201].

The formation of mixed hydrates can also be caused by the presence of air, when nitrogen or oxygen is incorporated into the small cavities. In all probability, the hydrates of various naturally-occurring gases (primarily natural gas), the constituents of which form hydrates in the H_I and H_{II} systems, should likewise be classified here; in addition to larger gas constituents, natural gas contains significant quantities of methane, another gas with small molecules.

An extensive survey of inclusion clathrate compounds with cavity structures was published by Solacolu and Solacolu [275] in 1973.

Chapter 3

STABILITY RELATIONS OF GAS HYDRATES

3.1. Conditions of formation and decomposition of gas hydrates

The agreeing findings and conclusions of researchers dealing with the formation, the range of existence and the properties of gas hydrates indicate that the initial conditions of gas hydrate formation are determined by the nature of the gas, the state of the water, the pressure and the temperature. The formation conditions are given by means of heterogeneous phase diagrams in p vs. T plots. This method, first used by Roozeboom [276] in the case of chlorine hydrate, has since been applied by many investigators to numerous other material systems. For example, the research results of Makogon [158] relating to the formation of methane hydrate are illustrated in Fig. 3.1.

In this figure, the curve $OFGH$ describes the pressure-dependence of the crystallization of pure water, while the curve AC is the vapour pressure curve of the pure hydrate-forming component, i.e. the temperature-dependence of its vapour pressure. The curve describing the pressure-dependence of the temperature of hydrate formation has several characteristic sections; its course is determined by the thermodynamic properties and phase states of the components. The curve section AB describes the conditions of formation of the hydrate from gaseous methane and ice at temperatures lower than 0 °C. For this section, the inequality $dT/dp > 0$ holds; thus, in the event of an increase in the temperature, the external pressure (i.e. the pressure of the hydrate-forming component) must be raised for the hydrate to be formed. Section BC describes the hydrate formation from liquid water in the presence of the gaseous hydrate-forming component. Here too, the slope of the curve demonstrates the necessity of the inequality $dT/dp > 0$. At point C, the vapour pressure of the hydrate is equal to that of the hydrate-forming component. At a temperature above that corresponding to this point, the hydrate-forming component too is in the liquid state, and therefore the slope of curve Cd is determined primarily by the molar volume change occurring during hydrate formation. This can be described thermodynamically by the equation:

$$\frac{dT}{dp} = T\frac{\Delta V_m}{\Delta H_{m,form}}$$

(3.1)

where $\Delta H_{m,form.}$ is the heat of formation of the gas hydrate at the given temperature, T and

$$\Delta V_m = V_m - V_{m,H} \tag{3.2}$$

The change in molar volume is consequently the difference between the molar volume (V_m) of the hydrate former at the equilibrium values of p and T, and that $(V_{m,H})$ relating to the hydrate state.

From Equation (3.1) it follows that, when $\Delta V_m > 0$, the section Cd has a positive slope, and when $\Delta V_m < 0$, a negative one. For all gases where the critical temper-

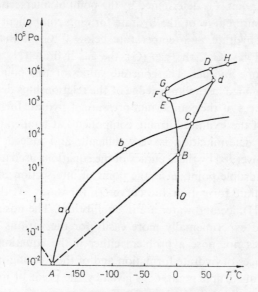

Fig. 3.1. Equilibrium diagram for formation
of CH₄ hydrate [158]

ature is higher than that of hydrate formation, the molar volume relating to the hydrate state, and hence the value of ΔV_m, can be determined in the knowledge of the densities at the given p and T pairs.

Makogon [158] found experimentally that, at not too high pressures, curve Cd has a positive slope in the cases of CO_2, C_2H_6 and H_2S, similarly to CH_4, and a negative slope in the cases of C_3H_8 and $CHCl_3$. The findings were confirmed by the molar volume change calculations.

In Fig. 3.1, section Ed defines those p vs. T conditions at which $dT/dp = 0$; for hydrate formation at a pressure corresponding to point d, therefore, the molar volumes of the gas are equal in the free and hydrate states.

If $p < p_d$, then $dT/dp > 0$, and if $p > p_d$, then $dT/dp < 0$. Accordingly, the temperature at point d is critical for the hydrate of given composition. The pressures of various gases relating to point d vary considerably, with orders of magnitude of 10^5–10^7 Pa. On increase of the pressure to a value above point d, the temperature of hydrate formation decreases for all types of gases.

In the heterogeneous state-diagrams of the different gas–water systems, a number of quadruple points can be found, corresponding to four-phase equilibrium. For gases where the temperature of hydrate formation is lower than the critical temperature, e.g. methane, ethane, propane, hydrogen sulphide and carbon dioxide, four such points can exist: A, B, C and D.

The position of point A is determined by the point of intersection of the equilibrium vapour pressure curve of the hydrate-forming component and that of formation of the gas hydrate, at a temperature below 0 °C. At this point, the solid (G_s) or condensed gas (G_c), the gas (G), the gas hydrate (H) and ice are all in equilibrium with one another. The concrete values of this point for the various gases can be determined in the knowledge of the relationships describing the temperature-dependences of the equilibrium pressure of hydrate formation and of the vapour pressure of the hydrate-forming component, at temperatures below 0 °C. The experimental determination is very difficult, and indeed, the accuracy of determination is governed by the accuracy of the equations used in the calculations.

The second quadruple point, B, is the point of intersection of the equilibrium curve ABC of hydrate formation and curve OF for water. At this point, the gas (G), the hydrate (H), ice and water are in equilibrium. The position of this point can be determined experimentally more easily for the various gas systems, but its calculation does not pose a problem either if the equations describing the pressure-dependences of hydrate formation and of the freezing-point of water are known. The third quadruple point, C, is given by the point of intersection of the vapour pressure curves of the hydrate-forming component and of the hydrate, at temperatures above 0 °C. At this point, gas (G), condensed gas (G_c), gas hydrate (H) and water are simultaneously present in equilibrium. The temperature of point C therefore determines the pressure above which the hydrate is formed in the presence not of the gaseous, but of the liquid hydrate-forming component. At this point, a sharp break occurs in the curve describing the hydrate vapour pressures if (in accordance with the above) the inequality $dT/dp < 0$ holds (e.g. in the cases of butane and chloroform).

The fourth quadruple point, D, is the point of intersection of the equilibrium curve Cd of hydrate formation and the high-pressure freezing-point curve GH for water. At this point, ice, water, condensed gas and hydrate are in equilibrium.

In spite of the fact that the various authors consistently refer to these quadruple points as critical points, properly speaking only point D is a true critical point, because only at pressures higher than the pressure corresponding to this can hydrate formation not occur at any temperature.

Along the individual curve sections of the diagram, three phases are in equilibrium.

In the construction of such complete heterogeneous phase-diagrams for gas–water systems, it is generally neglected by authors that the freezing-point of water changes to some extent as a consequence of the dissolution of gases in the water.

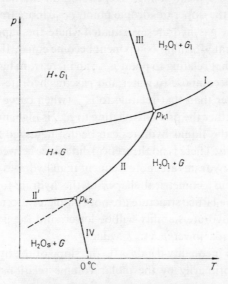

Fig. 3.2. Schematic representation of heterogeneous phase equilibrium conditions connected with hydrate formation [66]

Similarly, few papers have been published on the evaluation of hydrate structure changes at very low temperatures or at very high pressures. From both theoretical and practical points of view, the attention of researchers and specialists has mostly been drawn to hydrate formation processes in the temperature range 0–30 °C, and this is therefore the best-known area in the literature.

The schematic representation of the phase equilibrium relations connected with the conditions of hydrate formation can be seen in Fig. 3.2; this is similar to the p vs. T diagram for methane hydrate.

The individual curve sections in the state-diagram have the following meanings:

I. temperature-dependence of the vapour pressure of the hydrate-forming component saturated with water;

II. the vapour pressure curve of the hydrate at temperatures >0 °C;

II′. the vapour pressure curve of the hydrate at temperature <0 °C;

III. the melting-point curve of the hydrate;

IV. the depression of the freezing-point of the water as a result of dissolution of the hydrate-forming component.

It is evident from the figure that the gas hydrate can only exist at p vs. T values relating to areas lying to the left of curves II′, II and III, which meet at the points p_k (incorrectly termed critical points). Depending on the nature of the hydrate-forming component, the upper quadruple point, $p_{k,1}$, more frequently interpreted in connection with the gas hydrates, is situated where the vapour pressures of the hydrate and the hydrate-forming component become equal. If the pressure of the system is lower than that relating to point $p_{k,1}$, an increase in temperature will cause existing hydrates to decompose to water and gaseous hydrate-forming component at a temperature lower than that relating to $p_{k,1}$ (when curve II is crossed).

Depending on whether the pressure relating to $p_{k,1}$ is higher or lower than the atmospheric pressure, the liquid hydrates can be distinguished from the gas hydrates on a structural basis. There is no theoretical difference between these two hydrate types, though liquid hydrates are more frequent than hydrates of type H_{II} because of the dimensions and geometrical shapes of the hydrate-forming molecules.

If the cavities in the lattice structure are not completely occupied during hydrate formation, then the hydrate stability will be lower, and the position of point $p_{k,1}$ will be shifted towards lower p vs. T values.

On the basis of the inequality $dT/dp \gtreqless 0$, the slope of the hydrate melting curve (*III*) is determined primarily by the molar volume relationships, which depend on the nature of the material, but the developing structure type too has a considerable influence. If heating causes a hydrate to decompose into two liquid phases at a pressure above that of the decomposition-point, then for structure type H_I the melting curve generally displays a positive slope (because of the volume increase), and thus the decomposition temperature too rises; for structure type H_{II}, however, hydrate decomposition is accompanied by a volume decrease, of at most 2.5%, caused mainly by the unoccupied cavities, and is associated with a decrease in the decomposition temperature.

In the case of double or mixed hydrates, the small cavities in the H_{II} hydrate structure are also filled, and therefore the course of the melting curve is either independent of temperature, or has a negative slope, depending on the nature of the gas molecules and the degree of occupancy.

Hence, the melting-point of a gas hydrate at a given pressure is incongruent, and depends on the composition and on the degree of occupancy of the cavities. At pressures lower than that at $p_{k,1}$, the decomposition-point is strongly pressure-dependent, but it is much less so at higher pressures. With these double or mixed hydrates, the position of the decomposition-point is also influenced by whether the two or more constituents liberated on decomposition mix with each other or not. This is a very important factor if some gas constituent is in the liquid state under the conditions employed.

On the above basis, in comparative investigations of the conditions of hydrate formation and decomposition for the various gas sytems, it is indispensable to characterize the stabilities of the gas hydrates by means of some absolute or relative ratio. Such ratios may involve the p vs. T pairs fixing the positions of the quadruple points $p_{k,1}$ and $p_{k,2}$. The parameters of the "upper critical decomposition point", $p_{k,1}$, give the relative stability, while the pressure of hydrate dissociation at 0 °C, or the temperature at which the vapour pressure of the gas hydrate attains a value of 101 325 Pa are taken as the absolute stability.

The probability of formation of a hydrate is the greater, the higher its stability, i.e. the higher the temperature at which the vapour pressure of the hydrate reaches 101 325 Pa, or the lower its dissociation pressure at 0 °C.

The degree of stability of a gas hydrate, and consequently its dissociation temperature, are influenced considerably by the molecular size and geometrical shape of the hydrate-forming components. From the point of view of hydrate formation, a very small molecular diameter is not advantageous; thus, presumably because of the very low polarizability, independent hydrates of hydrogen, helium and neon are not known (their diameters, determined by various methods, being 0.267–0.277 nm, 0.188–0.267 nm and 0.239–0.354 nm, respectively). The other noble gases, however, are known to form hydrates. At 0 °C, the dissociation pressure of the hydrate of argon, with an atomic diameter of 0.38 nm, is 1.064×10^7 Pa, that of the hydrate of krypton, with a diameter of 0.40 nm, is 1.47×10^6 Pa, and that of the most stable of these hydrates, that of xenon, with a diameter of 0.44 nm, is only 1.52×10^5 Pa.

Of the hydrates of hydrocarbons, those of propane and isobutane are the most stable. These gases have molecular diameters of about 0.65 nm, and at 0 °C the dissociation pressures of their hydrates are approximately 1×10^5 Pa. The molecules of ethane, ethylene and acetylene cannot be considered spherical; the lengths of these molecules are about 0.55 nm, and the dissociation pressures of their hydrates at 0 °C are about 6×10^5 Pa. The decomposition pressure of the hydrate of methane, with a diameter of 0.41 nm, is around 2.7×10^6 Pa at 0 °C. It may be seen that the relative stability of the hydrates decreases with decreasing molecular size.

The production in a pure state of hydrates of hydrocarbons higher than butane has not yet been achieved under laboratory conditions. The sizes or spatial geometries of these hydrocarbon molecules presumably do not permit their incorporation into even the larger cavities of the water lattice, and even if their presence can be detected in the natural-gas hydrates, in all probability they are only entrapped between the crystal grains formed.

With regard to the tetrahedral bonding of the carbon atom, various models are suitable to illustrate the possible spatial structures of the paraffins. Such models are shown in Fig. 3.3 [277].

It can be seen from the figure that, with the obvious exception of ethane, the centres of the carbon atoms do not lie in a straight line, but at best are in a common plane. As a consequence of the free rotation around the axes of the single C—C bonds within the molecules, various spatial conformations can come about. The energy demands of free rotation are covered by the thermal motion of the molecules.

In models better approximating to reality, the sizes of the atoms (the van der Waals radii) and the spatially-oriented linkages of the electron orbitals forming the covalent bonds must also be taken into consideration. Such Stuart–Briegleb models of the paraffins in question are similarly shown in Fig. 3.3 (the lower row for reach of the individual hydrocarbons) [277].

In the electronic theory of atomic structure and atomic bonding, the energy content of a molecule varies with its conformation, the stabilities of the various conformations therefore differing. In a molecular aggregation in a given state, however, a large proportion of the molecules assume the most stable spatial form by means of rotation, thereby being in the lowest energy state. This favourable conformation is determined by the interaction of atoms not directly linked with one another in the molecule.

In ethane, two conformations can exist in which the relative positions of the hydrogen atoms bonded to the carbon atoms display the greatest difference. It may occur that the hydrogen atoms attached to one carbon atom are as near as possible to the hydrogen atoms on the other carbon atom; in the other extreme case, however, the hydrogen atoms of the two CH_3 groups are at the greatest possible distance from one another. A schematic representation of these two possibilities can be seen in Fig. 3.4(a) [277].

From theoretical organic chemical considerations, the most stable conformation of ethane is that involving the hydrogen atoms in the staggered position, for the van der Waals repulsive forces are then minimum; the eclipsed position is unstable, however, because of the greater potential energy. The rotation within the molecule means that transitions can occur between these two extreme conformations.

In the case of *n*-butane, four extreme conformations are possible. These are represented schematically in Fig. 3.4(b) [277].

The spatial conformation of *n*-butane with the minimum energy is the anti-periplanar one, because the relatively bulky methyl groups are then farthest from each other, while the hydrogen atoms are in the staggered state, both to one another and to the methyl groups. In the synperiplanar conformation, the two

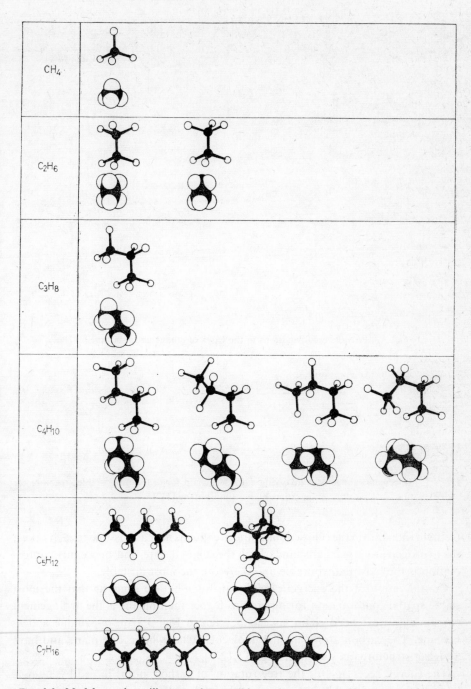

Fig. 3.3. Models used to illustrate the possible steric structures of paraffins, and the Stuart–Briegleb models [277]

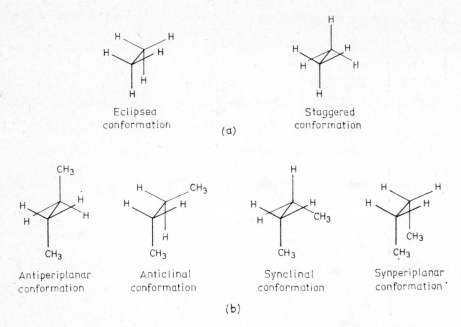

Fig. 3.4. Possible conformations in the cases of ethane and n-butane [277]

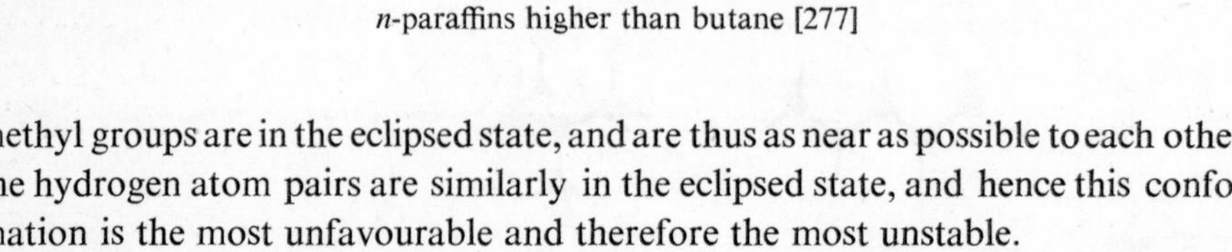

Fig. 3.5. Molecular arrangement involving extended chain form, the most stable structure for
n-paraffins higher than butane [277]

methyl groups are in the eclipsed state, and are thus as near as possible to each other;
the hydrogen atom pairs are similarly in the eclipsed state, and hence this conformation is the most unfavourable and therefore the most unstable.

Consideration of the energetic minimum for n-butane suggests that the most
stable spatial conformation for n-paraffins higher than butane is the arrangement
containing an extended chain, as can be seen in Fig. 3.3 for n-pentane and n-
heptane. The carbon atoms in such molecular chains lie in one plane and have
a zigzag structure, as shown in Fig. 3.5 [277].

The energy necessary for the free rotation of carbon atoms linked by a single
bond is amply covered by the energy of thermal motion, even at room tempera-

ture; in paraffins in the liquid or gaseous state, therefore (though the molecules strive to attain the conformation with the lowest energy state), spatial structures richer in energy (rotational isomers) can occur at room temperature.

In addition to rotational isomerism, in hydrocarbons containing four or more carbon atoms branching of the carbon chain may lead to structural isomers. The number of isomers thus formed increases with the number of carbon atoms: only two structurally isomeric open-chain butanes are known, compared with three pentanes, five hexanes and nine heptanes. The structural and rotational isomerism naturally influence not only the molecular size, but also the physical and chemical properties of the compound.

These structural considerations mean that it is not likely that paraffins higher than butane can be incorporated into even the larger cavities in the water lattice to yield gas hydrates. This appears to be verified by the fact that paraffin homologues higher than butane do not form gas hydrates at temperatures above 0 °C. At lower temperatures, even at greater pressures, there is a possibility for these hydrocarbons to be incorporated only if lattice defects occur.

The schematic molecular forms and longest dimensions of some known hydrate-forming components are shown in Fig. 3.6 [66].

The size of the benzene molecule (a diameter of 0.69 nm) would seem to permit it to form a liquid hydrate, but nevertheless in the pure state it does not form a hydrate. The probable cause of this is that this aromatic molecule possesses an abnormal field of force as compared to non-aromatic ones, and in the liquid state its molar volume is too large as well (90 cm^3/mol). The CCl_4 molecule, with a diameter close to 0.46 nm, has an even larger molar volume ($\sim$97 cm^3/mol). In the pure state, this compound does not form a hydrate either, but in the presence of a very small amount of guest species with low dimensions, e.g. air, it can form a mixed hydrate.

The relationships between the molecular sizes and stabilities of a number of substances and their gas hydrates can be observed in Fig. 3.7.

The temperature-dependence of the vapour pressure of a hydrate-forming material is an important factor influencing the position of the upper quadruple point. Temperatures relating to pressures in the range 1.01–60.6$\times$10^5 Pa for several guest molecule types, together with their critical data, are shown in Fig. 3.8 (based on the publications of Perry [278] and Stuhl [279]), and also in Table 3.1. (Supplementary data on the larger paraffins which do not form hydrates are also given in the table and the figure.)

In Table 3.1 the various gas hydrate-forming materials are arranged in order of increasing boiling-points, and the stabilities of their hydrates increase in this order too. The relationship between the dissociation temperatures of the hydrates and the boiling-points of the hydrate-forming components can be seen in Fig.

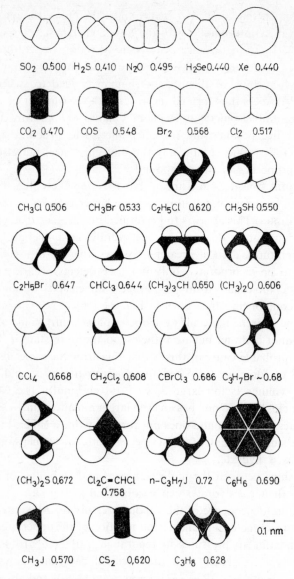

Fig. 3.6. Schematic pictures of molecules of several hydrate-forming substances, and the longest dimensions in nm [66]

3.9; the points do not lie on one line, but are situated within a band tending to the right as it proceeds upwards. The stabilities of the hydrates increase in the same direction with the rise of the band, while at the same time the gas hydrates

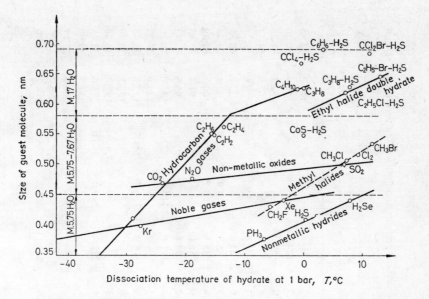

Fig. 3.7. Correlations between molecular size and the stability of gas hydrates

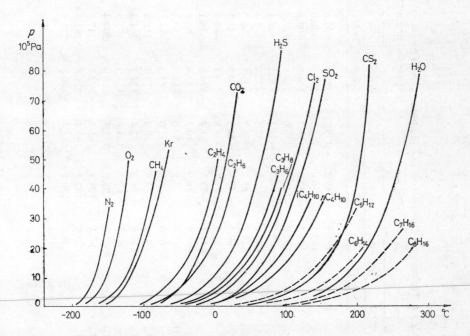

Fig. 3.8. Temperature-dependence of the equilibrium vapour pressure for several hydrate-forming substances [278, 279]

Table 3.1. Temperature-dependences of equilibrium vapour pressures, and also critical pressures and critical temperatures of several hydrate-forming materials [278]

p (10^5 Pa)	1.01	2.02	5.05	10.1	20.2	30.3	40.4	50.5	60.6	T_{cr} (°C)	p_{cr} (10^5 Pa)
					T (°C)						
N_2	−195.8	−189.2	−179.1	−169.8	−157.6	−148.3				−147.2	35.85
O_2	−183.1	−176.0	−164.5	−153.2	−140.0	−130.7	−124.1			−118.9	50.20
CH_4	−161.5	−152.3	−138.3	−124.8	−108.5	−96.3	−86.3			−82.1	46.26
Kr	−152.0	−143.5	−130.0	−118.0	−101.7	−88.8	−78.4	−66.5		−63.0	54.54
C_2H_4	−103.7	−90.8	−71.1	−52.8	−29.1	−14.2	−1.5	8.9		9.6	51.21
C_2H_6	−88.6	−75.0	−52.8	−32.0	−6.4	10.0	23.6			32.3	48.68
CO_2	−78.2	−69.1	−56.7	−39.5	−18.9	−5.3	5.9	14.9	22.4	31.1	73.73
H_2S	−60.4	−45.9	−22.3	−0.4	25.5	41.9	55.8	66.7	76.3	100.3	89.79
C_3H_6	−47.7	−31.4	−4.8	19.8	49.5	70.0	85.0			91.4	45.85
C_3H_8	−42.1	−25.6	1.4	26.9	58.1	78.7	94.8			96.8	42.42
Cl_2	−33.8	−16.9	10.3	35.6	65.0	84.8	101.6	115.2	127.1	144.0	76.76
iso-C_4H_{10}	−11.7	7.5	39.0	66.8	99.5	120.5				134.0	37.37
SO_2	−10.0	6.3	32.0	55.5	83.8	102.6	118.0	130.2	141.7	157.2	78.48
n-C_4H_{10}	−0.5	18.8	50.0	79.5	116.0	140.6				152.8	36.36
n-C_5H_{12}	36.1	58.0	92.4	124.7	164.3	191.3				197.2	33.33
CS_2	46.5	69.1	104.8	136.3	175.5	201.5	222.8	240.0	256.0	273.0	73.63
n-C_6H_{14}	68.7	93.0	131.7	166.0	209.4					234.8	29.90
n-C_7H_{16}	98.4	124.8	165.7	202.8	247.5					266.8	27.17
n-C_8H_{18}	125.6	152.7	196.2	235.8	281.4					296.2	24.95
H_2O	100.0	120.1	152.4	180.5	213.1	234.6	251.1	264.7	276.5	374.2	220.20

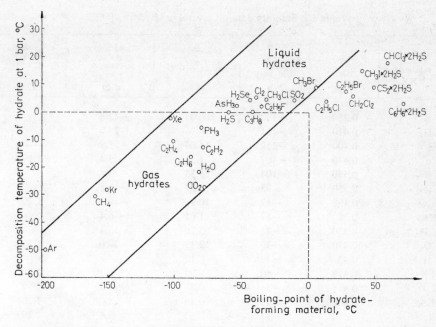

Fig. 3.9. Correlation between dissociation temperatures of hydrates and boiling-points of hydrate-forming substances [71]

and liquid hydrates are well separated from each other. The increase in the decomposition temperature of the hydrate with the increase in the boiling-point of the hydrate-forming component is connected with the increasing polarizability of the molecules, and with the intermolecular van der Waals interactions too.

Tables 3.2 and 3.3 present stability data on some simple gas hydrates, simple liquid hydrates and double hydrates, with the boiling-points and molecular sizes of the hydrate-forming components.

Among the conditions of formation of gas hydrates, the solubility in water of the gas plays an important role, since the dissolved gas particles form the nuclei which initiate the processes of hydrate precipitation and crystal growth. The solubilities of the various rare or quasi-rare gases in water increase with decreasing temperature or increasing pressure. The quantity of dissolved gas can be determined by means of the following equation:

$$RT \ln \frac{f_i}{x_i} = RT \ln k_i + V_i (p - p_1^*) \qquad (3.3)$$

where f_i is the fugacity of the gas component i, x_i is the mole fraction of the gas component dissolved in the water, k_i is the Henry absorption coefficient of the observed gas, p is the pressure of the gas phase, p_1^* is the vapour pressure of water

Table 3.2. Stability data on several gas hydrates [66]

Hydrate-forming substance X	$d=2r$ (nm)	Boiling point of X (°C)	Dissociation pressure of hydrate at 0 °C (10^5 Pa)	Dissociation temperature of hydrate at 10^5 Pa (° C)	Lattice constant of unit cell (nm)
Ar	0.380	−190	106.05	−42.8	—
CH_4	0.410	−161	26.26	−29.0	—
Kr	0.400	−152	14.64	−27.8	—
Xe	0.440	−107	1.51	−3.4	1.197
C_2H_4	0.560	−104	5.55	−13.4	—
C_2H_6	0.550	−93	5.25	−15.8	—
N_2O	0.495	−89	10.1	−19.3	1.203
PH_3	0.38	−87	1.61	−6.4	—
C_2H_2	0.550	−81	5.76	−15.4	—
CO_2	0.470	−79	12.48	−24.0	1.204
CH_3F	0.43	−78	2.12	—	—
H_2S	0.410	−60	0.97	+0.35	1.200
AsH_3	0.415	−55	0.81	+1.8	—
H_2Se	0.440	−42	0.46	+8.0	1.206
Cl_2	0.517	−34	0.33	+9.6	1.203
C_2H_5F	0.500	−32	0.70	+3.7	—
CH_3Cl	0.506	−24	0.41	+7.5	1.200
SO_2	0.500	−10	0.39	+7.0	1.194
CH_3Br	0.533	+4	0.25	+11.1	1.209
Br_2	0.568	+59	0.06	—	1.201

Table 3.3. Main stability data on several simple liquid hydrates and double hydrates [66]

Hydrate-forming substance X	$d=2r$ (nm)	Boiling point of X (°C)	Simple hydrate $X \cdot 17H_2O$ Dissociation temperature at 0 °C (10^5 Pa)	Simple hydrate $X \cdot 17H_2O$ Lattice constant of unit cell (nm)	Double hydrate $X \cdot 2H_2S \cdot 17H_2O$ Dissociation temperature at 10^5 Pa (°C)	Double hydrate $X \cdot 2H_2S \cdot 17H_2O$ Lattice constant of unit cell (nm)	Double hydrate $X \cdot 2H_2Se \cdot 17H_2O$ Dissociation temperature at 10^5 Pa (°C)	Double hydrate $X \cdot 2H_2Se \cdot 17H_2O$ Lattice constant of unit cell (nm)
COS	0.548	−30	—	1.20	0	1.73	—	—
C_3H_8	0.628	−46	1.01	—	+8	1.74	—	—
CH_3Br	0.533	+5	—	1.20	—	—	—	—
C_2H_5Cl	0.62	+13	0.27	1.730	+7.2	1.726	+10	1.741
C_2H_5Br	0.47	+38	0.21	—	+13.0	1.726	—	—
$CHCl_3$	0.644	+61	0.07	1.73	+16.3	1.729	+15	1.745
CCl_4	0.668	+77	—	—	+17.3	1.746	+19	1.760
C_6H_6	0.69	+80	—	—	+3.5	1.748	—	—
CCl_3Br	0.686	+104	—	—	+11.3	1.757	—	—

at the given temperature, and V_i is the partial molar volume of the gas component dissolved in the water.

With the increase of pressure, the solubility can be increased only up to a certain limit, however, if the associated equilibrium p and T values for hydrate formation are reached at a given temperature, then as a consequence of hydrate precipitation, the solubility of the gas in the water will decrease, while at the same

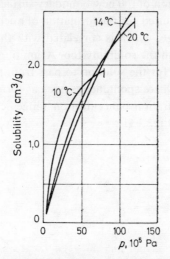

Fig. 3.10. Pressure- and
temperature-dependences of
solubility of methane in water [158]

time the gas content of the hydrate crystals will rise very rapidly. Figure 3.10 shows the pressure- and temperature-dependences of the solubility of gaseous methane in water.

From a consideration of the theories of crystallization, it is evident that a fundamental kinetic condition for the formation of crystalline gas hydrates is that a critical nucleus size should be formed during crystal formation, for until this size is reached the solid structure is thermodynamically unstable, and the crystal nuclei decompose as a result of thermal fluctuation. The critical nucleus size is influenced by the heat of crystallization (Q), the extent of supercooling ($T_p - T_o$) and the specific surface energy of the material (σ):

$$r_{cr} = \frac{2\sigma T_p}{Q(T_p - T_o)} \qquad (3.4)$$

T_p is the equilibrium temperature of crystallization, and T_o the temperature of the crystallization process.

At a given equilibrium temperature and pressure, the gas–water system is stable
in an unsaturated state. A phase change occurs only if the system comes into a
metastable supersaturated state because of change in the solubility or other physicochemical parameters. Though crystal deposition from the metastable state
is not ensured by the thermodynamic conditions, there is already a possibility
for the formation of nuclei. For this, it is necessary that the system should per-
form work for the formation of the new boundary surface between the two phases.
The amount of work required for the formation of nuclei is determined primarily
by the critical particle size, which is correlated with the degree of saturation of
the system, and hence with the solubility too. A limiting supersaturation concentration must be exceeded for the solution to pass from the metastable state to the
labile equilibrium state, where spontaneous nucleus formation is already ensured.
Saturation and supersaturation concentration ranges characteristic of solutions
in general are shown in Fig. 3.11.

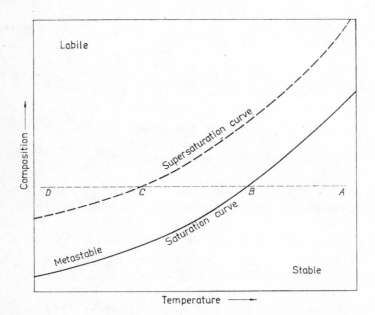

Fig. 3.11. Schematic diagram of the connection between supersaturation
and the formation of nuclei

In the stable solution range, neither nucleus formation nor nucleus growth is
possible. If a solution with the composition corresponding to point *A* is cooled,
the system comes into the metastable equilibrium state (point *B*), where the phase
change may begin spontaneously if the possibility of work performance necessary

for the formation of the interface of the two phases is given. The calculations of Makogon [158] demonstrate that the work of nucleus formation is equal to about one-third of the surface free energy of the nucleus; if the nuclei are assumed to be spheres with an approximate radius of r, this work can be determined via the following relationship:

$$W = \frac{4}{3}\pi r^2 \sigma \tag{3.5}$$

If a solution with the composition corresponding to point A in Fig. 3.11 is cooled, the thermodynamic possibility of crystallization develops only at point C, when

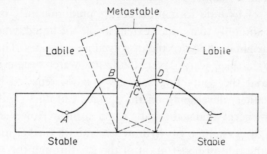

Fig. 3.12. Miers diagram illustrating the role of the boundary between metastability and lability [281]

a certain supersaturation is attained. The process may be further retarded, however, if the crystalline hydrate dissolves well in water, so that the formation of a macroscopic nucleus size needs a more intensive supersaturation or supercooling (point D).

It emerged from the investigations by Houghton [280] that the position of the metastable–labile boundary is significantly influenced by secondary factors, such as the viscosity, the flow velocity, the intensity of mixing, the cooling rate, the presence of nucleus-forming impurities, etc.; in certain cases, the nucleus formation occurs suddenly when even a slight supersaturation is reached.

The role of the metastable–labile zone boundary is well illustrated by the Miers diagram [281] in Fig. 3.12. The points A and E denote the lowest level of the centre of gravity of a brick-shaped parallelepiped. If the brick is stood on its shorter side, point C is the metastable energy level. During displacement from this position, the energy level of the centre of gravity of the brick follows curve CBA or CDE. Although the metastable energy level C is higher than the stable energy level A or E, the brick will be displaced from the metastable energy level only if excess energy corresponding to the centre of gravity positions B or D is provided.

However, as soon as the system has reached the critical energy level corresponding to points B or D, the subsequent spontaneous change will be directed not to a return to position C, but towards the positions A or E, respectively. This energy diagram proves that the metastable equilibrium is a true equilibrium, and for the displacement of the system from this state, a certain activation energy is needed, e.g. by means of further supersaturation.

Thermodynamic calculations relating to the formation of the crystal nucleus prove that there is only a very small probability of nucleus formation in the gas phase saturated with water vapour, because the molar volume of the gas there is very large, and hence there is very little probability of the water molecules grouping together in a manner corresponding to that found in the hydrate structure. The process of hydrate formation occurs predominantly on the gas–water contact interface, and only to a minor extent inside the liquid, where precipitation can begin in the neighbourhood of the dissolved gas particles. The forming nuclei are surrounded by a thick water layer. If the gas–water system is at rest during the experiment, and the contacting surfaces are not renewed by streaming or mixing, then the nuclei developing on the surface can do so only as a result of a larger activation energy under the static water surface. However, if the covering layer is sealed off by crystalline gas hydrate, the rate of further nucleus formation is determined by the rate of penetration of the gas through the solid surface, i.e. by the rate of diffusion.

While nucleus formation occurs with the greatest probability only on the gas–water interface, the further growth of the crystalline nuclei can proceed both on the boundary surface and within the gas or the water phase. The absorbed gas molecules are situated among the clusters formed; naturally, they can similarly be found in highest concentration on the gas–water contact surface. The surface water layer is therefore supersaturated with gas molecules, but at the same time this gas layer is likewise supersaturated with water. Thus, on the gas–water interface a transition layer is formed, in which both the gas and the water are supersaturated, so that the possibility exists for the nuclei to become stabilized, and for the nuclear growth to become more intensive. As the pressure is increased and the supercooling is enhanced, the structural ordering of the water becomes stronger and stronger, and this further increases the rate of hydrate formation.

If the gas sytem (e.g. that in a natural-gas pipeline) becomes supersaturated with respect to water only during the process, so that there is no gas–water contact surface, then in the period preceding condensation the water vapour molecules likewise assume clusters, with chains similar to those in ice. Depending on the thermodynamic conditions, these chains become interlocked with one another, and trap the gas molecules as nuclei in their cavities. Nuclear growth can begin on these nuclei if the conditions for this are provided.

At temperatures below 0 °C, i.e. in the presence of ice, it is assumed that the formation of hydrate nuclei occurs on the surface of the ice grains.

If a liquid hydrate-forming component is dispersed in a medium saturated with water, hydrate formation probably begins in the locally supersaturated zones, and the crystals then grow in an adsorption process.

As real metastable–stable state transitions are never fully reversible, the same is true for the processes of formation and decomposition of hydrates.

The factors determining the stability of water supersaturated with respect to gas (e.g. the temperature-dependence of the solubility of the dissolved component; the lattice structure developing in the water; the interactions between the dissolved gas and the solvent; the degree of dispersion; the conditions of ordering — relating to the degree of supercooling — of the appropriate number of individuals forming the stable crystal nucleus, in accordance with the desired lattice structure; the magnitude of the gas–water contact boundary surface; the mixing leading to the constant renewal of the contact surfaces; and the turbulence caused by streaming) do not play roles in the decomposition of gas hydrates.

The decomposition of gas hydrates (in actual fact their thermal dissociation) generally occurs not in one step, but via the following consecutive part-processes:

1. sublimation of host lattice.
2. decomposition of clathrate structure of host lattice;
3. diffusion of guest molecule through clathrate lattice structure;
4. desorption of guest molecule from surface of solid clathrate.

The activation energy necessary for the processes is required to overcome the lattice energy of the crystalline clathrate and the binding forces between the host lattice and the guest molecule.

The kinetically differing conditions of hydrate formation and decomposition result in the fact that, under isobaric conditions, the temperature of hydrate formation can be as much as 1–7 °C lower than the dissociation temperature.

Mention should be made of the survey by Malenkov [282] of the kinetics of formation of gas hydrates.

3.2. Effect of third component on stability of gas hydrates

The phase characteristics of a clathrate system originating from the interactions of a hydrate-forming material and water are significantly influenced by the presence of a third or even more components. This effect depends decisively on whether this additional component (or components) interacts with the hydrates-forming gaseous or possibly liquid component, or with the water as a clathrate-forming component; and, if the new component interacts with both the hydrate-forming substance and the water, then it is important with which the inter-action is stronger. The presence of the third component may result in very varied effects, and therefore the changes in the phase characteristics cannot be approximated to or described by a completely clear-cut and general picture. Concrete conclusions can be drawn only from investigations relating to single compounds or a given compound type. Besides the material properties, the effects of changes in the state parameters must also be taken into account. The theo-retical aspects of the phenomena emerging are very interesting, but their prac-tical importance is much greater: even in laboratory investigations, very small quantities of various impurities functioning as third or additional components are always present; this is even more the situation in the occurrence of gas hy-drates in industrial technological processes. Consequently, hydrate formation from gas mixtures, the problems arising in the production and transportation of natural gases, precipitation and fractionation tasks based on hydrate formation, the elimination of hydrate formation under given circumstances, or even the desalination of sea-water, all involve processes for which a knowledge of the individual hydrate characteristics of the given system, with its several compo-nents, is indispensably necessary.

The hydrate-formation conditions of a gas system with one or more components is thus changed to a greater or lesser extent by the presence of an additional "third" component. It can be stated in general that this effect depends on the gas composition, on the corresponding gas density, on the nature and quantity of the material changing the structural conditions in the water, and on the pressure prevailing in the system. In the presence of either electrolytes or non-polar solutes, the primary factors acting so as to alter the conditions of hydrate formation and dissociation are the structural changes depending mainly on the pressure, on the temperature and on the composition of the solute, and also the change in energy of the interactions between the molecules. The mechanism of the effects of these factors was dealt with in Section 2.3.

The conditions of gas hydrate formation, i.e. the stability of a gas hydrate under given conditions, are influenced mainly by other gases (auxiliary gases) present in addition to the hydrate-forming component, by electrolytes originally present in the water or otherwise added, by organic compounds introduced to change the conditions of hydrate formation (chiefly mono-, di- and trihydric alcohols), as well as by apolar phases added or condensed there (usually liquid hydrocarbons).

Research results connected with this topic prove that any concentration of electrolytes dissolved in the water will lower the temperature of hydrate formation at a given pressure; in low concentrations, alcohols increase the temperature of hydrate formation, but as the concentration is increased this temperature falls. In the latter case it can be assumed that the structural cavities in the water are partly filled (e.g. by CH_3 groups in the case of methanol), and accordingly the ice-like ordering of the hydrogen chains is enhanced in the environment of the organic molecules. At higher alcohol concentrations, the clathrate-forming aggregates break down, whereby the possibility of hydrate formation decreases in the same way as when the water structure is disturbed by electrolytes. From the results of examinations of the heat of mixing, the adiabatic compression, viscosity measurements and other investigations, it can be concluded that the methanol–water system with the most closely packed solution structure can be observed when there are four molecules of water per molecule of methanol. In solutions more concentrated than this, the structure begins to break down and the looser fitting decreases the possibility of formation of the clathrate structure.

This inhibitory effect of electrolytes and alcohols is very important in the processes of natural-gas production and transportation, but it can be used to advantage in other technological processes too.

A number of authors have studied hydrate inhibitors acting especially by disturbing the structure of water, but the first detailed review evaluation is that by Bond and Russell [82]. For the very strong hydrate-forming gas hydrogen sulphide, they determined how $CaCl_2$, NaCl, methanol, ethanol, ethylene glycol, diethylene glycol, sucrose and dextrose influence the temperature of hydrate formation, as functions of the inhibitor concentration and the gas pressure.

They found that the use of equivalent quantities of NaCl, $CaCl_2$, methanol and ethanol led in each case to similar decreases in the temperature of formation of H_2S hydrate; sucrose and dextrose are practically ineffective; the effects of the glycols lie between those of the salt solutions and those of the sugar solutions. As a consequence of the higher solubility of $CaCl_2$ than that of NaCl, the effect attainable with the former is greater than that with the most economical-seeming NaCl.

The most results connected with detailed investigations of the inhibition of the formation of gas hydrates by means of electrolytes and alcohols emerged in particular from the work of Makogon and Sarkisyants [157, 158].

The application of brines with salt contents greater than 100 g dm^{-3} for the inhibition of hydrates was developed by Lutoshkin and his co-workers [283]. Milev [284] reports the inhibition methods employed in the gas transmission pipeline between the Soviet Union and Bulgaria. From comparative investigations of the systems of solutions of methanol, diethylene glycol and $CaCl_2$ with the liquid hydrocarbons used in the experiments, here too it emerged that methanol has the greatest effectivity.

Soviet researchers achieved satisfactory results from the joint application of methanol and $CaCl_2$, this salt also having a high effectivity. In the case of a natural gas, Krasnov and Klimenok [285] examined the pressure-dependence of the inhibitory effect with pure methanol between the pressure limits $2-7 \times 10^6$ Pa, and with the $CaCl_2-H_2O$ system at pressures of $1-20 \times 10^6$ Pa. Their findings confirm that the phenomenon of inhibition can be interpreted as the dissolution of the crystalline hydrates.

In high-yield gas wells in the Soviet Union, a mixture of 10 vol. % methanol and 30 vol. % aqueous $CaCl_2$ in a proportion of 1:9 is frequently applied to prevent hydrate formation. Such a procedure is reported by Arshinov and his co-workers [286, 287]. They made use of laboratory experiments and produced mathematical correlations for determination of the density and dynamic viscosity of the mixture at given temperature, and to establish the necessary quantity and concentration of the inhibitor; in the knowledge of these parameters, conclusions could be drawn as to the lengths of operation periods during which hydrates are not formed. This inhibitor composition likewise proved optimum for gas production in the neighbourhood of the Mesoiak ice-zone, as reported by Sumets [288].

As the application of the salt–alcohol inhibitor spread, attention turned to the effects of the mixture on rocks and on the cement-based rings of bottom holes. From investigations in which the inhibitor was added in the well, Tsarev and Mordovskaya [289] found the best mixture to be that containing 10% $CH_3OH+20\%$ $CaCl_2+70\%$ H_2O+traces of HCl.

From the aspect of lowering the inhibitor losses, Burnykh and his co-workers [290] obtained good results by using alcohol mixtures, e.g. ethylene glycol–methanol or isopropanol–methanol in various proportions.

It should not be left out of consideration that, when mixtures of brines or salt solutions and methanol are fed into pipelines, salt precipitation and deposition occur due to the salting-out effect. At a gas temperature of 80 °C or lower, this crystal precipitation can be prevented by the addition of a 0.5% aqueous solution of Na_3PO_4 in a quantity corresponding to 10–15% of the brine. Researches on this were carried out by Kolesnikova [291]. The conditions of application of further mixed inhibitors were reported by Koshelev and his co-workers [292], for example, who employed as hydrate inhibitor a 5–20 wt. % aqueous solution of the ether–aldehyde fraction (EAF) formed as a side-product in the production of synthetic ethanol. The EAF contained on average 63% ethanol, 33% diethyl ether and 3.0% acetaldehyde. In the temperature and pressure ranges 0.7–12 °C and $6-80 \times 10$ Pa, it was found that the relationship

$$\Delta T = kw^n \tag{3.6}$$

holds for the decrease of the dissociation temperature of the hydrate. Here, w is the concentration of the EAF. It was found that for ethane, ethylene and natural gases $k=0.53$, 0.37 and 0.47, and $n=1.04$, 1.14 and 1.06, respectively.

The effect of formaldehyde on the decomposition of the hydrates of natural gases was investigated and interpreted by Gavriya and her co-workers [293].

Important experiments were carried out by Khoroshilov and his co-workers [294] to inhibit hydrate formation from disturbing the production of gas wells and at the same time to eliminate corrosion effects, with 1:1 electrolyte–hydrocarbon emulsions. 0.5% NaCl solution mixed with 250 mg dm^{-3} methanol was used as electrolyte, and petroleum as hydrocarbon. The corrosion behaviours of the metallic materials of gas wells were examined in the temperature range 30–80 °C in inhibitor solutions with concentrations between 50 and 1000 mg dm^{-3}. An inhibitor concentration of 300–500 mg dm^{-3} provided 95–98% protection against corrosion. The hydrate-inhibiting effects of these solutions were investigated with natural gases, and it was found that there is only an insignificant difference in comparison with the inhibition attained with pure methanol. In the pressure interval 5–15×10^6 Pa, the difference in the decrease of the equilibrium temperature of the hydrate for the gas sytem investigated was 1±0.4 °C. The laboratory results were confirmed by field investigatons.

In gas wells opened up by drilling through permafrost layers, with a 1.5–2.5 cm thick pipeline insulation, applied as a polyurethane spray, Khoroshilov and his co-workers [295] were able to attain hydrate-free operation without any alcohol injection, even though the gas-collecting line worked at an external temperature of −50 °C. The heat transfer coefficient of the polyurethane applied was less than 0.209 kJ m^{-2} h^{-1}. This method can be used only in continuously operating wells. In the event of a longer shut-down or the reopening of a well, hydrate inhibition was ensured by means of methanol addition.

After this brief general introduction, let us consider in more detail the effects of the various additive types on the stability conditions of gas hydrates.

3.2.1. Effects of gases on stability

The literature dealing with questions of hydrate stability changes in the presence of auxiliary gases emphasizes the hydrate-stabilizing effects of primarily nitrogen and oxygen, and to a much lower extent carbon dioxide; in this respect, mention was made earlier of the work of von Stackelberg and Meinhold [67].

In the Technical University for Heavy Industries in Miskolc, systematic investigations were made by Berecz and Balla-Achs [296] to clarify how the stability of methane hydrate is influenced by the carbon dioxide present in natural gases, and by the accompanying nitrogen content. Studies were performed with 0–100 vol.% CO_2 and 0–70 vol.% N_2.

Data obtained on the dissociation pressures of gas hydrates formed in the CH_4–CO_2–H_2O system are listed in Table 3.4. The results of the measurements

Table 3.4. Dissociation temperatures of hydrates of gaseous CH_4–CO_2 mixtures as functions of composition and pressure [296]

Gas composition		p_{diss} (10^5 Pa)									
		20.2	30.3	40.4	50.5	60.6	70.7	80.8	101	121.2	151.5
CH_4 (vol. %)	CO_2 (vol. %)	Temperature of hydrate dissociation (°C)									
100	—	−2.7	1.3	3.8	5.8	7.5	8.4	9.2	10.3	11.0	11.6
90	10	−2.0	1.8	4.3	6.0	6.9	7.4	7.9	8.6	9.2	9.7
80	20	−1.5	2.3	4.6	6.1	7.1	7.7	8.2	8.8	9.4	9.9
70	30	−0.8	2.8	4.8	6.4	7.3	8.2	8.7	9.7	10.1	10.5
60	40	−0.3	3.4	5.4	6.9	8.2	9.3	9.8	10.8	11.5	11.9
50	50	0.2	3.7	6.1	7.8	9.3	10.5	11.4	12.7	13.4	13.8
40	60	0.8	4.0	6.7	9.1	9.8	11.2	13.0	14.7	15.3	15.7
30	70	−2.2	3.0	7.0	9.8	11.8	13.3	14.1	15.6	16.5	17.4
20	80	−5.0	−1.8	4.4	8.3	10.4	11.8	12.7	15.1	16.3	17.7
10	90	−5.0	−0.3	3.2	5.9	7.8	9.2	10.3	11.7	12.5	13.1
—	100	3.8	6.6	9.0	9.7	9.85	9.95	10.0	10.2	10.4	10.5

verify the stabilizing effect of carbon dioxide on methane hydrate throughout the whole pressure interval examined. This effect is the highest at CO_2 content of 50–70 vol.%. At pressures below 5×10^6 Pa, the stabilizing effect progressively increases with increase of the CO_2 concentration, but at pressures above 5×10^6 Pa in the CO_2 concentration interval 5–30% the stability of the hydrate is practically unchanged. It even decreases slightly, and only at higher carbon dioxide contents than this does the hydrate stability increase to such an extent that the dissociation temperature exceeds those of the hydrates of the two pure gas constituents. In mixtures containing more than 70% carbon dioxide, the stability of the mixed hydrates is lowered considerably by the condensed phase formed as a consequence of retograde condensation. Figure 3.13 depicts isobars characterizing the variation in the stability of the hydrates with the composition.

The authors [296] similarly carried out a systematic series of investigations in an attempt to elucidate the conditions of formation and dissociation of hydrates in the CH_4–N_2–H_2O system as functions of the composition and pressure of the gas mixture. No stabilizing effect on the hydrate of methane was observed in gas mixtures containing 1–20% N_2. Slight increases in the nitrogen content progressively lowered the dissociation temperature of the mixed hydrate, that is the latter became more unstable. The experimental data relating to the conditions of dissociation of mixed hydrates formed from gas mixtures containing more than 20% nitrogen involved an appreciable experimental error and were poorly reproducible; in many cases there was extensive scattering. This phenomenon can be seen in Fig. 3.14, which confirms that the stabilizing effect of nitrogen on methane hydrate is not a clear-cut one.

The examinations revealed that the mixed hydrates formed do not have a uniform composition, and that periodical changes (break-points) can be observed in the interdependent p and T pairs in the process of decomposition of the hydrates during heating-up. In several cases, the existence of pure methane hydrate could be detected; indeed, it occurred on a number of occasions that, on the complete decomposition of the solid hydrate, the total pressure of the system exceeded the starting pressure. This phenomenon can only be interpreted if the solubilities of the gas components in water are taken into account. It can be

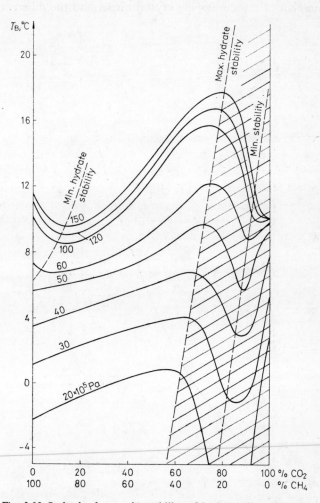

Fig. 3.13. Isobaric changes in stability of hydrates formed in the CH_4–CO_2–H_2O system as a function of the composition [296]

proved experimentally that the solubilities of paraffins in water are enhanced by carbon dioxide, hydrogen sulphide and unsaturated hydrocarbons; in contrast, the presence of nitrogen, hydrogen or helium decreases the solubilities. In the process of hydrate formation in the CH_4–N_2–H_2O system, the solubility of nitrogen in water is increased by a reduction of the temperature, gaseous methane being expelled continuously from the solution as a consequence. Depending on the effects of secondary factors acting during the process (cooling rate, renewal of the water surface due to mixing, etc.), the composition of the hydrate formed cannot be uniform because of the arrangement of the dissolved gas particles in the environment of the developing crystal nuclei and the different nature of the

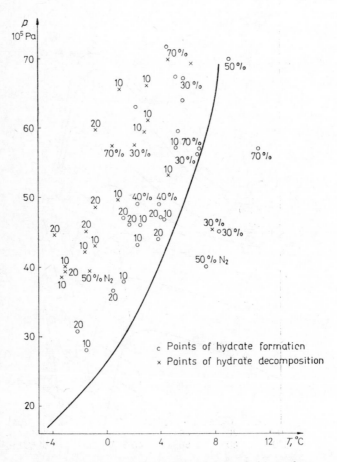

Fig. 3.14. Scattering of experimental data relating to the conditions of dissociation of CH_4–N_2 mixed hydrates [296]

basic structure in contact with the gas mixture. However, from Fig. 3.14 it can be stated that, in the presence of nitrogen, the formation of methane hydrate always occurs at a lower temperature than in the pure state; further, the dissociation temperature can increase only at a nitrogen content greater than 20 vol.%, and with a higher probability, the lower the pressure at which the hydrate dissociates. At a pressure above 7×10^6 Pa, however, this stabilizing effect can no longer be detected.

This result of the experimental series is in contrast with the literature experience that nitrogen, as a gas with very small molecules, stabilizes gas hydrates. This fact too confirms that individual investigations of the interactions and comprehensive information are needed to permit an unambiguous decision as to which conditions are suitable for a given gas system to undergo gas hydrate formation. Nitrogen itself forms a very unstable hydrate, and in all probability this effect prevails at high concentration for the mixed hydrates. Its stabilizing effect can nevertheless be clear-cut only in those gas systems where the predominant component is of type H_{II}, thus containing molecules incorporated only into the large cavities of the clathrate structure; the nitrogen therefore increases the cohesion force effects within the lattice by being able to fill the small cavities, which would otherwise remain vacant.

3.2.2. Effects of electrolytes on stability

As mentioned above, the first detailed investigations on the inhibition of formation of gas hydrate of hydrogen sulphide with various condensed-phase additives were made by Bond and Russell [82]. The upper four-phase equilibrium-point of hydrogen sulphide hydrate is found at 29.4 °C. At this temperature, the vapour pressure of H_2S (which is approximately the same as that of the hydrate system) is 2.33×10^6 Pa. The experimental results are given in Table 3.5.

It was found that, in systems free from inhibitors, the relative quantities of water and the hydrate do not influence the accuracy of measurement. Graphical plots of the experimental results can be seen in Fig. 3.15.

For all solutions, the equilibrium p vs. T curves intersect the vapour pressure curve of liquid H_2S. The shift in position of the point of intersection (T_m) depends on the nature and concentration of the inhibitor. If T_m is the maximum temperature at which the solid hydrate can be in equilibrium with the given solution, then the decrease in T_m referred to the mol dm^{-3} or wt. % values of the solution can be determined from the experimental data. These ΔT_m data are listed in Table 3.6.

As NaCl is the cheapest of the listed materials, this was chosen by the authors as the basis of comparison of the inhibitions. Of the salts, NaCl and $CaCl_2$ dis-

Table 3.5. Main data on inhibition of formation of H₂S hydrate according to
Bond and Russel [82]

Inhibitor		Hydrate-formation parameters		Maximum temperature at which H₂S hydrate can still be found in the given solution
Compound	Weight %	T (°C)	p (10^5 Pa)	T_m (°C)
—	—	10.0	3.13	
		18.0	7.27	
		26.5	15.35	29.5
		29.5	22.93	
		15.0	4.85	
		20.8	9.89	
		28.3	22.32	
NaCl	10.0	1.7	2.12	
	10.0	13.9	6.67	21.7
	10.0	21.7	19.19	
	26.4 (sat)	−4.0	4.34	
	26.4	3.0	6.87	
	26.4	5.0	10.50	6.8
	26.4	7.0	13.33	
	26.4	7.0	14.85	
CaCl₂	10.0	1.7	1.72	
	10.0	15.3	7.47	22.2
	10.0	22.2	19.39	
	21.1	−2.0	3.74	
	21.1	4.0	6.66	
	21.1	8.0	10.40	10.8
	21.1	11.0	15.15	
	21.1	11.1	16.06	
	36.0 (sat)	−7.8	9.09	−7.8
Methanol	16.5	0.0	2.83	
	16.5	10.0	7.47	17.2
	16.5	16.9	15.35	
Ethanol	16.5	7.5	4.04	
	16.5	14.7	9.09	19.3
	16.5	18.6	15.15	
Ethylene glycol	48.5			3.3
	72.8			−35.0
Diethylene glycol	47.5			14.4
	71.2			−1.1
	85.5			−9.0
	95.0			−40.0
Dextrose	50.0	11.7	7.07	
	50.0	16.1	10.20	19.4
	50.0	19.4	17.97	

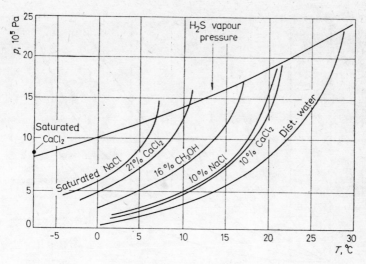

Fig. 3.15. Effects of inhibitors on the formation of H$_2$S hydrate [82]

played the highest inhibitory effects, the temperature of formation of H$_2$S hydrate decreasing to roughly the same extents in their presence.

The results of the hydrate inhibition experiments of Makogon and Sarkisyants [157], relating to pipeline gases, are presented in Fig. 3.16. These show the degrees to which the temperature of hydrate formation is lowered by the additives used, depending on their concentrations.

On the basis of Fig. 3.16, it can be stated that one of the most effective hydrate inhibitors is calcium chloride but, because of its corrosive action, it cannot be applied in practical natural-gas production and transportation. In connection

Table 3.6. Decrease in temperature of point T_m on inhibition of formation of H$_2$S hydrate [82]

Inhibitor	$\Delta T_{m, abs}$ (°C)	$\Delta T_{m, rel}$ (°C)	$\Delta T_{m, abs}$ (°C)	$\Delta T_{m, rel}$ (°C)
	relating to solution with composition expressed in mol/dm^3		relating to solution with composition expressed in mass units	
NaCl	4.3	1.00	8.8	1.00
CaCl$_2$	7.4	1.71	8.0	0.91
Methanol	2.5	0.57	9.5	1.08
Ethanol	2.9	0.68	7.8	0.89
Ethylene glycol	3.1	0.73	6.1	0.69
Diethylene glycol	3.1	0.73	3.6	0.41
Dextrose	3.1	0.71	2.0	0.23

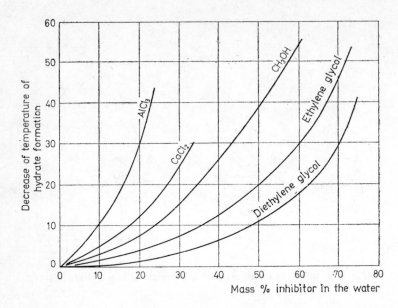

Fig. 3.16. Effects of inhibitors in decreasing hydrate formation, as a function of the composition [157]

with this, Makogon and Sarkisyants [157] made the noteworthy finding that the corrosive action of the very effective 30–35% $CaCl_2$ solution can be diminished by a factor of about 15 times by the addition of 0.5% $NaNO_2$, and by a factor of 40 times in the presence of 1.5% $NaNO_2$.

The variation of the "critical-point" of formation of propane hydrate (the upper four-phase equilibrium-point) in the presence of 0–30% NaCl was investigated and evaluated by Kostyuk [297]. He discovered that there is a relationship between the freezing-point of the salt solution and the extent to which the NaCl inhibits the formation of propane hydrate. The results were summarized in that, independently of the salt concentration, the difference between the temperatures of the freezing-point of the solution and of the formation of the hydrate is always 4.5–6.5 °C.

The mechanism of action of the various hydrate inhibitors must be connected with their decreasing effect on the vapour pressure of water, and thus indirectly with the freezing-point of the solution too.

In their experiment with natural gases, Aliyev and his co-workers [144] strived to elucidate the correlation between the freezing-point of the inhibitor and the temperature of hydrate formation. As inhibitors, $CaCl_2$, methanol, ethylene glycol and diethylene glycol were applied, and the following relationship was established:

$$t_f = -ak^2 \tag{3.7}$$

where t_f is the freezing-point of the solution in °C, k is the inhibitor concentration in wt. %, and a is a constant depending on the nature of the inhibitor.

At the same time, between the temperature of hydrate formation and the inhibitor concentration, the following relationship was found to hold:

$$t_H = -bk^2 + c \qquad (3.8)$$

where t_H is the temperature of hydrate formation in °C, k is the inhibitor concentration in wt. %, and b is a coefficient taking into account the nature of the inhibitor; it follows from the equation that c is the (pressure-dependent) temperature of hydrate formation in an inhibitor-free medium, where $k = 0$.

Combination of Equations (3.7) and (3.8) yields

$$t_H = \frac{b}{a} t_f + c \qquad (3.9)$$

i.e. a direct relationship between the temperature of hydrate formation and the freezing-point of the inhibitor. Table 3.7 gives the values of a and b found by Aliyev and his co-workers [144] for the inhibitors they used.

Table 3.7. The constants of Equation (3.9) for several inhibitors [144]

Inhibitor	a	b
$CaCl_2$	0.0465	0.0300
Methanol	0.0280	0.0195
Ethylene glycol	0.0150	0.0090
Diethylene glycol	0.0100	0.0050

The data from experiments made with a natural gas with a relative density of 0.59 are shown graphically in Figs. 3.17 and 3.18. The former illustrates the relationships between the inhibitor concentration and the temperature of hydrate formation in the pressure interval $1-5 \times 10^6$ Pa, while in Fig. 3.18 the temperature of hydrate formation is plotted against the freezing-point of the inhibitor.

Andryushchenko and Vasilchenko [145] tried to replace the hydrate inhibitors most widely used in practice by applying cheaper, but less effective additives, taking into account primarily the loosening action of these electrolyte solutions on the structure of water. By means of thermodynamic calculations, they proved that the chlorides and bromides of the alkaline earth metals, the alkali metals, the ammonium ion and the metals of the zinc group are the most effective inhibitors. With simple gas hydrates, the sulphates of the alkali metals, iron and nickel can also be used. Similarly good results were achieved with the nitrates of calcium, magnesium and aluminium. Of these salts, the chlorides of sodium, ammonium,

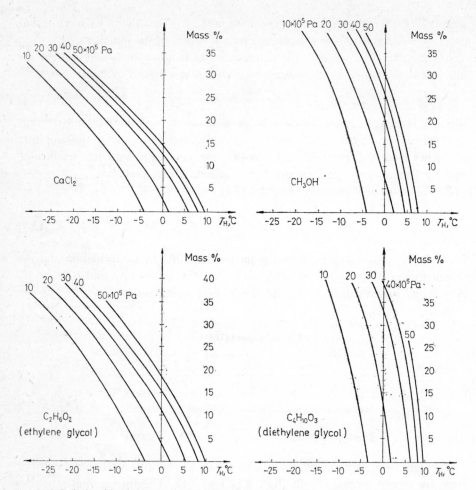

Fig. 3.17. Relationships between the inhibitor concentration (four different inhibitors) and the temperature of formation of the hydrate at various pressures for natural gas with a relative density of 0.59 [144]

calcium and magnesium are available relatively cheaply. The use of ammonium chloride under operating conditions is not advisable, as it is very aggressive from the aspect of corrosion, but the other compounds do not result in a dangerous degree of corrosion of carbon-steel pipelines in the presence of natural gases.

Ammonium salts are generally very good inhibitors, but their applicability is limited by the fact that the carbonate or hydrocarbonate is formed from the carbon dioxide very frequently present in the gaseous medium, and these products give rise to the clogging of the gas pipelines.

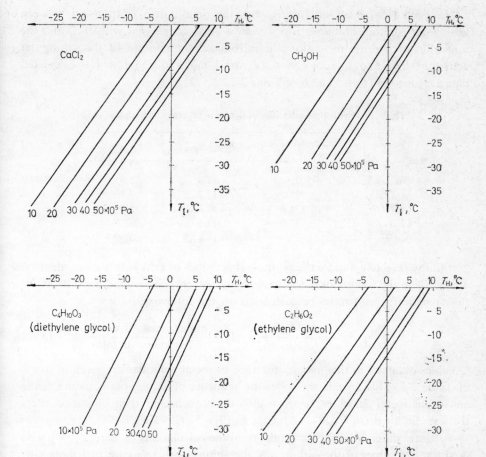

Fig. 3.18. Relationships between the temperature of hydrate formation and the freezing-point of the inhibitor (the same inhibitors as in Fig. 3.17) at various pressures for natural gas with a relative density of 0.59 [144]

The same authors [145] found that, for all salt solutions, a critical concentration can be determined, at which there is no possibility of hydrate formation, no matter how high the pressure or how low the temperature employed in the experiment. This is a concentration (the "limit of full hydration"; cf. Section 2.3.2) at which all the water molecules are bonded in the hydrate sheats of the salt ions in the electrolyte solution, so that the possibility of formation of a clathrate structure is completely eliminated. The values of the critical concentration which they gave for a number of electrolytes are to be seen in Table 3.8.

Makogon [158] dealt intensively with the quantitative relationship between salting-out (or the solubility, which is in a close connection with this) and the inhibition of hydrate formation; he introduced the concept of the salting-out activity, referred to 1 gram cation (A_p) or 1 mol cation (A_m). His calculated data are given in rows 5 and 6 of Table 2.2.

Table 3.8. Critical composition values for several electrolytes [145]

Salt	Critical composition (wt. %)	n_{H_2O}/n_{salt}	n_{H_2O}/n_{ion}
$CaCl_2$	26.0	17.8	5.93
$MgCl_2$	23.0	13.7	4.56
$Ca(NO_3)_2$	34.0	17.4	5.80
$LiCl$	17.0	11.5	5.75
$NaCl$	22.0	11.8	5.90

If it is wished to compare the salting-out activities of salts with a given identical anion, then the salting-out activities $A_{M_aX_b}$ of the corresponding salt M_aX_b (referred to 1 gram) must be calculated on the following basis:

$$A_{M_aX_b} = A_p \frac{a \cdot M_M}{a \cdot M_M + b \cdot M_X} = A_p \frac{a \cdot \text{molar mass of cation}}{\text{molar mass of salt}} \qquad (3.10)$$

Values obtained in this manner for various metal chlorides are given in row 8 of Table 2.2. This reveals the following sequence of salting-out activities of the anhydrous metal chlorides, which is also the sequence of their inhibition effectivities: $LiCl > BeCl_2 > AlCl_3 > MgCl_2 > CaCl_2 > NaCl > KCl$.

A slightly different sequence results from the calculations by Berecz [298], who used the z/r^2 values of the cations and the anions, with reference to 1 mole salt:

$$A'_{M_aX_b} = \frac{n_M \dfrac{z_M}{r_M^2} + n_X \dfrac{z_X}{r_X^2}}{n_M + n_X} \qquad (3.11)$$

$$A''_{M_aX_b} = \frac{n_M \dfrac{z_M}{r_M^2} + n_X \dfrac{z_X}{r_X^2}}{(n_M + n_X) M_{salt}} \qquad (3.12)$$

Equation (3.12) refers to 1 g salt, M_{salt} being the relative molecular mass of the salt.

Berecz found the following sequence for the effectivity of inhibition (see line 10 of Table 2.2):

$$BeCl_2 > LiCl > AlCl_3 > MgCl_2 > NaCl > CaCl_2 > KCl.$$

The values and sequence resulting from the latter calculations appear to be more probable, since these equations better take into account the surface charge densities and field strengths of the ions, which overall is one of the main factors in the positive hydration effect, or in other words the structure-maker effect, and hence in the perturbation of the water structure by the electrolyte.

It can be ascribed to the positive hydration effect of $CaCl_2$ that in a low-pressure gas pipeline, $CaCl_2$ is much more effective than alcohols in preventing the formation of ice and hydrates. A nomogram has been designed by Atonov and Soldatov [299], by means of which the freezing-point of a 30% $CaCl_2$ solution can be determined as a function of T and p.

3.2.3. Effects of alcohols on stability

In the natural-gas industry, alcohols (generally methanol and glycols) are mostly used as inhibitors to prevent hydrate formation. The first calculations in this respect were those of Hammerschmidt [300, 301], Poldermann [302] and Pieroen [95]. On the basis of their work, the decrease in the temperature of formation of a gas hydrate, which is dependent on the inhibitor concentration, can be calculated to a good approximation via the following formula:

$$\Delta T = \frac{Ks}{M(100-s)} \tag{3.13}$$

where ΔT is the decrease in temperature of the formation of the gas hydrate in °C, K is a constant depending on the nature of the inhibitor and on the heat of

Table 3.9. Values of constant K of Equation (3.13)
for several inhibitors

Inhibitor	K [300, 301]	K [158]
Methanol, ethanol and isopropanol	2335	1228
Ethylene glycol	2700	2195
Diethylene glycol	4000	2425
Triethylene glycol	5400	—
Propylene glycol	—	2195

formation of the hydrate, s is the concentration of the inhibitor in wt.% in the liquid phase, and M is the relative molecular mass of the inhibitor. Values of constant K calculated from the data of Hammerschmidt [300, 301] and of Makogon [158] are contained in Table 3.9. One cause of the differences between the data of these authors is the fact that the calculations of Hammerschmidt were made in °F, and those of Makogon in °C.

For the determination of the quantity of alcohol to be added to the pipeline, it is not sufficient to carry out calculations referring only to the gas phase; attention must also be paid to the fact that a certain proportion of the alcohol passes

Table 3.10. Calculation of required amount of alcohol inhibitor to be introduced into pipeline, according to Hammerschmidt [301]

1 Effective operating pressure of gas (10^5 Pa)	49.69
2 Dew-point and temperature of saturation of gas with water (°C)	15.6
3 Minimum temperature developing in production or transportation process (°C)	4.4
4 Water vapour content of gas at temperature and pressure of saturation (kg/10^6 m³ gas)	372
5 Water vapour content of gas at minimum operating temperature (kg/10^6 m³)	186
6 Change in water vapour content, calculated from the two foregoing data (kg/10^6 m³)	186
7 Temperature of gas hydrate formation at operating pressure (°C)	14.4
8 Desired freezing-point depression of solution (°C)	10.0
9 Methanol content of liquid phase necessary for this depression (mass %)	19.9
10 Phase equilibrium quotient of methanol as function of p and T	18.7
11 Concentration of methanol in gas phase (kg/10^6 m³)	373
12 Concentration of methanol in aqueous phase (kg/10^6 m³)	46.2
13 Rate of methanol injection (kg/10^6 m³)	419

into the gas phase, e.g. the vapour pressure of methanol is relatively high. Since the inhibition effect is particularly strongly pressure-dependent in the case of alcohols, the preplanning of the alcohol injection requires a knowledge of the hydrate characteristics of the gas flowing.

Following Hammerschmidt [301], the course of such a calculation for a natural gas with a relative density of 0.675 can be seen in Table 3.10.

Of the values in Table 3.10, the first three give the starting data; the fourth can be calculated from data 1 and 2 via Fig. 3.19 (d); the fifth can similarly be calculated via Fig. 3.19 (d), from data 1 and 3; the sixth is the difference between data 4 and 5; the seventh can be calculated from datum 1 via Fig. 3.19 (a); the eighth is the difference between data 7 and 3; the ninth can be calculated via Fig. 3.19 (b) from datum 8; the tenth can be calculated from data 1 and 3 via Fig. 3.19 (c); the eleventh is given by the product of data 9 and 10; the twelfth is given by the product of data 9 and 10, divided by 100, datum 9 being subtracted from the result; and the thirteenth is the sum of data 11 and 12.

Extending the work of Hammerschmidt, a relationship is reported by Makogon and Sarkisyants [157] by means of which the quantity of alcohol necessary under

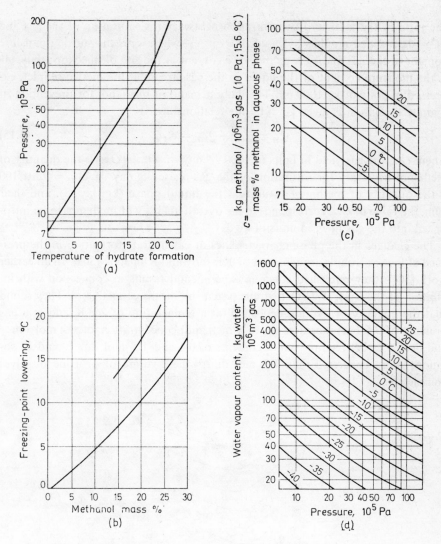

Fig. 3.19. Auxiliary diagrams used to calculate the individual fundamental data required for planning of the amount of alcohol to be added to inhibit hydrate formation, according to Hammerschmidt [301]

given conditions can be calculated:

$$e_M = x\left(\alpha + \frac{\Delta e}{100}\right) \tag{3.14}$$

where e_M is the specific methanol demand (kg/1000 m³), x is the mass percentage of methanol in the aqueous phase (obtained from graphical data), $\alpha = e_{MG}/x$ is

7

the ratio of the methanol concentration necessary for saturation of the gas and of x (determined graphically), and $\Delta e = e_n - e_k$, where e_n is the starting moisture content of the gas and e_k is its moisture content at the site of injection; thus, Δe is the moisture content of the liquid in the gas stream at the methanol inlet, expressed in units of kg/1000 m³. The daily demand of methanol for prevention of hydrate formation is given [157] by the relationship:

$$Q_M = Q_{MG} + Q_{MF} = e_M Q \tag{3.15}$$

where Q is the gas yield in units of 1000 m³ day^{-1}, $Q_{MG} = Qx\alpha$ is the quantity of methanol necessary for saturation of the gas phase (kg day^{-1}), $Q_{MF} = Q\Delta ex/100$ is the quantity of methanol saturating the liquid phase (kg day^{-1}), and their sum, the value of Q_M, corresponds to the overall quantity of methanol to be introduced into the gas flow, in units of kg day^{-1}.

The changes in the phase characteristics of gas hydrate formation in the presence of lower alcohols are reported by Krasnov and Klimenok [303]. In an earlier work [304], they published concrete experimental results in connection with inhibition of the $C_3H_8.17H_2O$ hydrate system by means of methanol. They found that the quantity of hydrate deposited (at a temperature of 263 K and at a gas pressure of 3.5 bar) increases up to a methanol concentration of 0.4 mol dm^{-3} and decreases between methanol contents of 0.4 and 1.5 mol dm^{-3}, and in solutions more concreted than 1.5 mol dm^{-3} the hydrate of propane dissolves completely.

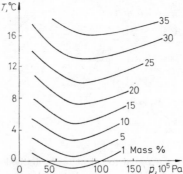

Fig. 3.20. Effects of methanol content and pressure on the temperature of formation of methane hydrate [158]

From their investigations of hydrate formation by natural gases in the presence of ice [305], they demonstrated that hydrate formation is increased here too if alcohol solutions of low concentrations are used. A good inhibitory effect could be achieved at $-5\,°C$ only with a methanol solution more concentrated than 27 mole %.

Table 3.11. Effects of methanol content on formation and dissociation temperatures of CH$_4$ hydrate [306]

Gas pressure	Methanol content											
	0%		5 vol. %		10 vol. %		15 vol. %		20 vol. %		25 vol. %	
(10^5 Pa)	T_{form}	T_{diss}	T_{form}	T_{diss}	T_{form}	T_{diss}	T_{form}	T_{diss}	T_{form}	T_{diss}	T_{form}	T_{diss}
15.15	—	−6.0	—	—	—	—	—	—	—	—	—	—
20.20	—	−2.7	—	−7.7	—	—	—	—	—	—	—	—
25.25	—	−0.5	—	−5.0	—	—	—	—	—	—	—	—
30.30	−5.8	1.3	—	−2.7	—	−5.8	—	—	—	—	—	—
35.35	−4.3	2.6	—	−0.8	—	−4.0	—	−6.8	—	—	—	—
40.40	−3.0	3.8	—	0.8	—	−2.6	—	−5.2	—	—	—	—
45.45	−1.8	5.0	−4.8	2.4	−6.5	−1.0	—	−3.6	—	−6.5	—	—
50.50	−0.8	5.8	−3.0	3.7	−4.5	0.1	—	−2.5	—	−5.3	—	—
55.55	0.0	6.7	−1.7	5.0	−3.3	1.4	−5.8	−1.4	—	−4.1	—	—
60.60	0.9	7.5	−0.6	6.3	−2.3	2.4	−4.8	−0.3	—	−3.1	—	−5.9
65.65	1.7	8.0	0.5	7.3	−1.5	3.2	−4.0	0.4	—	−2.3	—	−5.1
70.70	2.2	8.4	1.7	8.0	−0.7	3.9	−3.2	1.0	—	−1.7	—	−4.4
80.80	3.2	9.2	3.7	9.3	1.2	5.1	−1.8	2.2	—	−0.7	—	−3.2
90.90	3.8	9.8	5.1	9.9	2.5	5.8	−0.7	2.9	—	0.1	—	−2.3
101.00	4.3	10.3	5.8	10.0	3.2	6.3	0.1	3.4	—	0.6	—	−1.9
Freezing-point of solution	0 °C		−2.5 °C		−4.9 °C		−7.8 °C		−11 °C		−16 °C	

Table 3.12. Effects of methanol content on formation and dissociation temperature of CO_2 hydrate [306]

p_{CO_2}		Methanol content											
		0%		5 vol. %		10 vol. %		15 vol. %		20 vol. %		25 vol. %	
		T_{form}	T_{diss}	T_{form}	T_{diss}	T_{form}	T_{diss}	T_{form}	T_{diss}	T_{form}	T_{diss}	T_{form}	T_{diss}
(10^5 Pa)	°C												
15.15		−4.2	2.0	—	−3.0	—	−5.4	—	—	—	—	—	—
20.20		−2.6	3.8	—	−0.8	—	−2.8	—	−5.2	—	−7.6	—	—
25.25		−1.4	5.0	—	0.9	—	−1.3	—	−3.8	—	−6.4	—	−9.0
30.30		−0.5	6.6	−3.9	2.4	−5.6	0.1	—	−2.4	—	−4.7	—	−7.1
35.35		0.4	7.7	−2.8	3.2	−4.2	0.8	—	−1.6	—	−3.8	—	−6.0
40.40		1.0	9.0	−2.1	3.9	−3.5	1.4	—	−0.4	—	−2.5	—	−4.7
45.45		1.4	9.6	−1.6	4.3	−2.9	1.5	—	0.2	—	−1.9	—	−4.1
50.50		1.5	9.7	−1.3	4.5	−2.6	1.6	—	0.3	—	—	—	—
55.55		1.6	9.8	−1.6	4.6	−2.5	1.6	—	0.4	—	—	—	—
60.60		1.65	9.83	−1.7	4.7	−2.4	1.6	—	0.5	—	—	—	—
Freezing-point of solution		0.0 °C		−2.5 °C		−4.9 °C		−4.9 °C		−11.0 °C		−16 °C	

In a later work, Makogon [158] investigated in detail the gas pressure-dependence of the ability of methanol to bring about inhibition. Figure 3.20 illustrates the effects of the methanol content and the pressure on the temperature of formation of methane hydrate, i.e. on the corresponding $\Delta T\,°C$ value. It is evident from the figure that, at low methanol concentration ($\leqslant 3\%$), the phenomenon is observed (not only for methane, but also for natural gas) that, in the pressure interval $3.75–11.56 \times 10^6$ Pa, the temperature of formation of the hydrate of the gas does not decrease, but instead increases (i.e. ΔT is smaller than zero); thus, the hydrate is stabilized here. With these experiments, Makogon [158] wished to confirm that the effect of the inhibitor depends on the solubility of the gas in water, and that it can be interpreted by the variation of the solution structure forming due to the interaction of the gas–water system.

The experimental data of Berecz and Balla-Achs [306] too prove that, as the pressure is increased, the inhibitory effect of methanol decreases until a limit is reached. Their experiments were made with regard to the conditions of formation and dissociation of the hydrates of a dry natural gas containing 98% methane and of pure gaseous carbon dioxide. These data are shown in Tables 3.11 and 3.12. The temperature decreases characteristic of the dissociation of these gas hydrates are illustrated in Figs. 3.21 and 3.22.

It is obvious from the figures that, for the natural gas consisting mainly of methane, with a methanol concentration lower than 4 wt.% the inhibition is

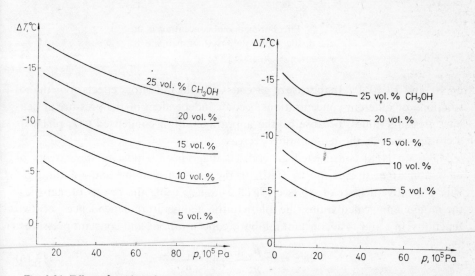

Fig. 3.21. Effect of methanol concentration on the dissociation temperature of methane hydrate [306]

Fig. 3.22. Effect of methanol concentration on the dissociation temperature of CO_2 hydrate [306]

the poorest in the pressure range $7-10 \times 10^6$ Pa; indeed, a slight degree of stabilization occurs here, which is more apparent in Fig. 3.23.

In the inhibition of carbon dioxide hydrates with methanol, it was found [306] that the decreases in the temperatures of both hydrate formation and hydrate decomposition in the presence of methanol are smaller than those for methane hydrate, but the effect mentioned is minimum here too at pressures between

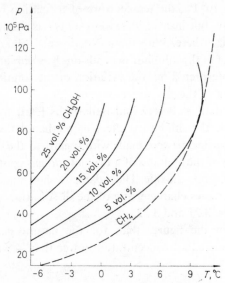

Fig. 3.23. Effect of methanol concentration on
the dissociation equilibrium of methane
hydrate [306]

2×10^6 and 3×10^6 Pa. In the case of gaseous CO_2, the inhibitory effects of methanol solutions of different concentrations become independent of the pressure at definite pressure values, because of the appearance of a condensed CO_2 phase as a consequence of the temperature decrease.

Makogon [158] constructed generalized diagrams to indicate the extents of inhibition by methanol and by $CaCl_2$ in the cases of methane and a natural gas with a relative density of 0.6 (see Fig. 3.24(a–d)); using the results of statistical thermodynamic calculations, he plotted the temperature-dependences of two functions ($\ln z$ and y) at constant inhibitor concentrations and constant pressures:

$$\ln z = \ln \frac{p_{w,H}}{p_{w,o}} = \ln \frac{p_w - p_s}{p_{w,o}} \tag{3.16}$$

where p_w is the vapour pressure of the condensed water (in Torr), $p_{w,o}$ is the vapour pressure of the water above the hypothetical empty host lattice, p_s is

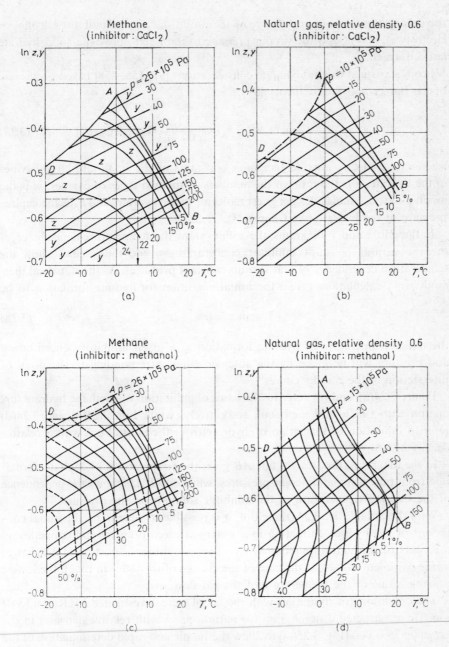

Fig. 3.24. Diagrams used to determine initial conditions of gas hydrate formation in various systems, according to Makogon [158]; (a) and (b): CaCl$_2$ inhibitor, (c) and (d): methanol inhibitor

the decrease in the vapour pressure of the water, due to the inhibitor (alcohol or electrolyte), and $p_{w,H}$ is the vapour pressure of the water above the solid hydrate lattice formed.

From statistical thermodynamic considerations [98, 158, 458] based on the use of the Barrer and Stuart theory [99]:

$$y = \frac{n}{(1+n)m}\ln{(1+C_{A1}p_A)} + \frac{1}{(1+n)m}\ln{(1+C_{A2}p_A)} \qquad (3.17)$$

where C_{A1} and C_{A2} are the Langmuir constants of the small and large cavities in the hydrate, n is the ratio of the numbers of the two types of cavities, m is the number of water molecules per guest molecule in the hydrate, and p_A is the vapour pressure of the hydrate (see Chapter 4).

In these $\ln z$ and y diagrams, the z lines show the temperature-dependence of $\ln z$ at various constant inhibitor concentrations, and the y lines show the temperature-dependence of y at various constant pressures. As the statistical thermodynamic calculations reveal the initial condition for hydrate formation to be

$$\ln z = y \qquad (3.18)$$

the initial temperature of hydrate formation at a given inhibitor concentration and pressure will be given by the abscissa value corresponding to the point of intersection of the z and y curves.

Figure 3.24(a) and (c) refer to the cases of inhibition of methane hydrate formation with $CaCl_2$ and methanol, respectively, while Fig. 3.24(b) and 3.24(d) refer to inhibition of formation of hydrate from the natural gas with a relative density of 0.6.

In the diagrams, the straight line AB gives the temperatures of hydrate formation from pure water at different pressures, while curve AD shows the dependence of the freezing temperature of the inhibitor solution on the pressure.

With the aid of Fig. 3.24, therefore, it is possible to determine the initial conditions of gas hydrate formation in a system at given pressure and temperature, the inhibitor concentration required, the change in this as a function of the change in the parameters (even in the case of the change of the latter in time), the lowest possible value of the temperature of the gas flow, etc.

The conditions of inhibition with methanol were investigated by Rosen [307] too. He constructed a nomogram for natural gases with relative densities in the range 0.575–1.000 (Fig. 3.25(c)) to allow the simple and rapid determination of the rate of injection needed for the desired depression of the temperature of hydrate formation. He reports that this quick method of determination can be well applied in practice, with an accuracy within $\pm 10\%$.

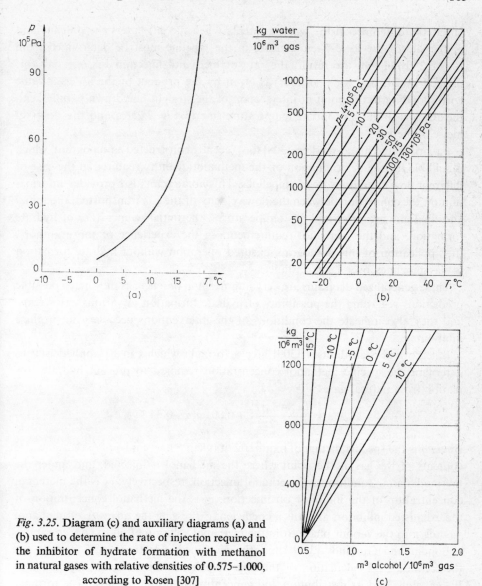

Fig. 3.25. Diagram (c) and auxiliary diagrams (a) and (b) used to determine the rate of injection required in the inhibitor of hydrate formation with methanol in natural gases with relative densities of 0.575–1.000, according to Rosen [307]

For calculations with the nomogram of Rosen (Fig. 3.25(c)), as starting data it is necessary to know the temperature of hydrate formation at the appropriate pressure on the basis of Fig. 3.25(a) or Fig. 3.24 (Makogon) the minimum temperature in the gas pipeline, and the decrease of temperature on hydrate formation, and it is also necessary to calculate the quantity of water able to condense. For the latter, the difference in the water contents when the gas enters the pipeline and at the site of the lowest temperature in the pipeline must be determined from Fig. 3.25(b). From the result, the rate of methanol injection for one day (or, divided by 24, that for one hour) is given by the product of the abscissa value corresponding to the point of intersection of the straight line drawn parallel with the abscissa to the appropriate temperature line in Fig. 3.25(c) and the flow velocity of the gas.

Favoritov and his co-workers [308] too, recently constructed a nomogram, which is suitable for the determination of the methanol quantity required in the case of different flow conditions in gas pipelines. This nomogram also provides information on the connection between the flow velocity of the gas transported, the initial and final gas pressures and gas temperatures, the initial temperature of hydrate formation, and the methanol requirement. In the experience of those authors, the application of this nomogram ensured operation without hydrate formation in gas pipelines over large distances.

These generalized diagrams are very important in practice, as they make possible predictions regarding the possibility of hydrate formation in various gas sytems, and they also indicate the conditions of the interventions necessary to produce inhibition.

Buleiko and Starodubtsev [309] give a formula which can be applied well to determine the specific methanol concentration required to prevent hydrate formation from natural gases:

$$q = \frac{(w_1 - w_2) \, c_2}{c_1 - c_2} + 0.001 \, c_2 \alpha + 0.34 \tag{3.19}$$

where q is the specific methanol requirement (g cm^{-3}), w_1 and w_2 are the moisture contents of the gas at the point where the methanol is injected, and under the thermodynamic conditions of methanol injection, respectively, c_1 is the methanol concentration of the inhibitor on injection, c_2 is the methanol concentration of the exhausted inhibitor, and α is a coefficient expressing the quantity of methanol passing into the vapour phase from the methanol solution of given concentration.

Bondar and Guseinov [310] similarly developed a mathematical formula for the technological calculation of the flow velocity of the methanol injected into the gas flow, and its distribution and concentration in the gas flow. The formula was made suitable for the case when the gas system had originally contained a given quantity of methanol:

$$m_{\mathrm{a}} = bk + \frac{b}{a - b} \, (aq + bk - 1.25abk) + Rc \tag{3.20}$$

where a is the initial concentration of methanol (kg dm^{-3}), b is the concentration of methanol remaining in the solution (kg dm^{-3}), c is the quantity of methanol in the condensed hydrocarbon, m_a is the specific methanol requirement, q is the water content of the gas ($dm^3/1000\ m^3$), R is the quantity of condensate separating out from the gas, and k is a constant coefficient.

In a publication reviewing their most recent results, Pick and Simanek [311] give an account of the calculation method they employ at present in Czechoslovakia to determine the optimum quantity of methanol needed to prevent hydrate formation in gas pipelines. The method is based on determination of the equilibrium concentration of methanol in the gas phase, and at the same time ensures that the methanol concentration in the liquid phase should be optimum, overdosing being avoided. For determination of the equilibrium methanol concentration of the gas phase, they use the expression

$$Pyv = x\gamma p_M \tag{3.21}$$

where y is the mole fraction of methanol in the gas phase, x is the mole fraction of methanol in the liquid phase, P is the total pressure of the system in MPa, p_M is the vapour pressure of the methanol, in MPa, at the temperature of the system, v is the fugacity coefficient, and γ is the activity coefficient of the methanol. If the activity coefficient of the methanol solution is taken as 1.05, and the fugacity coefficient is determined by thermodynamic calculation, the equation can be solved for x.

The minimum methanol concentration necessary for inhibition, so as to guarantee the undisturbed transportation of gas in the pipelines, is determined via the equation

$$m = 0.014\ 29Qy \tag{3.22}$$

where m is the amount of methanol to be injected (in units of $kg\ h^{-2}$), while Q is the quantity of gas transported at the operating pressure ($m^3\ h^{-1}$). In the determination of this latter datum, it must be taken into consideration that y too is a function of temperature; in the employment of the correlation, therefore, a working diagram is needed in which the $y = f(T)$ data relating to the various operating pressures are plotted. According to Pick and Simanek [311], this calculation method can be used reliably if the composition of the pipeline gas can be regarded as practically constant.

Parkhomenko [312] reports an apparatus permitting the automatic control of the injection of methanol into gas pipelines.

Although the inhibition brought about by methanol is a good one, there are some difficulties in its application, because its removal from the gas flow and its regeneration are accompanied by many technological problems.

Dashdamirov and Rasulov [313] have made proposals regarding the removal of methanol from the gas flow. They carried out investigations on the adsorption of methanol in 1:1, 1:2 and 2:1 alcohol–water mixtures. They could eliminate hydrate precipitation in the gas separator and in the post-separator pipeline only if the alcohol/water ratio was equal to or larger than 2:1. At such alcohol concentrations, they could achieve nearly total desorption with silica gel NS—62 and with activated carbon SKT.

Miroshnikov and his co-workers [314] investigated the effects of different inhibitors on the hydrates of the hydrocarbon gases, and found the glycol ethers to be the most suitable ones.

The various glycols are also good inhibitors, but they play an important role mainly in the process of gas drying. Although they are more expensive than the inhibitors discussed above, they have the great advantage of very good regenerability. The technology of this regeneration is known and is widely employed. For the inhibition of hydrate formation in natural gases flowing in pipelines, ethylene glycol is the most appropriate; in separators and heat-exchangers, on the other hand, di- and triethylene glycols can be used more economically, because of their lower vapour pressures. Ethylene glycol also has better viscosity characteristics, and it dissolves to a lesser extent in hydrocarbons than does diethylene glycol; accordingly, ethylene glycol generally has the better effectivity in the hydrate inhibition process.

Inhibition by means of alcohols or other organic compounds is evidently not restricted only to natural gases.

The different solubilities in alcohol of the various gases is a feature which cannot be neglected in inhibition with alcohols. Yorizane and his co-workers [315] evaluated the solubility data relating to gaseous CO_2, H_2S, H_2 and H_2S+H_2 mixtures thermodynamically. They determined the Henry constants characteristic of the solubilities of these gases in methanol up to a pressure of 50 bar in the temperature range between -30 and $0\,°C$, together with other thermodynamic data, such as the values of the heats of dissolution referred to the infinitely dilute solutions.

Bond and Russell [82] examined the possibility of inhibiting the process of hydrate formation by gaseous H_2S not only with electrolytes, but also with alcohols, glycols, sucrose and dextrose. In their experience, the use of alcohols as inhibitors made the investigations difficult, because a considerable quantity of water was bound from the liquid phase during hydrate formation. Accordingly, the alcohol concentration continuously changed during the process, and thus the concentration of alcohol in equilibrium with the hydrate could not be determined exactly. However, their investigations succeeded in establishing that the temperature of formation of H_2S hydrate decreases to very similar extents in the presence of methanol or ethanol; dextrose is practically ineffective, while in the case of a sucrose solution the accuracy of the determination is influenced by the comparatively high viscosity of the solution. When ethylene glycols were used, the points representing the actual hydrate formation could be determined only with a large scatter, and therefore the data were not published. Nevertheless, it was stated that glycols influence the formation of H_2S hydrate to only a moderate extent.

Taking into consideration the easy regenerability of the glycols, their high efficiency as inhibitors to prevent hydrate formation from natural gases and the gases accompanying them, and the low inhibitor losses, Burnykh and his co-workers [316] recommended the introduction of a mixture of 70–80% ethylene glycol (or possibly propanol) and 30–20% methanol into gas pipelines.

3.2.4. Effects of a condensed apolar phase and of other compounds on stability.
Hydrates of liquid hydrocarbons. Stabilities of mixed hydrates

The values of the formation and decomposition parameters of gas hydrates are influenced considerably if some gas component forms a condensed phase under the conditions of hydrate formation.

In deriving a formula suitable for calculation of the inhibitor quantity necessary to prevent the formation of hydrates of natural gases, Kedzierski and Krzyzanowski [317] took into account that the quantity of alcohol to be injected is appreciably affected not only by the moisture content of the gas, but also by the accompanying hydrogen sulphide present, and especially by the quantity of heavy hydrocarbons.

The investigations of a number of authors uniformly demonstrate that the presence of liquid hydrocarbons definitely decreases the stability of the hydrate formed. The cause of this is presumably that the condensed gas forms a viscous, foamy, liquid mass with water, in which the continuity of the gas stream slows down or prevents the development of a closed clathrate structure. Such experimental results have been reported by Musayev [129], Musayev and Chernikhin [128] and Fomina [134].

Kuliyev and his co-workers [318] investigated the conditions of hydrate formation from mixtures of various petroleum fractions (gas condensate, solar oil, ligroin) at temperatures of 7–11 °C and pressures of 45–50 bar. By varying the water/liquid hydrocarbon ratio between 1:0.57 and 1:1.50, they found that the stabilities of hydrates of gas systems containing liquid hydrocarbons are 3–3.5 times lower than those of the pure gas hydrates.

The combined effect of a gas condensate and of light petroleums (added separately to the system) in changing the conditions of hydrate formation and decomposition was investigated by Dzhavadov [138]. He found that the hydrates produced in the presence of petroleum have a looser structure, and adhere less to the surface of the pipeline. Both the gas condensate and petroleum form a film on the surface of water; this prevents contact between the water and the gas, eliminating gas diffusion, and thereby decreasing the possibility of the growth of the hydrate nuclei.

Musayev and Kuliyev [319] checked the conditions of hydrate formation in low-temperature separators. They established that, by bubbling the gas through a non-polar liquid system (condensate, crude iol, solar oil), the separation of both gas hydrate and ice can be prevented if adequate working parameters are used. The most important of these parameters are the height of the liquid column, the maximum gas content of the liquid, its density, the manner of feeding in and leading off the liquid, the rate of flow of the gas, and the optimum working pressure. In the course of one of their experiments, the separator, filled with condensate and kept at a working pressure of 50 bar, was connected with a gas having an initial flow rate of 0.1 $m^3 h^{-1}$, which resulted in a linear flow velocity of 0.025 m s^{-1} in the separator. While the gas flow velocity did not exceed 0.4 m s^{-1}, there was practically no foam formation. At a velocity above this, foaming began rapidly and to such a degree that at 0.8 m s^{-1} the boundary surface between the gas and the liquid practically disappeared and thereafter no hydrate formation could be observed.

The action of liquid hydrocarbons in inhibiting hydrate formation is very important under operating conditions, and since the presence of the condensed hydrocarbon phase alone is not always sufficient to prevent hydrate formation, reports have appeared on evaluations of the combined effects of alcohols or glycols and the condensate. Very detailed information can be found, for instance, in the paper of Scauzillo [320]; this involves laboratory work in which the conditions of formation of the hydrates of three different hydrocarbons were studied in the presence of brine, distilled water, an absorbent oil additive and diethylene glycols, both separately and together.

Scauzillo found that the temperature of formation of the hydrate of the natural gas examined was lowered by the brine by 1.7 °C in comparison to the value measured in the presence of distilled water. The difference was even greater with various concentrations of ethylene glycol additives. Up to a glycol content of 42%, the freezing-point depressions coincided well with the values calculated on the basis of the empirical relationship given by Hammerschmidt [301]. In the presence of liquid hydrocarbons, a further decrease could be observed in the temperature of hydrate formation; this decrease was the larger, the greater the quantity of oil present, i.e. the smaller the gas/oil ratio. When glycol too was added, there was no change in the tendency of the liquid hydrocarbons to effect inhibition, but the concrete ΔT values were altered; Scauzillo justified this as being due to the dissolution of the glycol in the oil. He assumed that the effective concentration of glycol is diminished by the presence of oil, and that the effects of the condensed hydrocarbon and the glycol are therefore not manifested additively in the aqueous phase.

The possible hydrates of derivatives of liquid hydrocarbons have more complicated structures than those of the gaseous hydrocarbons, and hence the effect of a third component is still more difficult to interpret in a clear-cut way than described above.

In their review, Lippert and his co-workers [87] survey the investigations of such systems. The third components influencing the properties of liquid hydrates are classified into three main groups:

1. Materials not forming hydrates, including those consisting of molecules too large to be incorporated into the cavities in the host lattice, or which dissolve very well in water.

2. Hydrate-forming materials, the hydrates of which are formed near 0 °C at a pressure of 1 bar.

3. Potential hydrate-forming materials, the hydrates of which are formed at pressures higher than atmospheric.

The effect of ethanol on the hydrate of chloroform was studied by Sampson [321]. The investigations were extended by Lippert [322] to the application of isopropanol and acetone; for chloroform hydrate, he gave the mean decrease in the decomposition temperature for one mol of the third component as 6.7, 7.5 and 8.7 °C for ethanol, isopropanol and acetone, respectively.

As regards additives forming hydrates near 0 °C at atmospheric pressure, systems were examined which dissolve only in the non-aqueous component, e.g. the mixed hydrates of chloroform and methyl iodide [322]. In this system, the decomposition temperature always lay between the decomposition temperatures of the hydrates formed by the two pure constituents, and on this basis it was assumed that a mixed hydrate forming a solid solution may be produced.

According to Lippert [322], the condition for formation of a solid solution is that there should be no essential difference between the molecular dimensions of the components undergoing mixing and hydrate formation.

For the investigation of the conditions of decomposition of such mixed hydrates, cyclopentane was chosen as reference compound [322]. The decomposition temperature of cyclopentane hydrate under atmospheric conditions is 7.2 °C, which is the highest value for a liquid hydrate. Methyl iodide (which forms a hydrate) and methylcyclopentane and isopentane (which do not form hydrates) were used as the third component. The experiments gave the expected results: in the presence of the hydrate-forming methyl iodide, a mixed hydrate was formed, the decomposition-point of which lay between those of the hydrates of the two pure components; in the cases of the additives isopentane and methylcyclopentane, however, which dissolve in cyclopentane but not in water, the decomposition-point of cyclopentane hydrate was decreased by about 0.05 °C by 1 mole % of the third component.

It was concluded from these experiments that, on the formation of mixed hydrates, the dominating factor in the determination of the properties is the molecular dimensions of the components present in the system. If each component is capable of hydrate formation, then solid solutions of hydrates are formed corresponding to the mixing proportions, the decomposition-points of which lie between the decomposition-points of the hydrates of the two pure components.

If the molecules of one of the components are too large to form a hydrate, whereas those of the other one are capable of hydrate formation, characteristic decreases are observed in the extent of hydrate formation and in the decomposition-point of the hydrate. True double hydrates can be formed only if one molecule is large (a van der Waals diameter in the range 0.55–0.79 nm), while the other one is very small (e.g. H_2S). For an approximation to these latter cases, a uniform small decomposition-point depression of 0.05 °C per mole % was considered appropriate in the mixtures of cyclopentane with the components not forming hydrates.

The investigations by Maass and Boomer [323] revealed that ethylene oxide has an especially high affinity for water, and that it forms a hydrate of hydrocarbon type containing 6 or more water molecules, which melts at 10.7 °C. Lippert [322] too investigated how this potential hydrate-forming material, which dissolves in both aqueous and non-aqueous phases, acts on the hydrate of chloroform. His systematic studies demonstrated that the hydrates of ethylene oxide and chloroform are not of the same type, and that ethylene oxide added as a third component lowers the stability of chloroform hydrate.

In the University of Oklahoma, interesting supplementary experiments were planned for the application of hydrate formation to solve heavy water fractionation. In heavy water, the hydrogen-bonds are longer and stronger than in common water, and consequently larger and stronger cavities can be formed in the heavy water structure. In spite of the fact that the work of Godchot and his co-workers [324] did not reveal an essential difference between the stabilities of the deuterates and the hydrates of krypton and xenon, it was assumed by Lippert and his co-workers [87] that, in the case of less volatile hydrate-forming components, e.g. in liquid hydrates, much greater differences can be expected between the stabilities and other characteristic properties of deuterates and hydrates.

However, from a study of the conditions of formation of the hydrates of natural gases, Schöller [325] came to the conclusion that there too the deuterates had lower stability than the hydrates, though the gas contained about 1 % propane or higher (and consequently less volatile) hydrocarbon components. Urea and ethylenediaminetetraacetic acid too were effectively used as inhibitors to diminish the stabilities of hydrocarbon hydrates; further, on the addition of acetylace-tone, the formation of the gas hydrate was completely prevented.

The difference in stability of hydrates and deuterates was also dealt with by Ivanov and Tsarev [326]. Their investigations showed that the energy necessary to break the hydrogen-bonds between the water molecules is 618 J mol^{-1} higher for $CH_4.6D_2O$ than for $CH_4.6H_2O$, and 474.4 J mol^{-1} higher for $C_3H_8.8D_2O$ than for $C_3H_8.8H_2O$. Literature data were used to determine the energy dif-ference between the hydrates and the deuterates from the dissociation pressures of other hydrates and deuterates too.

The vapour pressures of the hydrates and deuterates of halogenated hydrocar-bons in the presence of solid and liquid water and heavy water were systematized by Tezikov and Stupin [327].

By means of NMR and low-temperature dielectric absorption investigations of clathrate- and deuterohydrates, Davidson and his co-workers [328] compared the reorientation properties of guest molecules of dipolar character (C_1, C_2, C_3 and iso-C_4 hydrocarbons, H_2S, and neopentane$+H_2S$).

The dielectric properties of hydrates can give information about two important aspects of their behaviour: the relaxation of the water molecules, which form a structure somewhat looser than that in ice, and the barriers hindering the reorien-tation and rotation of the gas molecules. The similarity observed by Davidson and Wilson [329, 330] between the low-frequency dielectric properties of the hy-drates of ethylene oxide and acetone may be significant, although their structures are different. Ethylene oxide forms a gas hydrate (H_I), whereas acetone yields a liquid hydrate (H_{II}). Such similarity in behaviour can be expected in the study of other hydrates too.

As is apparent from the foregoing, in the accurate low-frequency dielectric measurements on these systems certain experimental difficulties arose. Although the preliminary studies of other hydrates with H_I structure suggested that the dielectric behaviour is generally similar to that of ethylene oxide hydrate, the exact relaxation rates of the water molecules depend on the nature of occluded gas molecules.

Davidson, Ripmeester and their co-workers [331–337] have carried out interest-ing NMR investigations of many different hydrates in recent years, and have given detailed information about the results of these experimental and theoretical studies.

Chapter 4

THERMODYNAMICS OF GAS HYDRATES

4.1. Chemical thermodynamic investigations of gas hydrates

In the first few decades after the discovery of gas hydrates, the primary consideration was the determination of the compositions of the various hydrate systems, i.e. the molar proportions of gas and water. Since this was not merely a simple analytical chemistry task, the work had to be combined with physicochemical considerations and chemical thermodynamic calculation methods leading to the most exact answer possible.

Following the discovery of Trouton's rule, it was used by de Forcrand in the general form:

$$\frac{Q}{T} = K \tag{4.1}$$

(where Q is the corresponding heat of the phase transition) for determination of the molecular mass [338], the heat of evaporation and the heat of melting [339], and the minimum bond energy of gas hydrates [340, 341]. In these latter investigations, he reported concrete values of the heat of formation (Q') of the gas hydrates from several gas–water systems, and he stated that a heat of formation value differing from this can be obtained (Q'') when the gas–ice equilibrium develops.

As a result of his calculations, it became quite clear that the difference between the two heats of formation corresponds to the heat of freezing of the n mol of water bound to the gas molecules during hydrate formation. As de Forcrand determined the value of the heat of freezing of one mol of water to be 1430 cal, he proposed the following relationship to establish the quantity of water bound in the hydrate:

$$\frac{Q''-Q'}{1430} = n \tag{4.2}$$

The value of Q'' can be determined directly, or it can be calculated from the dissociation pressure via the Clausius–Clapeyron equation, while Q can be obtained from the relationship $Q=30T'$ (where T' is the temperature at which the vapour

Table 4.1. Thermodynamic data and compositions of gas hydrates [32, 340, 341]

Gas	T (K)	T' (K)	Q' (kJ)	Q'' (kJ)	Calculated composition (according to de Forcrand [340, 341])	Probable formula (according to Villard [32])
Ar	86.0	229.2	55.59	28.71	$Ar.4.5H_2O$	$Ar.4H_2O$ or $Ar.5H_2O$
CH_4	109.0	244	68.34	30.59	$CH_4.6.3H_2O$	$CH_4.6H_2O$
CO_2	194.8	251.8	67.55	31.56	$CO_2.6H_2O$	$CO_2.6H_2O$
N_2O	185.0	253.7	68.09	31.81	$N_2O.6H_2O$	$N_2O.6H_2O$
C_2H_2	18.80	257.6	66.54	32.31	$C_2H_2.5.7H_2O$	$C_2H_2.6H_2O$
C_2H_6	18.80	257.2	74.03	32.23	$C_2H_6.7H_2O$	$C_2H_6.7H_2O$
C_2H_4	169.0	259.6	76.66	32.43	$C_2H_4.7.4H_2O$	$C_2H_4.7H_2O$
H_2S	211.0	273.3	68.30	34.27	$H_2S.5.7H_2O$	$H_2S.6H_2O$
C_2H_5F	241.0	276.7	84.10	34.69	$C_2H_5F.8.27H_2O$	$C_2H_5F.8H_2O$
SO_2	263.0	280.0	82.89	35.11	$SO_2.8H_2O$	$SO_2.8H_2O$
CH_3Cl	250.0	280.5	78.71	35.15	$CH_3Cl.7.2H_2O$	$CH_3Cl.7H_2O$
Cl_2	283.4	282.6	76.74	35.44	$Cl_2.6.91H_2O$	$Cl_2.7H_2O$
Br_2	332.0	>273	(?)	(?)	$Br_2.10H_2O$	$Br_2.10H_2O$

pressure of the hydrate reaches a value of 101 325 Pa); accordingly, all the data were available for the indirect analysis of the gas hydrates.

The results of de Forcrand are summarized in Table 4.1, where T is the boiling-point of the gas, T' is the temperature at which the vapour pressure of the hydrate reaches the value 101 325 Pa, Q' is the heat of formation of the hydrate in the water–gas system, and Q'' is the heat of formation of the hydrate in the ice–gas system.

The sequence of the hydrate-forming gases in this table is the sequence of stabilities of the hydrates.

For quite a long period, this first fundamental calculation method turned the attention of the researchers towards the improvement of the means of determining heats of formation and dissociation. In the various calculation methods, the Clausius–Clapeyron relationship referring to the temperature-dependence of the vapour pressure is used almost without exception. If the parameters of formation or decomposition of the hydrates are known, and if the concrete pressure and temperature data are plotted in the form $\log p$ vs. $1/T$, a linear relationship results:

$$\log p = A - \frac{B}{T} \qquad (4.3)$$

where B is the slope.

In the sense of the Clausius–Clapeyron equation:

$$\frac{dp}{dT} = \frac{\Delta H}{T \Delta V} \qquad (4.4)$$

where ΔH is the quantity of heat absorbed or released in the process, ΔV is the volume change during the reaction, and dp/dT is the variation of the equilibrium pressure with respect to the temperature. By rearrangement of Equation (4.4):

$$\Delta H = T \Delta V \frac{dp}{dT} \qquad (4.5)$$

Differentiation of Equation (4.3) leads to:

$$0.4343 \frac{dp}{p} = -\frac{B}{T^2} dT \qquad (4.6)$$

Hence:

$$\frac{dp}{dT} = -\frac{Bp}{0.4343 T^2} \qquad (4.7)$$

Substitution of Equation (4.7) into Equation (4.5) gives

$$\Delta H = -\frac{T \Delta V B p}{0.4343 T^2} = -\frac{B \Delta V p}{0.4343 T} \qquad (4.8)$$

By means of this equation, no matter what units are used, the heat change accompanying the reaction can be calculated.

The work of Frost and Deaton [51] to determine the heats of formation and decomposition of the hydrates of natural gases and their pure components had the main aim of establishing the actual compositions of the hydrates. The data found in their investigations are given in Table 4.2. Their results agreed very well with the composition data obtained by direct analysis.

Table 4.2. Thermodynamic data and compositions of hydrates of natural gas and its constituents [51]

Gases	V_m (dm^3 mol^{-1}) (at 0 °C and $1.013\,25 \times 10^5$ Pa)	Heat of formation				Calculated hydrate composition
		above 0 °C		below 0 °C		
		J mol^{-1}	cal mol^{-1}	J mol^{-1}	cal mol^{-1}	
Methane	22.21	63.54	15.17	20.44	4.88	$CH_4 . 7.18H_2O$
Ethane	21.99	75.19	17.95	25.64	6.12	$C_2H_6 . 8.25H_2O$
Propane	21.66	134.29	32.07	26.54	8.34	$C_3H_8 . 17.95H_2O$
Natural gas (NG)	22.13	78.55	18.76	24.40	5.83	$NG . 9H_2O$

Attention was drawn by Roberts and his co-workers [342] to the fact that the ΔV in the Clausius–Clapeyron equation (i.e. the volume change accompanying the phase change) practically agrees with the molar volume of the gas (the deviation is at most 2.5%, if the quantity of the condensed phase is negligible). We can then write

$$\Delta V = \Delta V' = ZRT/p \qquad (4.9)$$

where Z is the compression or deviation factor, the determination of which was carried out experimentally in the cases of CH_4 and C_2H_6. In the knowledge of the compression factor, it was found that both gas hydrates had compositions corresponding to the heptahydrate.

These fundamental and important research results relating to the compositions of the hydrates must be criticized, however, since neither the solubility in water of the hydrate-forming component nor the presence of water vapour in the gas were taken into account.

The determination of such correction data is undoubtedly very difficult, but in any event it must be based on considerations of the solubility and heat of dissolution data. The improvement of calculations in this direction became possible only later, following the work of Glew [114], and then in the later 1960's as a consequence of statistical thermodynamic calculations. In recent years, Kobayashi and his co-workers [343–346] have carried out considerable experimental work to determine the water contents of several gases (predominantly natural gases) in equilibrium with their hydrates.

Ng Heng-Joo and Robinson [347] similarly determined the water content of the gas phase in equilibrium with the hydrate by utilizing statistical thermodynamic data, and on this basis put forward a proposal for the prediction of the possibility of hydrate formation. In their calculations they determined the pressure and temperature-dependences of the fugacity of the water above the empty hydrate lattice. In reality, however, an empty hydrate lattice does not exist, for the lattice is stabilized by the inclusion of the gas molecules in the cavities. The properties of the empty lattice can therefore be determined only indirectly, but they do serve as a suitable reference basis. In the course of their determinations, Ng Heng-Joo and Robinson made use of the state equation of Peng and Robinson [348] and, by means of a critical evaluation, further developed the earlier results of Sloan and his co-workers [344]. Their calculated and experimentally found data agreed with an accuracy of nearly 3%, but this difference does not exceed the accuracy of the determination of the experimental data.

The precipitation of gas hydrates in the crystalline state is accompanied by the process of solidification of the components, and their heats of formation are correlated with this heat of solidification. By assuming that there are no specific chemical bonding forces between water and the hydrate-forming component, von Stackelberg [69] calculated the vapour pressures of the hydrates formed, using the boiling-points of the hydrate-forming components.

As the formation of hydrates is a surface phenomenon too, because it is a process beginning at the boundary of two phases, Eucken [349] started from the fundamental relationship applied to physical adsorption:

$$W_{ads} = \sqrt{L_1 L_2} \tag{4.10}$$

where W_{ads} is the bond energy of adsorption, L_1 and L_2 are the heats of evaporation of the absorbed material (gas) and the adsorbent (water), respectively. In Equation (4.10), using the value $L_2 = 800$ cal mol$^{-1} = 33\,494$ J mol^{-1} based on the data of Eucken, and $L_1 = 21 T_b$ from Trouton's rule:

$$W_{ads} = -400 \sqrt{T_b} \text{ cal mol}^{-1} = -1674.7 \sqrt{T_b} \text{ J mol}^{-1} \tag{4.11}$$

Eucken concluded that this value of W_{ads} corresponds to the heat of reaction, ΔH, of the following process:

$$X(g) + n\,H_2O(s) \rightarrow X.nH_2O(s) \tag{4.12}$$

if it is assumed that the water lattice of the gas hydrate does not differ energetically from that of ice, and further that there is no entropy difference between the two kinds of solid water structure. With these assumptions, the entropy of the reaction corresponding to Equation (4.12) is equal to the entropy necessary for the solidification of 1 mol of gas. On the basis of Trouton's rule, the entropy of liquefaction of the gases, $\Delta S_1 = 21$ cal $= 87.92$ J, and hence a value of $\Delta S = -21$ cal $= -87.92$ J emerges for the solidification. However, this was diminished by von Stackelberg [69] by a value of $\Delta S_2 = 1$ cal $= 4.187$ J, as the gas constituents still possess a considerable entropy within the hydrate structure, permitting their rotational movement:

$$\left.\begin{aligned}\Delta S &= -(21+1) = -22 \text{ cal K}^{-1} \\ \text{or} \qquad \Delta S &= -(87.92+4.19) = -92.11 \text{ J K}^{-1}\end{aligned}\right\} \qquad (4.13)$$

Accordingly, setting out from the basic relationship

$$\Delta G = \Delta H - T\Delta S \qquad (4.14)$$

and substituting the above values, the free enthalpy of formation of the hydrate at 0 °C is therefore

$$\left.\begin{aligned}\Delta G &= -400 \sqrt{T_b} + 273 \cdot 22 = -400 \sqrt{T_b} + 6000 \text{ cal mol}^{-1} \\ \text{or} \qquad \Delta G &= -1675 \sqrt{T_b} + 273 \cdot 92.1 = -1675 \sqrt{T_b} + 25,142 \text{ J mol}^{-1}\end{aligned}\right\} \qquad (4.15)$$

if the initial gas pressure was 1×10^5 Pa.

It is also true, however, that using for the standard state the value $p_H^o = 10^5$ Pa $= 1$ bar

$$\Delta G = RT \ln p_H = 1250 \log p_H \qquad (4.16)$$

where p_H is the vapour pressure (measured in bar) of the hydrate at 0 °C.

Combination of the two equations leads to the relationship

$$-400 \sqrt{T_b} + 6000 = 1250 \log p_H \qquad (4.17)$$

from which

$$\log p_H = 4.8 - 0.32 \sqrt{T_b} \qquad (4.18)$$

When ΔG is expressed in units of J mol^{-1}, Equation (4.18) assumes the form

$$\log p_H = 20.11 - 1.34 \sqrt{T_b} \qquad (4.19)$$

Equation (4.18) is thus a suitable relationship for calculation of the hydrate vapour pressure to be expected under given starting conditions if the boiling-point of the hydrate-forming component is known.

For temperatures other than 0 °C, if the phenomenon of physical adsorption is taken into consideration, similarly based on von Stackelberg [69], the following relationship can be obtained:

$$4.57T \log p_H = -400 \sqrt{T_b} - 1437n + 22T - 1437nT/273 \qquad (4.20)$$

where n is the amount of substance (expressed in mol) of water bound to one mol of the hydrate-forming component.

The results of calculations with the above relationship for a temperature of
0 °C can be seen in Table 4.3. The quotient in the final column of the table proves
that the difference between the measured and calculated values of the vapour
pressure depends primarily on the dimensions and shape of the hydrate-forming
molecules, and in this respect on their ability to occupy the cavities in the water
lattice.

Table 4.3. Calculated and measured p_H data [69]

Hydrate-forming substance	Boiling point (K)	p_H calculated (10^5 Pa)	p_H measured (10^5 Pa)	$\dfrac{p_{H,\ measured}}{p_{H,\ calculated}}$
Ar	87	66.66	106.05	1.6
CH$_4$	112	26.26	26.26	1.0
Kr	122	19.19	14.64	0.7
Xe	165	4.85	1.51	0.3
C$_2$H$_4$	171	4.24	5.55	1.3
C$_2$H$_6$	180	3.23	5.25	1.6
N$_2$O	184	2.93	10.10	3.4
C$_2$H$_2$	189	2.52	5.75	2.3
CO$_2$	194	2.22	12.42	5.6
H$_2$S	213	1.31	0.91	0.7
C$_3$H$_8$	228	0.91	1.01	1.1
Cl$_2$	239	0.72	0.33	0.5
CH$_3$Cl	249	0.57	0.41	0.7
SO$_2$	263	0.41	0.39	1.0
CH$_3$Br	277	0.30	0.25	0.8
C$_2$H$_5$Cl	286	0.25	0.26	1.0
Br$_2$	332	0.09	0.06	0.7
CHCl$_3$	334	0.09	0.06	0.7

The data in Table 4.3 show that the calculation method described previously
is suitable for the determination, with a good approximation, of the hydrate vapour pressure to be expected, but it also demonstrates that the absolute stability of
a gas hydrate is in a simple correlation with the boiling-point of the hydrate-
forming component.

If the vapour pressure of the hydrate (p_H) and the vapour pressure of the hydrate-
forming component (p_g) are equal, we reach the upper quadruple-point (critical-
point) in the phase diagram of the gas hydrate system; this is suitable for the
determination of the relative stability of a gas hydrate, for if $p_g < p_H$, there is
generally no possibility for the hydrate to be formed, only the liquid hydrate-
forming component resulting.

From the Clausius–Clapeyron equation and Trouton's rule, the vapour pressure
of the hydrate-forming component at 0 °C can be described by the equation:

$$\log p_g = 4.60 - 0.0169\ T_b \qquad (4.21)$$

With the increase of the boiling-point, therefore, the vapour pressure of this component decreases faster than the hydrate vapour pressure, which may be seen in Equation (4.18) to be correlated only with the value of $\sqrt{T_b}$.

In the determination of the critical-point of a hydrate, the following assumption must hold:

$$\log p_H = \log p_g \tag{4.22}$$

and consequently the equality

$$4.8 - 0.32\sqrt{T_b} = 4.60 - 0.0169 T_b \tag{4.23}$$

holds only if $T_b = 333.15\,\mathrm{K} = 60\,°C$. This verifies the earlier-mentioned experimental experience that gas hydrates are formed only by those gases with boiling-points not in excess of $60\,°C$ under atmospheric conditions.

As regards substances which form hydrates of type H_I, the compound with the highest boiling-point is bromine ($T_b = 59\,°C$), while the corresponding compound for hydrates of type H_{II} is chloroform ($T_b = 61\,°C$). Although the molecular size of iodine (0.63 nm) suggests that it would be capable of hydrate formation, it is nevertheless unable to do so, for its boiling-point is $183\,°C$.

In contrast with the above conclusion, however, certain compounds with boiling-points higher than $60\,°C$ are known to form hydrates: CCl_4 ($T_b = 77\,°C$), $C_2H_4Cl_2$ ($T_b = 84\,°C$), CH_3CCl_3 ($T_b = 114\,°C$) and C_2H_5I ($T_b = 72\,°C$). The molecular sizes of these compounds are 0.67–0.68 nm, values just low enough to permit their incorporation into the cavities in the water lattice. Forces not taken into consideration in the previous calculations must be assumed to play a role here in the course of hydrate formation.

In accordance with the systematization by von Stackelberg [69], based on calorimetric and vapour pressure measurement data, depending on the amount of water bound in the hydrate, the heat of formation of a gas hydrate as in the equation

$$X(g) + nH_2O(l) = X.nH_2O(s)$$

can have the following values:

if $n \approx 6$ (ideal H_I type), $\Delta H = -(58.6 - 71.2)\ \mathrm{kJ\ mol^{-1}}$;
if $n \approx 8$ (e.g. Br_2, CH_3Br), $\Delta H = -(79.5 - 83.7)\ \mathrm{kJ\ mol^{-1}}$;
if $n \approx 17$ (liquid hydrate), $\Delta H = -(121.4 - 134.0)\ \mathrm{kJ\ mol^{-1}}$.

These values are obtained to a good approximation from the heats of solidification of the starting materials in the above equation, as can be seen from Table 4.4.

From similar considerations, von Stackelberg and Frühbuss [68] determined values of 272–314 kJ mol^{-1} for the heats of formation of double hydrates produced in the presence of hydrogen sulphide. They observed that the incorporation of H_2S molecules increases the entropy decrease accompanying the double hydrate formation, as a consequence of which the incorporation of hydrate-forming components into the clathrate structure (if permitted by the molecular size) is only possible up to a maximum boiling-point value of $T_b = 383.15\ K = 110\ °C$.

Table 4.4. Heats of fusion and evaporation relating to heats of formation of gas hydrates [69]

	$n=6$	$n=8$	$n=17$
Heat of fusion of water (n . 6.006 kJ mol^{-1})	36.03	48.07	102.2
Heat of fusion of X ($\sim$ 4.2 kJ mol^{-1})	4.2	4.2	4.2
Heat of evaporation of X (kJ mol^{-1})	12.5–20.9	25.08	20.9–29.26
Total heat change (kJ mol^{-1})	52.6–61.0	77.33	127.1–135.4

In the case of mixed hydrates, the interpretation of the heat change accompanying the process of formation is more complicated. Though a number of references are to be found in this respect, mainly based on the work of von Stackelberg and Meinhold [67], the opinion of Parent [350] must be accepted that the heats of formation reported by the various authors should always be accepted with reservation. The reason for this is that the accuracy of the experimental p vs. T pairs (the most important fundamental data in the calculation) depends to a considerable extent of the purity of the material used, on the experimental conditions, on the number of points recorded during the investigation, on the means of constructing the straight line fitting the scattered experimental points in the log p vs. $1/T$ diagram, and on the accuracy of the compression (deviation) factor Z used in the calculations.

The literature heat of formation data naturally take into account a state of total occupancy of the cavities. This can be achieved in laboratory practice only with very careful work, but there is no evidence that such a degree of filling can actually be attained during technological processes. In the event of incomplete occupancy, however, the gas hydrates formed are richer in water than the theoretical, and this is a circumstance of importance in the calculation of the heat of formation data.

Efforts to eliminate these disturbing factors and to determine the heat of formation as one of the characteristic physicochemical parameters of gas hydrates are still actively being made by researchers dealing with this theme; accordingly, in almost every publication in which theoretical questions are discussed, in addition to reports of equilibrium p vs. T data, references can be found to the heat of formation.

At the same time, from the 1950's up to the present, numerous theoretical considerations relating to statistical thermodynamic calculations have been published in connection with the gas hydrates. These are able to penetrate more deeply than the earlier calculations based only on classical thermodynamics, but of course they by no means invalidate the use in the future of such classical theoretical investigations.

Byk and Fomina [120] found that determination of the relationship dT/dp (characteristic of the phase equilibrium in the process of formation of gas hydrates) from the Clausius–Clapeyron equation is only possible if the solubility in water of the gas in question can be neglected. Later, working together with Koshelev [351], they gave the dissociation pressures of the hydrates of the constituents of natural gases as functions of the decomposition temperature, taking into consideration the general equation

$$\log p = A - \frac{B}{T} \tag{4.24}$$

The dissociation pressures of the hydrates of the various gas constituents were described under the water–hydrate–gas or ice–hydrate–gas equilibrium conditions. Their equations are to be found in Table 4.5.

Other investigations by Byk and Fomina [352] indicated that the heat changes accompanying the process $X(g) + nH_2O(l \text{ or } s) \rightarrow X \cdot nH_2O(s)$ can be divided into three part-steps:

(1) Evaporation of the water or ice;
(2) Formation of a metastable crystalline H_2O lattice;
(3) Stable hydrate formation with the inclusion of the hydrate-forming component.

If the above considerations are taken into account, the thermal effect of hydrate formation is

$$Q_1 = -\lambda_{H_2O(l)} + L^\circ + Q_{ads} \tag{4.25}$$

or

$$Q_2 = \lambda_{H_2O(s)} + L^\circ + Q_{ads} \tag{4.26}$$

where $\lambda_{H_2O(l)}$ and $\lambda_{H_2O(s)}$ are the heats of crystallization and of fusion of water and ice, respectively, Q_1 and Q_2 are the heats of formation of hydrate from the gas and from water or ice, respectively, L° is the sublimation heat accompanying

Table 4.5. Expressions describing temperature-dependence of dissociation pressures of hydrates of natural gas components [351]

Component	log p (bar)	Temperature interval (°C)
	Water–hydrate–gas	
Methane	14.7068–3630.7849 T^{-1}	0–17
Methane	18.3221–4678.7140 T^{-1}	17–34
Ethane	16.6345–4348.3 T^{-1}	0–14.5
Ethylene	14.5531–3772.882 T^{-1}	0–9
Propane	26.4081–7149.1062 T^{-1}	0–4
Propylene	27.9057–7446 T^{-1}	0–2
Isobutane	29.33 –7240 T^{-1}	0–1.8
Isobutylene	27.9298–7629 T^{-1}	0–4.2
Hydrogen sulphide	13.9648–3826.3495 T^{-1}	0–29.5
Carbon dioxide	16.8885–4323.5675 T^{-1}	0–10
Nitrogen	14.1293–3257 T^{-1}	0–18
	Ice–hydrate–gas	
Methane	5.6414–1154.6078 T^{-1}	0–11
Ethane	6.9206–1694.8599 T^{-1}	0–10
Ethylene	5.9703–1425.8209 T^{-1}	0–30
Propane	5.4242–1417.93 T^{-1}	0–12
Propylene	4.9627–1168 T^{-1}	0–15
Isobutane	9.106 –1688 T^{-1}	0–11
Hydrogen sulphide	4.8592–1334.1919 T^{-1}	0–23
Carbon dioxide	13.4238–3369.1245 T^{-1}	0–6
Nitrogen	5.5598– 927 T^{-1}	0–4.5

the formation of the metastable water lattice (which depends on the lattice type), and Q_{ads} is the adsorption heat of the hydrate-forming gas. From literature heats of formation of hydrates, the fundamental data $L_I^\circ = 12.3$ kcal mol^{-1} (51.50 kJ mol^{-1}) and $L_{II}^\circ = 12.18$ kcal mol^{-1} (51.00 kJ mol^{-1}), and the heats of evaporation of water and ice, they determined the adsorption heats of the various hydrate-forming components. $CH_4.6.54H_2O$ and $C_3H_8.17.95H_2O$ were found to be stable at 0 °C. For $CH_4.6.54H_2O$, the adsorption heats for CH_4 in the cases of water and ice had the values 4.5 kcal mol^{-1} (18.84 kJ mol^{-1}) and 4.6 kcal mol^{-1} (19.26 kJ mol^{-1}), respectively, while the corresponding values for C_3H_8 were 6.5 kcal mol^{-1} (27.21 kJ mol^{-1}) and 2.6. kcal mol^{-1} (27.63 kJ mol^{-1}), respectively. Thus, the values calculated via the previous two equations were the same within 2.5%. The numerical values of the adsorption heats confirm that the hydrate-forming components are linked to the water lattice by van der Waals forces.

The investigations by Gritsenko and his co-workers [353, 354] led to the conclusion that the Clausius–Clapeyron equation cannot be applied to multiphase, multicomponent thermodynamic systems. This finding was justified by the fact that, according to the Clausius–Clapeyron equation, the function $\Delta H = f(T)$ displays a maximum as the temperature decreases, which cannot be interpreted physically. A new equation was constructed for the determination of the heats of formation of hydrates, according to which the decrease of temperature is accompanied by the decrease of ΔH. For the case of a three-component system, the equation has the following form:

$$Q^{1,2} = \left[n(V^2 - V^1) - \frac{n}{n+1}\left(\frac{\partial V}{\partial x_1}\right)^1 + x_2^1\left(\frac{\partial V}{\partial x_2}\right)^2 - \right.$$

$$\left. - \frac{n}{n+1}RT\left(\frac{\partial \ln x_1}{\partial p}\right)^1 - \frac{n}{n+1}RT\left(\frac{\partial \ln x_2}{\partial p}\right)^1 \right] T\frac{dp}{dT} \qquad (4.27)$$

Indices 1 and 2 in the equation refer to the liquid and solid solution (hydrate) phases, respectively.

The work of Iskenderov and Musayev [355] drew attention to the fact that the latent heats of formation of hydrocarbon hydrates increase with the increase of the molar mass, and therefore the heats of formation of the hydrates of natural gases depend on the gas composition.

Let us next consider investigations of the properties of gas hydrates by means of classical thermodynamic methods, chiefly from the aspect of thermodynamic equilibrium.

If it is assumed that the gas is a perfect gas or an ideal gas mixture, and that the liquid mixture possibly in equilibrium with it is also a mixture with ideal behaviour, under isothermal conditions the partial pressures p_i of the components in the gas phase will follow Dalton's law:

$$p_i = y_i p \qquad (4.28)$$

where y_i is the mole fraction of the solute component in the gas phase, and p is the total pressure in the gas volume. They will also follow Raoult's law:

$$p_i = x_i p_i^* \qquad (4.29)$$

where x_i is the mole fraction of the given component in the liquid phase, and p_i^* is the vapour pressure of the pure component. It is evident that, at a given temperature, both laws must hold for the partial pressure of the given component. Hence:

$$x_i p_i^* = y_i p \qquad (4.30)$$

From this:

$$\frac{y_i}{x_i} = \frac{p_i^*}{p} = K \qquad (4.31)$$

This is the evaporation ratio (or distribution coefficient, or phase equilibrium constant), which is equal to the equilibrium constant of the (liquid–vapour equilibrium) process:

$$X_{(liquid)} \rightleftharpoons X_{(vapour)} \tag{4.32}$$

where ideal behaviour is assumed:

$$K = \frac{a_{i\,(vapour)}}{a_{i\,(liquid)}} = \frac{y_i}{x_i} = \frac{p_i^*}{p} \tag{4.33}$$

This equilibrium constant K plays a very important role in the physicochemical evaluation of hydrate systems, for in a description of the processes it is necessary to know how, at a given pressure, the composition of the phase depends on the composition of the condensed phase in equilibrium with it.

For the general reaction of gas hydrate formation:

$$X(g) + nH_2O(l) = M.nH_2O(s) \tag{4.34}$$

the equilibrium constant is:

$$K = \frac{a_H}{a_g a_{H_2O}^n} \tag{4.35}$$

where a_H is the activity of the hydrate, a_g is the activity of the gas phase, and a_{H_2O} is the activity of the water. At the temperature in question, the activities of the solid and liquid phases may be taken as unity ($a_H = a_{H_2O} = 1$), since both are pure condensed phases, and thus the equilibrium constant of the process is the reciprocal of the gas activity. However, as the activity of the gas is equal numerically to the fugacity:

$$a_g = f_g \tag{4.36}$$

if a value of 1 bar is selected as the reference state f_g^θ, the value of a_{H_2O} will be unity only if the aqueous phase is a pure phase, but not if it is a mixed phase, e.g. a solution which contains a hydrate inhibitor. For this latter case, Equation (4.35) can be written in the form:

$$a_{H_2O}^n = \frac{1}{Kf_g} \tag{4.37}$$

On the basis of this equation, a number of authors have attempted to draw conclusions regarding the changes in activity of the water in processes of hydrate formation, primarily in the description of thermodynamic changes in the presence of inhibitors, and in the determination of hydrates formed under various conditions. The latter data can be calculated via the following considerations.

In the knowledge of the equilibrium pressure of the gas in the presence of water or some aqueous solution, its fugacity can be determined at the temperature and total pressure of the system. If this calculation is carried out for two experimental points, for example, where

$$a^n_{H_2O(1)} = \frac{1}{Kf_{g(1)}} \tag{4.38}$$

and

$$a^n_{H_2O(2)} = \frac{1}{Kf_{g(2)}} \tag{4.39}$$

we arrive at the relationship:

$$\frac{a^n_{H_2O(1)}}{a^n_{H_2O(2)}} = \frac{f_{g(2)}}{f_{g(1)}} = \left(\frac{a_{H_2O(1)}}{a_{H_2O(2)}}\right)^n \tag{4.40}$$

from which

$$\frac{\log f_{g(2)} - \log f_{g(1)}}{\log a_{H_2O(1)} - \log a_{H_2O(2)}} = n \tag{4.41}$$

If the fugacity of the gas in the equilibrium system is known, therefore, together with the activity of the water in the two different states, a good guide can be obtained as to the composition of the hydrate. With this method, Miller and Strong [50] found that the value of n is independent of the pressure and the temperature. In the determination of the composition of propane hydrate, 77 experimental data were obtained for the value of n. A value less than 5.6 was found in 17 cases, and a value larger than 6.4 in 5 cases. Of the 77 values, n=6.7 was the largest, and n=4.5 the smallest; the mean value was 5.8. Accordingly, the method was considered adequate for determination of the hydrate composition, as the number of small cavities in the hydrate structure formed here is 6.

According to the investigations of Wilms and van Haute [356], the calculation methods applied by Miller and Strong do not give satisfactory results in the case of low gas concentrations. For the determination of the compositions of gas hydrates, therefore, they set out from the above basis and made use of the theory of solid solutions to construct a general formula by means of which reproducible results were obtained for both hydrate types in the whole concentration range.

For the use of the equilibrium constant, primarily in the cases of the components of natural gases, Wilcox and his co-workers [357] constructed diagrams describing the equilibrium pressure vs. temperature relationships between the solid and gas phases in hydrate formation. These were based on the equation

$$K = \frac{y_i}{x_i} = \frac{y_{i(in\ vapour)}}{x_{i(in\ hydrate)}} \tag{4.42}$$

The correctness of their calculations was proved by practical applications. The diagrams for various gases are shown in Figs. 4.1–4.6.

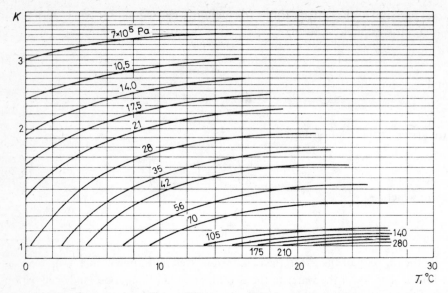

Fig. 4.1. Dependence of the equilibrium constant on pressure and temperature in
the case of methane hydrate [357]. Reprinted by permission from
Ind. Eng. Chem. **33**, 662 (1941). Copyright by the American Chemical Society

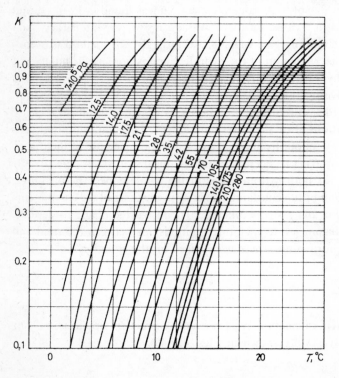

Fig. 4.2. Dependence of
the equilibrium constant
on pressure and temper-
ature in the case of
ethane hydrate [357].
Reprinted by permission
from *Ind. Eng. Chem.* **33**,
662 (1941). Copyright by
the American Chemical
Society

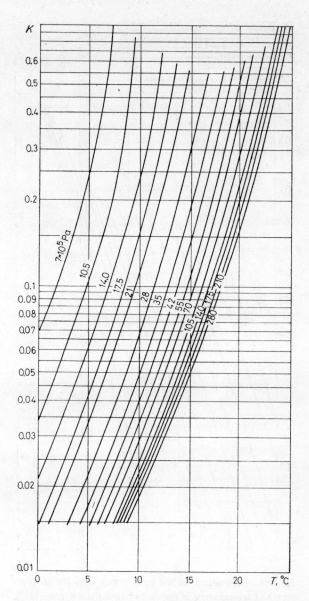

Fig. 4.3. Dependence of the equilibrium constant on pressure and temperature in the case of propane hydrate [357]. Reprinted by permission from *Ind. Eng. Chem.* **33**, 662 (1941). Copyright by the American Chemical Society

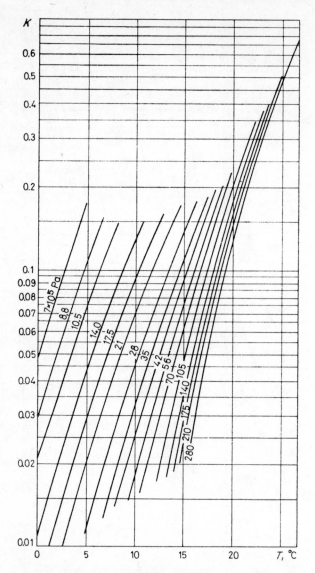

Fig. 4.4. Dependence of the equilibrium constant on pressure and temperature in the case of isobutane hydrate [357]. Reprinted by permission from *Ind. Eng. Chem.* **33**, 662 (1941). Copyright by the American Chemical Society

In the application of solid–vapour equilibrium constants in the case of gas mixtures, with the condition

$$\sum x_i = 1 \qquad (4.43)$$

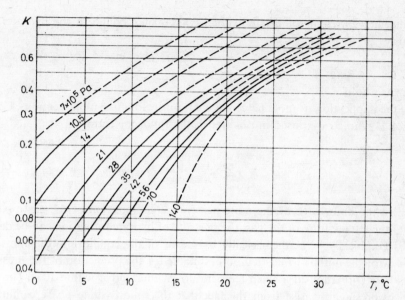

Fig. 4.5. Dependence of the equilibrium constant on pressure and temperature in the case of hydrogen sulphide hydrate [357]. Reprinted by permission from *Ind. Eng. Chem.* **33**, 662 (1941). Copyright by the American Chemical Society

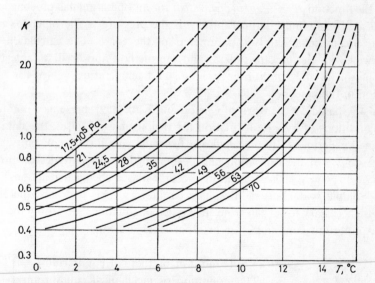

Fig. 4.6. Dependence of the equilibrium constant on pressure and temperature in the case of carbon dioxide hydrate [357]. Reprinted by permission from *Ind. Eng. Chem.* **33**, 662 (1941). Copyright by the American Chemical Society

for the mole fractions of the individual components bound in the hydrate, the equality:

$$\sum \frac{y_i}{K} = 1 \qquad (4.44)$$

must also hold.

By the application of mole percentage data instead of the mole fractions, Equation (4.44) can be well used in practice in the following form:

$$\sum_{i=1}^{n} \frac{y_i}{K_i} = 100 \qquad (4.45)$$

Thus, in the knowledge of the mole percentage composition of the gas mixture and of the temperature of the gas, the pressure of hydrate formation can be determined graphically or mathematically by means of the equilibrium diagrams. In the view of the authors, this method is suitable for the prediction of the pressure of hydrate formation at a given temperature, with an error within 10%. The differences presumably arise from the fact that the solid hydrate phase cannot be regarded as ideal, and the inaccuracy of the gas analyses is also reflected in the value of K.

A computer programme was prepared by Pick and Simanek [358] for calculation of the function $K = f(T, p, a_1, a_2, \ldots a_n)$ for an operating gas pipeline system, where T is the minimum temperature, p is the pressure in the pipeline, and a_1, $a_2, \ldots a_n$ are the characteristic parameters of the curve. The calculated values agreed well with the literature data, and the method appeared suitable for the determination of the equilibrium pressure and temperature values for hydrate formation in the pipeline.

Table 4.6, based on the work of Carson and Katz [359], presents a comparison of the conditions of hydrate formation for various natural gases, the data being determined experimentally or calculated from the equilibrium constants.

In the course of the formation and decomposition of gas hydrates, processes take place in which phase transitions occur without chemical reactions. In the thermodynamic evaluation of the phase equilibria of gas hydrates, therefore, it is likewise necessary to set out from the system of equations governing the general phase equilibrium conditions for heterogeneous systems:

$$\left.\begin{array}{l} T^{(1)} = T^{(2)} = \ldots = T^{(P)} \quad \text{(condition of thermal equilibrium)} \\ p^{(1)} = p^{(2)} = \ldots = p^{(P)} \quad \text{(condition of mechanical equilibrium)} \\ \left.\begin{array}{ccc} \mu_1^{(1)} = \mu_1^{(2)} = \ldots = \mu_1^{(P)} \\ \vdots \quad\ \vdots \qquad\quad \vdots \\ \mu_C^{(1)} = \mu_C^{(2)} = \ldots = \mu_C^{(P)} \end{array}\right\} \text{(conditions of chemical equilibrium)} \end{array}\right\} \quad (4.46)$$

Table 4.6. Comparison of conditions of formation of hydrates of various natural gases on the basis of data calculated from equilibrium constants and experimentally measured data [359]

Reference	Hydrate-formation parameters		
	Temperature (°C)	Measured pressure (10^5 Pa)	Calculated pressure (10^5 Pa)
Wilcox et al. [357]	4.4	11.03	11.03
	10.0	20.93	21.84
	15.5	42.42	47.86
	19.4	77.06	90.00
	23.9	212.10	255.93
Wilcox et al. [357]	4.4	16.33	11.11
	10.0	31.67	34.08
	15.5	66.10	71.90
	21.1	171.80	170.39
	23.3	282.80	>280
Wilcox et al. [357]	4.4	13.63	13.13
	10.0	27.43	26.08
	15.5	57.57	56.27
	21.1	153.42	135.34
	23.3	253.10	238.86
Deaton and Frost [360]	4.4	8.28	7.42
	10.0	16.96	15.77
Deaton and Frost [360]	4.4	9.89	10.6
	10.0	19.69	21.21
Deaton and Frost [360]	4.4	11.73	11.87
	10.0	22.97	24.24
Deaton and Frost [360]	1.6	8.83	8.90
	7.2	17.53	17.12
	12.8	35.70	35.85
Deaton and Frost [360]	4.4	12.86	12.62
	10.0	25.80	25.75
Deaton and Frost [360] (Hammerschmidt [41])	4.4	14.84	12.82
	10.0	29.79	25.80
Deaton and Frost [360]	4.4	20.85	17.67
	10.0	41.71	35.00
	15.5	83.43	74.44

From this system, by subtracting from the number of all independent variables the number of independent equations possible between them, the Gibbs phase rule, describing the relationship between the number of phases (P) in equilibrium with one another, the number of components (C) and the number of degrees of freedom (F), is simply obtained:

$$P + F = C + 2 \qquad (4.47)$$

The isothermal–isobaric transitions of substances, accompanying the phase changes, involve alterations in the chemical potentials of the components and the free enthalpies of the phases. Turning now to the isothermal–isobaric equation for two phases, let the chemical potential of the i-th component in one phase be $\mu_i^{(1)}$, and in the other phase $\mu_i^{(2)}$, and let dn_i mol of substance pass from phase (1) to phase (2) during the phase transition. In this process, the free enthalpy of phase (1) is diminished by an amount

$$dG^{(1)} = -\mu_i^{(1)} \, dn_i \qquad (4.48)$$

while that of phase (2) is increased by an amount:

$$dG^{(2)} = \mu_i^{(2)} \, dn_i \qquad (4.49)$$

Hence, the overall free enthalpy change for the system is:

$$dG = dG^{(1)} + dG^{(2)} = \mu_i^{(2)} \, dn_i - \mu_i^{(1)} \, dn_i = (\mu_i^{(2)} - \mu_i^{(1)}) \, dn_i \qquad (4.50)$$

In the event of the equilibrium of the two phases:

$$\mu_i^{(2)} = \mu_i^{(1)} \qquad (4.51)$$

from which, for the overall system, $dG = 0$.

If Equality (4.51) does not hold, then, in accordance with the fundamental laws of thermodynamics, the phase change occurs in the direction for which $dG < 0$, i.e. $\mu_i^{(1)} > \mu_i^{(2)}$.

The effects of changes in pressure and temperature can similarly be determined on the basis of the Clausius–Clapeyron equation.

From their investigations at high pressure, McLeod and Campbell [112] devised an empirical relationship, based on the above considerations, for prediction of the temperature of hydrate formation. Starting from Condition (4.51), in the case of chemical equilibrium the partial molar free enthalpy (i.e. the chemical potential) of the i-th hydrate-forming component must be the same in the two phases. For monovariant equilibrium, the following relationship must be satisfied:

$$\mu_i^{(1)} + d\mu_i^{(1)} = \mu_i^{(2)} + d\mu_i^{(2)} \qquad (4.52)$$

Since

$$d\mu_i = V_i \, dp - S_i \, dT \qquad (4.53)$$

it follows that

$$V_i^{(1)} \, dp - S_i^{(1)} \, dT = V_i^{(2)} \, dp - S_i^{(2)} \, dT \qquad (4.54)$$

and in the case of equilibrium:

$$\frac{dp}{dT} = \frac{S_i^{(1)} - S_i^{(2)}}{V_i^{(1)} - V_i^{(2)}} = \frac{H_i^{(1)} - H_i^{(2)}}{T(V_i^{(1)} - V_i^{(2)})} = \frac{\Delta H_i}{T \Delta V_i} \qquad (4.55)$$

McLeod and Campbell state that for a multicomponent gas mixture

$$\frac{dp}{dT} = \frac{\Delta H_1}{T \Delta V_1} = \frac{\Delta H_2}{T \Delta V_2} = \frac{\Delta H_3}{T \Delta V_3} = \ldots \text{etc.} \qquad (4.56)$$

and

$$T \frac{dp}{dT} = \frac{x_1 \Delta H_1 + x_2 \Delta H_2 + x_3 \Delta H_3 + \ldots}{x_1 \Delta V_1 + x_2 \Delta V_2 + x_3 \Delta V_3 + \ldots} = \frac{\Delta H}{\Delta V} \qquad (4.57)$$

For the value of ΔV, the gas volume V can acceptably be substituted. Then, by using the gas law and taking into consideration the relationship:

$$pV = ZRT \qquad (4.58)$$

(where Z is the deviation factor), they obtain the known relationship:

$$\frac{dp}{dT} = \frac{p \Delta H}{ZRT^2} \qquad (4.59)$$

or:

$$\frac{d \ln p}{dT} = \frac{\Delta H}{ZRT^2} \qquad (4.60)$$

By combining ΔH, Z and R in this equation into one constant, they obtain:

$$\frac{d \ln p}{dT} = \frac{c}{T^2} \qquad (4.61)$$

The function $d \ln p/dT$, which is the slope of the equilibrium hydrate vapour pressure curve, can be determined from the $\log p$ vs. $1/T$ diagram. With this slope denoted by m, they derived the empirical equation:

$$c = mT^2 \qquad (4.62)$$

The experiments of McLeod and Campbell [112] indicated that this relationship can be applied equally to gas mixtures and to their individual gas components. They carried out calculations on several natural gas samples, and the factors c for the individual gas constituents were determined as a function of pressure. In the literature their data are given in Rankine degrees (R); by reduction to K, the values of the empirical constant for the natural gases and their constituents

Table 4.7. The values of constant c of Equation (4.63)
for natural gas components, according to McLeod
and Campbell [112]

p (10^5 Pa)	c			
	CH_4	C_2H_6	C_3H_8	iso-C_4H_{10}
424.2	18 933	20 806	28 382	30 696
494.9	19 096	20 848	28 709	30 913
565.6	19 246	20 932	28 764	30 935
636.3	19 367	21 094	29 182	31 109
707.0	19 489	21 105	29 200	30 935

can be determined via the equation:

$$c = 0.2143\,T^2 \tag{4.63}$$

These values are listed in Table 4.7.

With the use of Equation (4.63) in the case of a $CH_4+C_2H_6$ gas mixture with a C_2H_6 content of 19.1%, at a pressure of 4.256×10^7 Pa:

$$c = 0.809c_{CH_4}+0.191c_{C_2H_6} = 0.809\times18\,933+0.191\times20\,806 = 19\,290.74 \tag{4.64}$$

From this, the temperature of hydrate formation at the same pressure is:

$$T = \sqrt{\frac{19\,290.74}{0.2143}} = 300.03\ K = 26.88\,°C$$

McLeod and Campbell found that the non-hydrocarbon constituents of natural gases (e.g. CO_2, N_2 and H_2S) have only a negligible influence on the temperature of hydrate formation at such high pressures.

Bukhgalter [142] also used the above relationship, applying it to a natural gas, and he observed the good agreement of the calculated and measured data. At a given pressure, there was a difference of only 0.4 °C in the temperature of hydrate formation.

For prediction of the conditions of formation of gas hydrates, the literature also contains an empirical formula in which the general equation of hydrate formation was established by means of Hildebrand's rule [140] and Trouton's rule, and by utilizing the data of Deaton and Frost [361], starting from the fundamental equation:

$$\frac{dp}{dT} = \alpha\left(\frac{p}{T}\right)^{\beta} \tag{4.65}$$

The authors concerned, Yorizane and Nishimoto [139], wrote the fundamental equation in the form:

$$p^{(1-\beta)} = \alpha T^{(1-\beta)}+c \tag{4.66}$$

and determined the values of α, β and c for various hydrates in the presence of liquid and solid water. They concluded that the relationship between α and c exhibits a regular variation for the different hydrocarbons (as can be seen in Fig. 4.7), while a value of -0.1563 was found appropriate for β. In the equation, the pressure was given in psi and the temperature in R (Rankine degrees), and their

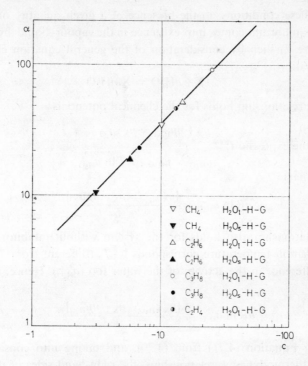

Fig. 4.7. Connection between the α and c values
of Equation (4.66) for various hydrates [139]

calculations on this basis for ethane, for example, did not lead to a difference greater than $\pm 1.7\%$ for the pressure of hydrate formation at the given temperature, in the case of either ice or liquid water.

In the presence of inhibitors, the conditions of formation of gas hydrates change, and this naturally influences the heat of formation and the composition of the hydrate too. The thermodynamic considerations relating to the change in the three-phase equilibrium temperature at a given pressure in the presence of a third component were reviewed in great detail by Pieroen [95]. In his work, he set out from the following assumptions:

(a) The pressure of the system is constant;

(b) The third component does not dissociate, and it does not itself form a hydrate;

(c) The composition of the gas phase or the liquid phase rich in the hydrate-forming component X is constant;

(d) The composition of the hydrate is constant.

If, under these conditions, in the presence of a given quantity of inhibitor, a three-phase equilibrium comes into existence in the vapour–liquid–hydrate system at temperature T, then by consideration of the general equation characterizing hydrate formation:

$$X + nH_2O \rightleftharpoons X.nH_2O \tag{4.67}$$

the following relationship holds for the chemical potentials:

$$\mu_X + n\mu_{H_2O} = \mu_{X.nH_2O} \tag{4.68}$$

Introducing the expression:

$$\mu_{H_2O} = \mu_{H_2O}^\varnothing + RT \ln a_{H_2O} \tag{4.69}$$

it is evident that:

$$nR \ln a_{H_2O} = \frac{[\mu_{X.nH_2O} - \mu_X - n\mu_{H_2O}^\varnothing]_T}{T} \tag{4.70}$$

The same relationship also holds for the system without inhibitor, but in this case the formation temperature corresponds to T_0 (different from T), and there must be a difference in the activity of the water too (a'_{H_2O}). Hence, the resulting equation is

$$nR \ln a_{H_2O} = \frac{[\mu_{X.nH_2O} - \mu_X - n\mu_{H_2O}^\varnothing]_{T_0}}{T} \tag{4.71}$$

Subtracting Equation (4.71) from (4.70), and taking into consideration the fundamental thermodynamic relationships, the right-hand sides of the equations can be written in the following form:

$$\frac{(\mu)_T}{T} - \frac{(\mu)_{T_0}}{T_0} = -\int_{T_0}^{T} \frac{H}{T^2} \, dT \tag{4.72}$$

and thus:

$$nRT \ln \frac{a_{H_2O}}{a'_{H_2O}} = \Delta H^* \left[\frac{1}{T} - \frac{1}{T_0} \right] \tag{4.73}$$

where:

$$\Delta H^* = H_{X.nH_2O} - H_X - nH_{H_2O} \tag{4.74}$$

If it is assumed that the heat of formation is independent of the temperature and of the presence of the third component, this equation is approximately equal to the equation:

$$\Delta H = [H_{X.nH_2O} - H_X - nH_{H_2O}]_{T_0} \tag{4.75}$$

and the heat of formation ΔH^* is equal to ΔH, which is the heat of formation of 1 mol of the hydrate $X . nH_2O$ at a temperature T_0 (in the absence of the third component), expressed on the basis of the vapour–liquid equilibrium. This heat of formation is sometimes referred to as the "uncorrected" heat of formation in the literature, and it is this which can be determined from concrete experimental data by means of the Clausius–Clapeyron equation. In reality, the reason why this is not a correct heat of formation is that it does not refer to pure liquid water or to pure water-free gas components, since these components dissolve in each other to some extent. Most hydrate-forming materials dissolve only very slightly in water, and this difference can therefore be neglected in the subsequent calculations, the following equation being taken as holding:

$$nR \ln \frac{a_{H_2O}}{a'_{H_2O}} = \Delta H \left[\frac{1}{T} - \frac{1}{T_0} \right] \tag{4.76}$$

In general, the activity of the water does not differ essentially from its mole fraction, even in solutions containing relatively high amounts of inhibitor, and accordingly Equation (4.76) can be written in the form:

$$\ln \frac{x_{H_2O}}{x'_{H_2O}} = \frac{\Delta H}{nR} \left[\frac{1}{T} - \frac{1}{T_0} \right] \tag{4.77}$$

If the hydrate-forming component does dissolve very slightly in water, then by introducing the equality $x'_{H_2O}=1$, we obtain the equation:

$$\ln x_{H_2O} = \frac{\Delta H}{nR} \left[\frac{1}{T} - \frac{1}{T_0} \right] \tag{4.78}$$

On introduction of the temperature decrease $\Delta T = T_0 - T$, and the relationship $x_{inhib.} = 1 - x_{H_2O}$ for the mole fraction of the third component, then at very low mole fractions of inhibitor, where

$$\ln x_{H_2O} = \ln (1 - x_{inhib.}) \approx -x_{inhib.} \tag{4.79}$$

Equation (4.78) can be written in the form:

$$\Delta T = -\frac{nRT_0^2}{\Delta H} x_{inhib.} \tag{4.80}$$

The evaluation by Pieroen suggested that Equations (4.73), (4.77), (4.78) and (4.80) are quite exact; however, because of inexactitudes in the calculations or measurements, the accuracy of Equations (4.77) and (4.78) is influenced to a certain degree by the values of ΔT, ΔH and n.

In a gas–water system containing an inhibitor, the expected decrease of temperature of hydrate formation in the presence of s wt.% inhibitor, on the basis of Equation (4.80), is:

$$\Delta T = -\frac{nRT_0^2}{\Delta H}\frac{18s}{(100-s)\,M_{\text{inhib.}}}(1-x_{\text{inhib.}})\qquad(4.81)$$

or, if $x_{\text{inhib.}}$ is really very small:

$$\Delta T \approx -\frac{nRT_0^2}{\Delta H}\frac{18s}{(100-s)\,M_{\text{inhib.}}}\qquad(4.82)$$

The decrease in the temperatures of formation of hydrates of hydrocarbon gases in the presence of alcohol inhibitors, converted to °C, according to Hammerschmidt [301] is:

$$\Delta T = 1297\frac{s}{(100-s)\,M_{\text{inhib.}}}\qquad(4.83)$$

If this equation is compared with Equation (4.82), taking the values of the heats of formation and the average value $T_0 = 280$ K as basis, according to Pieroen we obtain:

$$\Delta T = 1240\frac{s}{(100-s)\,M_{\text{inhib.}}}\qquad(4.84)$$

When all the incertitudes in the derivation (chiefly in the values of n, T_0 and ΔH) are taken into consideration, the agreement between the theoretical Equation (4.84) and that obtained experimentally, (4.83) can be stated to be good. Thus, in the knowledge of the values of ΔH and n, there is a possibility for the determination of either the expected temperature of hydrate formation or the necessary inhibitor quantity.

In the course of their investigations of the mechanisms of action of hydrate inhibitors, Krasnov and Klimenok [362] carried out kinetic experiments with aqueous solutions of a hydration activator, an inert material and a hydrate inhibitor. The effects of temperature, pressure and the hydrocarbon concentration were compared. Their detailed results are not at our disposal.

4.2. Statistical thermodynamic investigations
of gas hydrates

In addition to thermodynamic calculations, further information relating to the structures of gas hydrates as sets of particles can be obtained on the basis of kinetic-statistical theories. These theories determine the main features of the system of matter via statistical consideration of the motion of the particles, a better ap-

proximation thereby being obtained for the changes, the mechanism of the transformations, and the factors influencing the rates of the processes.

Statistical thermodynamic investigations of gas hydrates started in the 1950's.
The most important works in this field are those of van der Waals [97, 363], van
der Waals and Platteeuw [98, 364], Barrer and Stuart [99], and McKoy and
Sinanoglu [103].

The two different structure types for the gas hydrates were described unambiguously by von Stackelberg and Müller [66], who provided appropriate information
concerning the cavity sizes and cavity arrangements in hydrates of types H_I and
H_{II}. The statistical thermodynamic investigations were carried out by the various
authors by taking into consideration these structure types.

Barrer and Stuart [99] developed their theory on the following models:

(1) Species X is present in both cavity types in the gas hydrate.

(2) Species X is present in only one type of cavity.

(3) One type of cavity contains species X, while the other type contains species Y.

(4) One type of cavity contains species X, while the other type contains species
Y and Z.

(5) Both types of cavity may contain both species X and species Y.

Consequently, for the investigation of the phenomena, a gas mixture is needed
which consists of species X, Y and Z; the activities of these species are a_X, a_Y
and a_Z, respectively, and the corresponding partial pressures are p_X, p_Y and p_Z.
These gas constituents are incorporated into the cavities in the host lattice. In
the interest of the simplification of the calculations, it must be assumed that

(a) at a given time, only one molecule of any of the species X, Y or Z can be
present in a cavity;

(b) gas constituents enclosed in neighbouring cavities do not interact at all
with one another;

(c) the water lattice does not become deformed as a consequence of the enclosure of the gas molecules.

However, these assumptions are only good approximations: by means of special investigations it can be demonstrated that two nitrogen molecules can be
accommodated at the same time in the larger cavities of the H_{II} structure, for
instance, even though nitrogen itself forms a clathrate phase of H_I type.

Assumption (b) can never be fulfilled completely in practice, for it is made im-
possible by the rotation of the gas molecules in the cavities, and consequently
by the field of force of the gas molecules. At the same time, this means that condition (c) cannot be fully correct either.

With consideration to all these points, the following calculation methods were applied by Barrer and Stuart [99] for the two-phase equilibrium systems.

If the clathrate compound contains $(N_1 + N_2).m$ water molecules, where N_1 is the number of cavities of one type, and N_2 that of the other type, in the equilibrium state the proportions $_X\theta_1$, $_Y\theta_1$, $_Z\theta_1$ and $_X\theta_2$, $_Y\theta_2$, $_Z\theta_2$, respectively, of the cavities are filled with X, Y and Z molecules. The free energy A of this clathrate unit, consisting of water molecules and X, Y and Z molecules, can be given on the basis of statistical thermodynamic considerations:

$$\exp\left(-\frac{A}{kT}\right) = \exp\left(-\frac{A^\circ}{kT}\right) \frac{N_1!}{N_{1X}\theta_1! \, N_{1Y}\theta_1! \, N_{1Z}\theta_1! \, N_1(1 - \sum_{X,Y,Z} {}_X\theta_1)!} \cdot$$

$$\cdot \frac{N_2!}{N_{2X}\theta_2! \, N_{2Y}\theta_2! \, N_{2Z}\theta_2! \, N_2(1 - \sum_{X,Y,Z} {}_X\theta_2)!} \prod_{X,Y,Z} [{}_Xj_1(T)]^{N_1 {}_X\theta_1} \cdot$$

$$\cdot \prod_{X,Y,Z} [{}_Xj_2(T)]^{N_2 {}_X\theta_2} \tag{4.85}$$

In this equation, A° is the free energy of the empty host lattice consisting of $(N_1 + N_2).m$ water molecules, while $_Xj_1(T)$ and $_Xj_2(T)$ are the partition functions of the X molecules enclosed in the first and second cavity types, respectively. The combinatorial factor expresses the possibility of the distribution of the X, Y and Z molecules in the N_1 and N_2 cavities.

For large numbers, the Stirling formula holds:

$$N! \approx N^N \cdot e^{-N} \tag{4.86}$$

or

$$\ln N \approx N \ln N - N$$

Hence:

$$\frac{A}{kT} = \frac{A^\circ}{kT} + N_1\left\{\sum_{X,Y,Z} {}_X\theta_1 \ln {}_X\theta_1 + \left(1 - \sum_{X,Y,Z} {}_X\theta_1\right) \ln\left(1 - \sum_{X,Y,Z} {}_X\theta_1\right)\right\} -$$

$$- N_1 \sum_{X,Y,Z} {}_X\theta_1 \ln {}_Xj_1(T) + N_2\left\{\sum_{X,Y,Z} {}_X\theta_2 \ln {}_X\theta_2 + \right.$$

$$+ \left(1 - \sum_{X,Y,Z} {}_X\theta_2\right) \ln\left(1 - \sum_{X,Y,Z} {}_X\theta_2\right)\right\} - N_2 \sum_{X,Y,Z} {}_X\theta_2 \ln {}_Xj_2(T) \tag{4.87}$$

For an infinitesimally small isothermal change in the composition at constant volume, we obtain the relationship:

$$dA = \sum_{X,Y,Z} \mu_X({}_X\theta_1 \, dN_1 + N_1 \, d_X\theta_1) +$$

$$+ \sum_{X,Y,Z} \mu_X({}_X\theta_2 \, dN_2 + N_2 \, d_X\theta_2) + \mu_{H_2O}(dN_1 + dN_2) m \tag{4.88}$$

where μ_X and μ_{H_2O} are the chemical potentials of component X and water, respectively.

If $_X\theta_1$, $_Y\theta_1$, $_Z\theta_1$ and $_X\theta_2$, $_Y\theta_2$, $_Z\theta_2$, i.e. the degrees of occupancy of the cavities of the two types, can be regarded as constant, and if the ratio N_1/N_2 is replaced by n:

$$\left(\frac{\partial A}{\partial N_2}\right)_{_X\theta_1,\,_X\theta_2,\,\ldots,\,n} = \mu_X(n_X\theta_1 + _X\theta_2) + \mu_Y(n_Y\theta_1 + _Y\theta_2) +$$

$$+ \mu_Z(n_Z\theta_1 + _Z\theta_2) + \mu_{H_2O}(n+1)m \tag{4.89}$$

With the conditions $N_1 = $ constant and $N_2 = $ constant, after simplification and taking Relationship (4.88) into consideration, Equation (4.87) leads to:

$$\mu_X = \frac{1}{N_1}\left(\frac{\partial A}{\partial _X\theta_1}\right)_{N_1,\,N_2,\,_X\theta_2,\,_Y\theta_1,\,_Y\theta_2,\,_Z\theta_1,\,_Z\theta_2} =$$

$$= kT \ln \frac{_X\theta_1}{(1 - \sum\limits_{X,Y,Z} {}_X\theta_1)_X j_1(T)} \tag{4.90}$$

and

$$\mu_X = \frac{1}{N_2}\left(\frac{\partial A}{\partial _X\theta_2}\right)_{N_1,\,N_2,\,_X\theta_1,\,_Y\theta_1,\,_Y\theta_2,\,_Z\theta_1,\,_Z\theta_2} =$$

$$= kT \ln \frac{_X\theta_2}{(1 - \sum\limits_{X,Y,Z} {}_X\theta_2)_X j_2(T)} \tag{4.91}$$

Equations (4.90) and (4.91) can evidently be written in an analogous manner for components Y and Z too.

With the aid of the resulting relationships, a possibility arises for the determination of the chemical potential of the water in the clathrate structure, given by the following equation:

$$\mu_{H_2O} = \frac{1}{(n+1)m}\left(\frac{\partial A^\circ}{\partial N_2}\right)_n + \frac{kTn}{(n+1)m} \ln\left(1 - \sum\limits_{X,Y,Z} {}_X\theta_1\right) +$$

$$+ \frac{kT}{(n+1)m} \ln\left(1 - \sum\limits_{X,Y,Z} {}_X\theta_2\right) \tag{4.92}$$

If $\theta = 0$, that is the water is in a metastable state because of the empty cavities in the water lattice, then the chemical potential is:

$$\mu_{H_2O}^\circ = \frac{1}{(n+1)m}\left(\frac{\partial A^\circ}{\partial N_2}\right)_n \tag{4.93}$$

In hydrate structure H_1, the 2 pentagonal dodecahedral cavities and the 6 tetrakaidecahedral cavities are surrounded by 46 water molecules; thus, $n = N_1/N_2 = = 3$, and $m = 5.75$. In hydrate structure H_{II}, 136 water molecules give rise to 16 pentagonal dodecahedral cavities and 8 hexakaidecahedral cavities; accordingly, $n = 2$ and $m = 5.67$.

Under isothermal conditions, according to the laws of chemical thermodynamics for the vapour and gaseous components:

$$\mu_{H_2O(g)} = \mu^\varnothing_{H_2O(g)} + RT \ln a_{H_2O(g)} \tag{4.94}$$

and

$$\mu_{X(g)} = \mu^\varnothing_{X(g)} + RT \ln a_{X(g)} \tag{4.95}$$

where index g denotes the gas or vapour phase in equilibrium with the crystalline hydrate phase, and $\mu^\varnothing$ indicates the standard chemical potentials. Since $\mu_{H_2O(g)} = = \mu_{H_2O}$ and $\mu_{X(g)} = \mu_X$ at equilibrium, under isothermal conditions we have:

$$\ln a_{H_2O(g)} \approx \ln p_{H_2O} = \frac{\mu^\varnothing_{H_2O} - \mu^\varnothing_{H_2O(g)}}{kT} +$$

$$+ \frac{n}{(n+1)m} \ln \left(1 - \sum_{X,Y,Z} {}_X \theta_1\right) + \frac{1}{(n+1)m} \ln \left(1 - \sum_{X,Y,Z} {}_X \theta_2\right) \tag{4.96}$$

and:

$$\ln a_{X(g)} \approx \ln p_X = \ln \frac{{}_X \theta_1}{\left(1 - \sum_{X,Y,Z} {}_X \theta_1\right) {}_X j_1(T)} - \frac{\mu^\varnothing_{X(g)}}{kT} =$$

$$= \ln \frac{{}_X \theta_2}{\left(1 - \sum_{X,Y,Z} {}_X \theta_2\right) {}_X j_2(T)} - \frac{\mu^\varnothing_{X(g)}}{kT} \tag{4.97}$$

Introducing the molar partition function, the partial pressure of the gas component is:

$$p_{X(g)} = kT \frac{\Phi_X(T) {}_X \theta_1}{{}_X j_1(T)\left(1 - \sum_{X,Y,Z} {}_X \theta_1\right)} = kT \frac{\Phi_X(T) {}_X \theta_2}{{}_X j_2(T)\left(1 - \sum_{X,Y,Z} {}_X \theta_2\right)} \tag{4.98}$$

where $\Phi_X(T)$ is the molar partition function of gaseous component X, the volume factor being neglected.

If both types of cavities are occupied by identical gas molecules, then by taking into consideration the degrees of occupancy of the cavities, the composition of the hydrate can be given by the relationship:

$$\frac{(N_1 {}_X \theta_1 + N_2 {}_X \theta_2)}{N_1 + N_2} X \cdot mH_2O \tag{4.99}$$

where

$$_X\theta_1 = \frac{_X K_1 p_X}{1 +\ _X K_1 p_X} \quad \text{and} \quad _X\theta_2 = \frac{_X K_2 p_X}{1 +\ _X K_2 p_X} \tag{4.100}$$

while the values of the equilibrium constants are

$$_X K_1 = \frac{1}{kT} \frac{_X j_1(T)}{\Phi_X(T)} \quad \text{and} \quad _X K_2 = \frac{1}{kT} \frac{_X j_2(T)}{\Phi_X(T)} \tag{4.101}$$

Consequently, if the cavities of one type are occupied only by species X, and the cavities of the other type are occupied only by species Y, then the composition of the hydrate is of the form

$$n_X \theta_1 X \cdot_Y \theta_2 Y \cdot m H_2 O$$

and the degrees of occupancy of the cavities can be expressed in accordance with this.

However, the clathrate–gas two-phase equilibrium system can only exist rarely, or under specially selected experimental conditions. Under natural conditions of formation, the possibility of a three-phase equilibrium is much more frequent.

A number of three-phase equilibria can occur as regards the hydrate-forming component X and water, e.g.

(a) $X(g) + m H_2 O(l) \rightleftharpoons X \cdot m H_2 O(s)$

(b) $X(g) + m H_2 O(s) \rightleftharpoons X \cdot m H_2 O(s)$

(c) $X(l) + m H_2 O(l) \rightleftharpoons X \cdot m H_2 O(s)$

(d) $X(s) + m H_2 O(s) \rightleftharpoons X \cdot m H_2 O(s)$

In the cases of (a) and (b), equilibrium between the hydrate-forming component X and the hydrate phase $X \cdot m H_2 O$ can exist only along the hydrate p vs. T curve in the phase diagram. In this section, the equilibrium conditions require that

$$\mu_{H_2O}^{(l)} = \mu_{H_2O} \quad \text{and} \quad \mu_{H_2O}^{(s)} = \mu_{H_2O} \tag{4.102}$$

With regard to this, it follows that

$$\mu_{H_2O}^{(l)} - \mu_{H_2O}^{\circ} = \frac{kTn}{(n+1)m} \ln(1 - _X\theta_1) + \frac{kT}{(n+1)m} \ln(1 - _X\theta_2) \tag{4.103}$$

and:

$$\mu_{H_2O}^{(s)} - \mu_{H_2O}^{\circ} = \frac{kTn}{(n+1)m} \ln(1 - _X\theta_1) + \frac{kT}{(n+1)m} \ln(1 - _X\theta_2) \tag{4.104}$$

From the chemical thermodynamic relationships, in accordance with the foregoing equations it holds along the vapour pressure curve of the hydrate that

$$\frac{1}{(n+1)m} \frac{\mathrm{d}}{\mathrm{d}T} \left[n \ln \left(1 - \sum_{\mathrm{X,Y,Z}} \mathrm{x}\theta_1\right) + \ln \left(1 - \sum_{\mathrm{X,Y,Z}} \mathrm{x}\theta_2\right) \right] =$$

$$= -\frac{\Delta H}{RT^2} + \frac{\Delta V}{RT} \frac{\mathrm{d}p}{\mathrm{d}T} \qquad (4.105)$$

where ΔH denotes the heat accompanying the transformation of the empty meta-stable clathrate crystal to liquid or to solid ice, if the volume change is ΔV, and $p = p_{H_2O} + p_X + p_Y + p_Z$ is the sum of the partial pressures. At constant pressure, the right-hand side of the equation will be equal to $-\Delta H/RT^2$, while at constant temperature and varying pressure it will have a value of $\Delta V/RT$. If the value of the term to be differentiated in the equation is constant, the right-hand side of the equation will be zero, and we then obtain the equality

$$\Delta H = T\Delta V \frac{\mathrm{d}p}{\mathrm{d}T} \qquad (4.106)$$

Barrer and Stuart [99] recommend the following data for calculation of the value of ΔV:

molar volume of liquid water	$18.0 \ \mathrm{cm^3 \ mol^{-1}}$
molar volume of ice	$19.6 \ \mathrm{cm^3 \ mol^{-1}}$
molar volume of clathrate of type I	$22.8 \ \mathrm{cm^3 \ mol^{-1}}$
molar volume of clathrate of type II	$23.0 \ \mathrm{cm^3 \ mol^{-1}}$

There is little probability that the degree of occupancy of the cavities would be constant throughout under conditions corresponding to the p vs. T curve, because the clathrate system is not fully stoichiometric, and its composition (as mentioned previously) can vary depending on the applied pressure and temperature. On the basis of the findings of Eucken [365], von Stackelberg [69] considers that ΔH is practically zero in the case of transformation of the empty clathrate crystal to ice. However, this cannot be accepted (although ΔH is really small), because the intercalation of the hydrate-forming component into the cavities in the empty clathrate crystals must be accompanied by a heat change (the intercalation heat); this can be determined via Equations (4.97) or (4.98) with the aid of the equations

$$\left(\frac{\partial \ln p_X}{\partial T}\right)_{\mathrm{x}\theta_1, \mathrm{y}\theta_2} = \frac{\Delta H_1}{RT^2} \qquad (4.107)$$

and

$$\left(\frac{\partial \ln p_X}{\partial T}\right)_{\mathrm{x}\theta_2, \mathrm{y}\theta_2} = \frac{\Delta H_2}{RT^2} \qquad (4.108)$$

where ΔH_1 and ΔH_2 are the intercalation heats relating to the two different types of cavities. These latter equations too hold under the condition that the degree of occupancy (θ) of the cavities is constant under equilibrium conditions, and therefore these equations can be accepted only as yielding approximate determinations.

More accurate data can be obtained by appropriate interpretation of the interactions of the dispersion and repulsion forces during the intercalation. In calculations of this kind, Barrer and Stuart [99] used the Lennard-Jones 12–6 potential in the form of the equation

$$\varepsilon = -A\left(\frac{1}{r^6} - \frac{r_0^6}{2r^{12}}\right) \tag{4.109}$$

where r is the actual distance of the water–gas molecule pair from the centre of the cavity, and r_0 is its equilibrium value. They employed the Kirkwood–Müller expression for the value of the constant A:

$$A = 6N_A mc^2 \frac{\alpha_1\alpha_2}{\alpha_1/\chi_1 + \alpha_2/\chi_2} \tag{4.110}$$

where N_A is the Avogadro number, m is the mass of the electron, c is the velocity of light, α_1 and α_2 are the polarizabilities and χ_1 and χ_2 are the diamagnetic susceptibilities of the individual molecules.

The value of the total potential for the overall volume of the crystal is given by integration of Equation (4.109). This method provides a practically accurate result, because the value of constant A has been confirmed theoretically. By integration, we obtain the equation

$$\Delta E = -\frac{\pi N A}{R^3} P(x) + \frac{\pi N A r_0^6}{5R^9} Q(x) \tag{4.111}$$

where N is the number of water molecules per mol of the clathrate phase ($N_{H_I} = 2.66 \times 10^{22}$ and $N_{H_{II}} = 2.62 \times 10^{22}$), and the values of $P(x)$ and $Q(x)$ are as follows:

$$P(x) = \frac{4x^6}{3(x^2-1)^3} \tag{4.112}$$

$$Q(x) = \frac{x^9}{2}\left[-\frac{1}{8(x+1)^8} + \frac{1}{9(x+1)^9} + \frac{1}{8(x-1)^8} + \frac{1}{9(x-1)^9}\right] \tag{4.113}$$

In these latter expressions, $x = R/a$, where R is the distance of the centre of the cavity from the wall, and a is the distance of the centre of the gas molecule enclosed in the cavity from the centre of the cavity.

10*

If the energies ΔE_i of Ar, Kr or Xe atoms assembled in such a system are plotted graphically for the various clathrate types and free diameters, as a function of the distance from the centre of the cavity, on the basis of Equation (4.111), then the sorption heats of the individual gases can be calculated via the equation

$$\Delta H_i = \Delta E_i + p\Delta V = \Delta E_i - RT \tag{4.114}$$

when ΔH_i corresponds practically to the most negative function value $\Delta E_i = f(a)$. If the distributions of the two types of cavities are the same (and for °C taking RT as 2286 J mol^{-1}), the values obtained are

  Ar: $\Delta H_i = -14.65$ kJ mol^{-1}
  Kr: $\Delta H_i = -21.77$ kJ mol^{-1}
  Xe: $\Delta H_i = -30.14$ kJ mol^{-1}

The sorption heats thus obtained are very close to the values resulting for the heats of intercalation from Equations (4.107) and (4.108); this provides good support for the fact that, even if the heat of formation of the empty clathrate crystal from ice, ΔH, is not quite zero, it is nevertheless very small.

Applying statistical thermodynamic calculations to the conditions of formation of gas hydrates, Barrer and Stuart [99] found that the intercalation of some hydrate-forming component into the structural cavities in water decreases the equilibrium vapour pressure of the water; consequently, the clathrate phase is formed when its vapour pressure becomes equal to or less than the vapour pressure of the liquid or solid water. For a hydrate-forming component X, the following relationship can be written

$$\ln p_{H_2O} = \ln p_{H_2O}^{\circ} + \frac{n}{(1+n)m} \ln (1 - {}_X\theta_1) +$$

$$+ \frac{1}{(1+n)m} \ln (1 - {}_X\theta_2) \tag{4.115}$$

or

$$\ln p_{H_2O}^{(ice)} = \ln p_{H_2O}^{\circ} + \frac{n}{(1+n)m} \ln (1 - {}_X\theta_1) +$$

$$+ \frac{1}{(1+n)m} \ln (1 - {}_X\theta_2) \tag{4.116}$$

Equations (4.115) and (4.116) are suitable at the same time for the determination of the critical values of ${}_X\theta_1$ and ${}_X\theta_2$, and hence for the indication of the critical composition at which the formation of solid gas hydrate can occur even in the event of a partial degree of occupancy of the cavities.

By taking into consideration Equations (4.100) and (4.101), these critical conditions can be characterized with the aid of the equilibrium constants by means of the relationship

$$\ln \frac{p_{\mathrm{H_2O}}^{\circ}}{p_{\mathrm{H_2O}}^{(\mathrm{ice})}} = \frac{n}{(1+n)m} \ln \left(1 + {}_{\mathrm{X}}K_1 p_{\mathrm{X}}\right) +$$

$$+ \frac{1}{(1+n)m} \ln \left(1 + {}_{\mathrm{X}}K_2 p_{\mathrm{X}}\right) \tag{4.117}$$

where p_{X} is the vapour pressure necessary for formation of the hydrate.

In the presence of liquid water and several hydrate-forming components, Equation (4.117) can be written in the form

$$\ln \frac{p_{\mathrm{H_2O}}^{(\mathrm{l})}}{p_{\mathrm{H_2O}}^{\circ}} = \frac{n}{(1+n)m} \ln \left(1 - \sum_{\mathrm{X,Y,Z,\ldots}} {}_{\mathrm{X}}\theta_1\right) +$$

$$+ \frac{1}{(1+n)m} \ln \left(1 - \sum_{\mathrm{X,Y,Z,\ldots}} {}_{\mathrm{X}}\theta_2\right) \tag{4.118}$$

By means of this relationship, the effects of "auxiliary" gases on hydrate stability can be interpreted too.

In reality, however, pure liquid water is never present in the process of hydrate formation; as a result of the solubility of the gases in water, or of the presence of an inhibitor added as third component, we are concerned rather with aqueous solutions. Accordingly, therefore, the following equation is the correct one:

$$\ln \frac{p_{\mathrm{H_2O}}^{(\mathrm{soln.})}}{p_{\mathrm{H_2O}}^{\circ}} = \frac{n}{(1+n)m} \ln \left(1 - \sum_{\mathrm{X,Y,Z,\ldots}} {}_{\mathrm{X}}\theta_1\right) +$$

$$+ \frac{1}{(1+n)m} \ln \left(1 - \sum_{\mathrm{X,Y,Z,\ldots}} {}_{\mathrm{X}}\theta_2\right) \tag{4.119}$$

In the sense of Raoult's law (4.22):

$$p_{\mathrm{H_2O}}^{(\mathrm{soln.})} = x_{\mathrm{H_2O}}^0 \, p_{\mathrm{H_2O}}^{(\mathrm{l})} \tag{4.120}$$

Hence, as a consequence of the decrease in the vapour pressure of the solution, the degree of occupancy of the cavities necessary for stability increases; this is equivalent to the fact that the pressure necessary for hydrate formation must be higher than in the case of pure water.

Byk and his co-workers [366] used the calculation method of Barrer and Stuart [99] to determine the vapour pressure of the water in the gas hydrate system. Assuming that the total quantity of water or ice was incorporated in the clathrate structure, they obtained

$$\ln \frac{p_{\mathrm{H_2O}}^{(\mathrm{hydr.})}}{p_{\mathrm{H_2O}}^{\circ}} = \psi', \quad \text{where} \quad \psi' = f/p_{(\mathrm{total})} \tag{4.121}$$

Hence:

$$\ln p_{\mathrm{H_2O}}^{(\mathrm{hydr.})} = \ln p_{\mathrm{H_2O}}^{\circ} + \psi' \tag{4.122}$$

For both hydrate types, an equation can be written in the form

$$\log p_{H_2O} = A - B \log T - C/T \qquad (4.123)$$

the constants of which were tabulated. By generalization of the results obtained for methane hydrate and for propane hydrate, their calculations were also made suitable for the determination of the partial pressure of the water vapour in hydrates of various natural gases. For the individual gas systems, data are given in 5 K intervals in the temperature range 223–283 K. For a natural gas with the composition 82.53% CH_4 + 5.99% C_2H_6 + 3.26% C_3H_8 + 0.86% C_4 + 0.23% CO_2 + + 7.19% N_2, the curves describing the pressures of the vapour above water and above the hydrate (calculated in this manner) intersect each other at a pressure of 34 bar and at a temperature of 11.8 °C. The experimental value of the temperature at which the hydrate was formed at this pressure was 11.3 °C. Though the difference between the measured and the calculated values is not a considerable one, the calculation method can be further refined. It was also established [367] that the ratio of the vapour pressure of the saturated water and the partial pressure of the water vapour in the hypothetically empty hydrate lattice varies linearly with the temperature, regardless of whether the determination is performed below 0 °C or above 0 °C, but the two straight lines are not coincident, and they do not intersect each other at a temperature of 0 °C. This finding is supported by the tabulated compilation of the data calculated for 10 refiner gases.

By programming the solution of the Barrer–Stuart equation, Koshelev and his co-workers [368] elaborated a convenient method for calculation of the hydrate composition and the equilibrium constants of formation of the hydrates of the components (determined by Wilcox and his co-workers [357]), and the dependences of these on temperature. Their calculated values for the vapour pressure of the hydrate agreed well with the experimental data of Frost and Deaton [369].

For the determination of the fugacity of the water vapour in equilibrium with a hydrocarbon hydrate, Kuliyev and his co-workers [370] later proposed the relationship

$$\log (f/y) = \frac{V(p-y)}{RT} \qquad (4.124)$$

where f is the fugacity of the water vapour, y is the partial pressure of the water vapour, p is the total pressure expressed in Torr (≈ 1.33 mbar), V is the molar volume of water in cm^3 mol^{-1}, and R is the gas constant. The relationship was claimed to give good results under a pressure of 4×10^6 Pa. The results of Koshelev and his co-workers [583] were utilized in this work.

The following relationships were recommended by van der Waals [97] for the determination of the absolute values of the equilibrium constants:

$$_xK_1 = \frac{2\pi a_1^3 G_1}{kT\exp\dfrac{\varepsilon_1}{kT}} \quad \text{and} \quad _xK_2 = \frac{2\pi a_2^3 G_2}{kT\exp\dfrac{\varepsilon_2}{kT}} \tag{4.125}$$

where, for small and large cavities, respectively, $a_1 = 0.40$ nm and $a_2 = 0.43$ nm denote the distance of the centre of the cavity from the centre of the water molecule forming the cavity wall, ε_1 and ε_2 are the sorption energies at the centre of the cavities, and G_1 and G_2 are integral values in the determination of which all essential parameters are taken into consideration. These can be determined on the basis of the Lennard-Jones 12–6 potential from the fundamental equation

$$\varepsilon(r) = 3\varepsilon^* \left[\left(\frac{\sigma^*}{r} \right)^{12} - \left(\frac{\sigma^*}{r} \right)^6 \right] \tag{4.126}$$

where ε^* is the maximum negative interaction energy of a water molecule pair at a distance of r, and σ^* is the distance of the pair in the situation when $\varepsilon(r) = 0$. This can be defined in terms of a volume value:

$$\sigma^* = (V^*)^{1/3} \tag{4.127}$$

when

$$a = \{\sqrt{2V_0}\}^{1/3} \tag{4.128}$$

If the pair distance is a, then the value of the potential is

$$\varepsilon(a) = \varepsilon^* \left[-2\left(\frac{V^*}{V_0} \right)^2 + \left(\frac{V^*}{V_0} \right)^4 \right] \tag{4.129}$$

In order that the interactions of water molecules not directly in contact with the cavity can be taken into account, Pople [371] suggested the use of a value of 2.4 in place of the factor 2.0 inside the brackets in Equation (4.129). By taking this into consideration, Barrer and Stuart [99] gave the following relationship to express the sorption energies relating to the two different cavity types in the hydrate structure of type H_I:

$$\varepsilon_1 \text{ (or } \varepsilon_2) = \Lambda^* \left[-2.4\left(\frac{V^*}{V_0} \right)^2 + \left(\frac{V^*}{V_0} \right)^4 \right] \tag{4.130}$$

The coefficient Λ^* in the equation can thus be calculated in the knowledge of ε_1, ε_2, V^* and V_0. These data are necessary for the determination of the integral G which figures in the above expressions for the equilibrium constants:

$$G_1 = \int_0^{0.30544} y^{1/2} \exp \frac{\Lambda^*}{kT} \left[2\left(\frac{V^*}{V_0} \right)^2 M(y) - \left(\frac{V^*}{V_0} \right)^4 L(y) \right] dy \tag{4.131}$$

This relationship of the integral was determined by Wentorf and his co-workers [372], who gave the function values $M(y)$ and $L(y)$ too.

The calculation data relating to three of the noble gases are to be found in Tables 4.8 and 4.9. On the basis of these data, the method of Barrer and Stuart yields the following critical compositions, necessary for hydrate formation to occur:

$0.7Ar.2.6Ar.23H_2O$, i.e. $Ar.6.97H_2O$
$0.8Kr.2.6Kr.23H_2O$, i.e. $Kr.6.76H_2O$
$0.9Xe.2.8Xe.23H_2O$, i.e. $Xe.6.22H_2O$

Table 4.8. Calculated data relating to noble gas hydrates of types H_I and H_{II} [99]

Type of hydrate	H_I		H_{II}	
Type of cavity	1	2	1	2
Diameter of empty cavity (nm)	0.52	0.59	0.48	0.69
Distance between centres of cavities (nm)	0.80	0.87	0.76	0.97
ε energy at centre of cavity (kJ mol^{-1})				
Ar	−10.45	−8.36	−12.12	−5.85
Kr	−17.97	−14.21	−20.48	−10.03
Xe	−28.01	−22.57	−30.51	−15.88
ε maximum energy in cavity (kJ mol^{-1})				
Ar	−12.54	−12.12	−12.96	−9.61
Kr	−20.06	−19.44	−20.90	−15.46
Xe	−29.26	−26.33	−30.51	−21.32

The experience of researchers dealing with gas hydrates demonstrates that the composition of the hydrate is not identical with that of the gas phase. Barrer and Stuart [99] put forward proposals for the determination of the fractionation factor in the case of the precipitation of gases as hydrates, by means of which the separation of the various components can be preplanned.

In their calculations, they started from a binary gas mixture $X+Y$, and the values of the fractionation factor η for the two different types of cavities in the hydrate were described by means of the following relationships:

$$\eta_1 = \left(\frac{Y}{X}\right)_{gas} \left(\frac{X}{Y}\right)_{hydrate} = \frac{x\,\theta_1}{y\,\theta_1} \frac{p_Y^m}{p_X^m}$$

$$\eta_2 = \frac{x\,\theta_2}{y\,\theta_2} \frac{p_X^m}{p_Y^m}$$

(4.132)

Table 4.9. Calculated data relating to noble gas hydrates [112]

Gas	H_I hydrate				Small cavities			Large cavities				$\dfrac{p^{ice}_{H_2O}}{p^o_{H_2O}}$ at 0 °C
	σ^* (nm)	$V^* \times 10^{24}$ (cm³)	$\dfrac{V^*}{V_0}$	$\dfrac{A^*}{kT}$	G_2	XK_2	$\dfrac{V^*}{V_0}$	$\dfrac{A^*}{kT}$	G_1	XK_1	
Ar	0.295	25.6	0.566	6.89	1.94×10^{-2}	2.03×10^{-8}	0.350	13.2	1.14×10^{-1}	6.1×10^{-8}	0.73
Kr	0.307	29.0	0.641	9.64	9.4×10^{-3}	2.66×10^{-7}	0.400	17.7	9.4×10^{-2}	5.9×10^{-7}	0.71
Xe	0.322	33.3	0.735	12.27	3.5×10^{-3}	8.3×10^{-6}	0.455	21.8	5.8×10^{-2}	1.59×10^{-7}	0.58

where p_X^m and p_Y^m are the equilibrium pressures in the gas mixture. From Equation (4.100):

$$\frac{_X \theta_1}{_Y \theta_1} = \frac{_X K_1}{_Y K_1} \frac{p_X^m}{p_Y^m} \quad \text{and} \quad \frac{_X \theta_2}{_Y \theta_2} = \frac{_X K_2}{_Y K_2} \frac{p_X^m}{p_Y^m} \tag{4.133}$$

Accordingly, on this basis the fractionation factors can be given in the form

$$\eta_1 = \frac{_X K_1}{_Y K_1} \quad \text{and} \quad \eta_2 = \frac{_X K_2}{_Y K_2} \tag{4.134}$$

These equations are suitable for the determination of the degree of separation, but their application in the solution of practical problems is cumbersome.

In the course of their calculations, van der Waals and Platteeuw [98] set out from practically the same assumptions as Barrer and Stuart [99] did. They derived the thermodynamic properties of clathrates by means of the statistical mechanical interpretation of a model in which ideal, three-dimensional localized adsorption is considered.

As regards the substance Q (water in the case of the gas hydrates) forming the fundamental lattice in the structure of a clathrate system, in their model they make a sharp distinction between the stable modification, which is in the liquid (Q^l) or crystalline (Q^α) state under the given conditions, and the metastable modification (Q^β), containing empty cavities. In the lattice substance Q, $1 \ldots n$ cavities of type i can exist, the number of these being v_i for 1 mol of Q. In gas hydrates of type H_I, the value of this parameter is $v_1 = 1/23$ and $v_2 = 3/23$ per mol water for the small and the large cavities, respectively, while for the H_{II} structure $v_1 = 2/17$ and $v_2 = 1/17$.

If a clathrate crystal at temperature T consists of N_Q molecules of water occupying a volume V, it is stabilized at equilibrium by the enclosing of the gas constituents A $\ldots$ J $\ldots$ M in the cavities. Let the absolute activities of the guest molecules be $\lambda_A \ldots \lambda_M$; the chemical potential of gas component J is then

$$\mu_J = kT \ln \lambda_J \tag{4.135}$$

The independent variables in the description of the system are

$$T, V, N_Q, \lambda_A \ldots \lambda_M$$

In the establishment of the generalized partition function, van der Waals and Platteeuw [98] used the following considerations and symbols:

A^β is the free energy of the empty host lattice at given values of T, V and N_Q; the combinatorial factor expresses the mode of distribution of the dissolved molecules $N_{Ai} \ldots N_{Mi}$ in the $v_i N_Q$ cavities of type i; and $h_{Ji}(T, V)$ is the partition function of the J molecules on their intercalation into the cavities of type i.

The equation of the partition function (PF) for a given system is

$$\text{PF} = \exp\left(-\frac{A^\beta}{kT}\right) \prod_i \left[\frac{(v_i N_Q)!}{(v_i N_Q - \sum_J N_{Ji})! \prod_J N_{Ji}!} \prod_J h_{Ji}^{N_{Ji}}\right] \tag{4.136}$$

From this, the grand partition function can be obtained if summation is performed with regard to the absolute activities of the intercalated components for all possible values of N_{Ji}:

$$\varXi = \exp\left(-\frac{A^\beta}{kT}\right) \sum_{N_{Ji}} \prod_i \left[\frac{(v_i N_Q)!}{(v_i N_Q - \sum_J N_{Ji})! \prod_J N_{Ji}!} \prod_J h_{Ji}^{N_{Ji}} \lambda_{Ji}^{N_{Ji}}\right] \tag{4.137}$$

then, applying the multinomial rule:

$$\varXi = \exp\left(-\frac{A^\beta}{kT}\right) \prod_i \left(1 + \sum_J h_{Ji}\lambda_J\right)^{v_i N_Q} \tag{4.138}$$

In Equation (4.138), h_{Ji} is the cell partition function, and $\varXi$ is the grand partition function for the dissolved component, but for the solvent it is only an ordinary partition function. On the basis of the relationship

$$\mathrm{d}(kT \ln \varXi) = S\,\mathrm{d}T + p\,\mathrm{d}V + \sum_K N_K \mathrm{d}\mu_K - \mu_Q \mathrm{d}N_Q \tag{4.139}$$

or introducing the absolute activities too and rearranging:

$$kT \mathrm{d}\ln \varXi = \left(\frac{U}{T}\right) \mathrm{d}T + p\,\mathrm{d}V + \sum_K kTN_K \mathrm{d}\lambda_K/\lambda_K - \mu_Q \mathrm{d}N_Q \tag{4.140}$$

In its given form, Equation (4.140) corresponds to the grand partition function derived by Rushbrooke [373], without the term $\mu_Q N_Q$, referring to the water, i.e. the solvent.

The hydrate composition can be determined directly from the equation

$$N_K = \sum_i N_{Ki} = \lambda_K (\partial \ln \varXi / \partial \lambda_K)_{T, V, N_Q, \lambda_J(J \neq K)} =$$
$$= \sum_i v_i N_Q h_{Ki} \lambda_K / (1 + \sum_J h_{Ji})\lambda_J \tag{4.141}$$

As N_K must be a linear and a homogeneous function of the number of cavities of the various types, it must hold that

$$N_{Ki} = v_i N_Q h_{Ki} \lambda_K / (1 + \sum_J h_{Ji})\lambda_J \tag{4.142}$$

from which the probability of finding a K molecule in the cavity of type i is

$$Y_{Ki} = N_{Ki}/v_i N_Q = h_{Ki}\lambda_K / (1 + \sum_J h_{Ji})\lambda_J \tag{4.143}$$

Based on the calculations of van der Waals and Platteeuw [98], the following formula arises for the expression of the chemical potential of the solvent:

$$\frac{\mu_Q}{kT} = -(\partial \ln \Xi/\partial N_Q)_{T,V,\lambda_J} = \frac{\mu_Q^\beta}{kT} - \sum_i v_i \ln\left(1 + \sum_J h_{Ji}\right)\lambda_J \qquad (4.144)$$

where $\mu_Q^\beta = \partial A^\beta/\partial N_Q$ is the chemical potential of the β-modification of water.

The internal energy of the hydrate too can be determined by the above procedure:

$$\frac{U}{kT^2} = \left(\frac{\partial \ln \Xi}{\partial T}\right)_{V,N_Q,\lambda_J} = \frac{U^\beta}{kT^2} + \sum_i v_i N_Q \frac{\sum_J \lambda_J \partial h_{Ji}/\partial T}{1 + \sum_J h_{Ji}\lambda_J} \qquad (4.145)$$

or, using Equation (4.143), the relationship involving the probability of occupation of the cavities:

$$\frac{U - U^\beta}{kT^2} = N_Q \sum_{J,i} v_i y_{Ji} \frac{\partial \ln h_{Ji}}{\partial T} \qquad (4.146)$$

If the vapours of the dissolved materials (gases) display ideal behaviour, then:

$$\lambda_K = p_K/[kT\Phi_K(T)] \qquad (4.147)$$

where $\Phi_K(T)$ is the molecular partition function of gas component K, the value of which is given by the relationship

$$\Phi_K(T) = (2\pi m_K kT/h^2)^{3/2} j_K(T) \qquad (4.148)$$

where m_K is the molecular mass of K, and $j_K(T)$ is the internal partition function of K, taking into account the rotation in the cavity.

If the vapour of the hydrate-forming component does not behave ideally, then the relationship can be used with the modification that the fugacity of component K must be substituted into the equation instead of the vapour pressure p_K.

Making use of the partition function, the probability that a cavity of type i is occupied by a molecule of species K is:

$$y_{Ki} = C_{Ki}p_K/\left(1 + \sum_J C_{Ji}p_J\right) \qquad (4.149)$$

where

$$C_{Ki} = \frac{1}{kT} \frac{h_{Ki}(T,V)}{\Phi_K(T)} = \frac{h_{Ki}(T,V)\lambda_K}{p_K} \qquad (4.150)$$

C_{Ki} is the Langmuir constant, and accordingly C_{Ji} too can be given. By taking this into consideration:

$$1 - \sum_K y_{Ki} = \left(1 + \sum_J h_{Ji}\lambda_J\right)^{-1} = \left(1 + \sum_J C_{Ji}p_J\right)^{-1} \qquad (4.151)$$

and thus

$$p_K = \frac{1}{C_{K1}}\left[\frac{y_{K1}}{1-\sum_J y_{J1}}\right] = \frac{1}{C_{K2}}\left[\frac{y_{K2}}{1-\sum_J y_{J2}}\right] \tag{4.152}$$

and by using Equation (4.144):

$$\mu_Q = \mu_Q^\beta + kT \sum_i v_i \ln\left(1-\sum_K y_{Ki}\right) \tag{4.153}$$

The derived relationships thus give good information about the connection between the chemical potential of the solvent water forming the host lattice of the hydrate, the equilibrium vapour pressure and the hydrate composition.

These relationships were used by van der Waals and Platteeuw [98] and in essence by Barrer and Stuart [99] too in calculations of the thermodynamic properties of gas hydrates.

Equation (4.153) can be interpreted in that gas hydrates behave as ideal dilute solutions, where the chemical potential of the solvent is independent of the properties of the solute, and its value can be determined in the knowledge of the concentrations of the hydrate-forming components dissolved (enclosed) in the various types of cavities.

Practical experience indicates that a condition of stability in the presence of the α-modification of the lattice-forming material is that

$$\mu_Q \leqslant \mu_Q^\alpha \tag{4.154}$$

and thus the potential difference between the two modifications, which is a function of the temperature and total pressure, is:

$$\Delta\mu(T, p) = \mu_Q^\beta - \mu_Q^\alpha \tag{4.155}$$

From Equation (4.153):

$$\sum_i v_i \ln\left(1-\sum_K y_{Ki}\right) \leqslant -\frac{\Delta\mu}{kT} \tag{4.156}$$

or by using Equation (4.149):

$$\sum_i v_i \ln\left(1+\sum_K C_{Ki} p_K\right) \geqslant \frac{\Delta\mu}{kT} \tag{4.157}$$

where, in the case of equality, p_K is the minimum vapour pressure of component K necessary for the clathrate to be more stable than phase Q^α in the univariant equilibrium Q_s^α–clathrate–gas, when two solids and a single gas phase are in equilibrium with one another. For this case, the equilibrium relationships referring to gas component A are:

$$y_{A1} = \frac{C_{A1} p_A}{1+C_{A1} p_A}, \dots, y_{An} = \frac{C_{An} p_A}{1+C_{An} p_A} \tag{4.158}$$

and:

$$\sum_i v_i \ln(1-y_{Ai}) = -\frac{\Delta\mu}{kT} \tag{4.159}$$

Using the value of y_{Ai}, the total A content of the hydrate is:

$$Y_A = \sum_i v_i y_{Ai} \tag{4.160}$$

In practice, the equilibrium liquid water (Q^L)–clathrate–gas is more important; for this, the above derivations are valid if the equality $\mu_Q^\alpha = \mu_Q^L$ can be accepted. Difficulties arise, however, from the fact that the determination of μ_Q^L and the difference $\mu_Q^\beta - \mu_Q^L$ is not at all easy; besides the values of p and T, they are also functions of the quantity of gas component dissolved in the water, which in turn is connected with the nature of the solute. This is likewise an obstacle to the determination of the equilibrium.

The gas hydrates are characteristic examples of the clathrates, in which both types of cavities are present at the same time. Following the calculation procedure of van der Waals and Platteeuw [98], we have concretely that

$$y_{A1} = \frac{C_{A1}p_A}{1+C_{A1}p_A}, \quad y_{A2} = \frac{C_{A2}p_A}{1+C_{A2}p_A} \tag{4.161}$$

$$v_1 \ln(1-y_{A1}) + v_2 \ln(1-y_{A2}) = -\frac{\Delta\mu}{kT} \tag{4.162}$$

and

$$Y_A = v_1 y_{A1} + v_2 y_{A2} \tag{4.163}$$

The concentration of the absorbed gas is not constant for a gas hydrate at a given temperature, however, but depends on the nature of the hydrate-forming gas.

The grand partition function derived in Equation (4.138) contains as an unknown the cell partition function h_{Ji}, which appears in the later relationships too, and therefore the possibility of determining this is very important. The method used by Lennard-Jones and Devonshire [101] in the study of liquids (which served for the quantitative determination of the partition functions of the dissolved molecules in the cavities) was found suitable by van der Waals and Platteeuw [98, 364] for the thermodynamic prediction of gas hydrates.

In the introductory part to the statistical mechanical calculations relating to the gas hydrates, we gave some of the necessary basic conditions primarily connected with the dimensions of the incorporated gas molecules and their arrangements in the cavities. For the interpretation of the energetic relations, a further two conditions must be stated:

(a) The dissolved molecules can rotate freely in the cavities, which can be considered spherically symmetrical;

(b) The potential energy of a dissolved molecule situated at a distance r from the centre of the cavity can be expressed by the spherically symmetrical potential $w(r)$ equation proposed by Lennard-Jones and Devonshire [101].

These conditions hold completely only for monoatomic gases, e.g. the noble gases, but they are virtually fulfilled for the relatively small, almost spherically symmetrical CH_4, CF_4 and SF_6, and the non-polar diatomic molecules too. In the case of non-spherically symmetrical, slightly elongated molecules, such as N_2 and O_2, the rotation of the molecules in the cavities is hindered, as they are oriented parallel to the cavity wall; as a consequence of this, the cavities are flattened to some extent and hence the rotation of the molecules becomes even less free. This latter effect results in an increase in the vapour pressure of the clathrate, or a decrease in the entropy, compared to the values calculated on the basis of free rotation [272].

With the method of Lennard-Jones and Devonshire [101], the average contribution to the potential energy of a molecule enclosed in a cavity, as a consequence of its interactions with the wall, can be given by the force law

$$\varphi(R) = 4\varepsilon\left[\left(\frac{\sigma}{R}\right)^{12} - \left(\frac{\sigma}{R}\right)^6\right] \tag{4.164}$$

where R is the distance between the dissolved molecule and the elementary part of the wall, while ε and σ are energy and distance parameters characteristic of the type of interaction. In the case of gas hydrates, the wall element of a cavity of type i is surrounded by z_i water molecules. σ is the distance at which the attraction and repulsion forces are in equilibrium, i.e. $\varphi(\sigma)=0$. The parameter ε is the value of $-\varphi(R)$ for the strongest attraction, which occurs at the value $R= =2^{1/6}\sigma$; and the van der Waals radius of one atom [212] is given by $R/2 =0.56\sigma$.

The interactions described by Equation (4.164) were summed by Lennard-Jones and Devonshire for the case when z molecules are distributed on the spherical surface, and on this basis they gave the function $(w)r$ for the value of the potential field originating within the sphere, in which all orientation possibilities were averaged. According to this, for a gas molecule at the centre of the cell:

$$w(0) = z\varepsilon(\alpha^{-4} - 2\alpha^{-2}) \tag{4.165}$$

and

$$w(r) - w(0) = z\varepsilon\left[\alpha^{-4}l\left(\frac{r^2}{a^2}\right) - 2\alpha^{-2}m\left(\frac{r^2}{a^2}\right)\right] \tag{4.166}$$

where l and m are algebraic functions of the dimensionless variable r^2/a^2. The parameter α expresses the cell radius a in terms of the effective molecular diameter σ:

$$\alpha = \frac{a^3}{\sigma^3\sqrt{2}} \tag{4.167}$$

By taking into consideration the energetic assumptions, the cell partition function can be written in the following form:

$$h = \Phi(T) \exp\left[-\frac{w(0)}{kT}\right] \int_{\text{cell}} \exp\left[-\frac{w(r)-w(0)}{kT}\right] 4\pi r^2 \, dr \tag{4.168}$$

The value of the partition function $\Phi(T)$ corresponds to the value determined in Equation (4.148), and expresses primarily the effects originating from the translational motion. The second part of the equation gives the potential energy $(w(0))$ of a dissolved molecule situated at the centre of the cell, referred to a perfect gas, while the quantity under the integral is the "free volume" of the molecule wandering in the cavity. For this latter quantity we may write:

$$\int \exp\left[-\frac{w(r)-w(0)}{kT}\right] 4\pi r^2 \, dr = 2\pi a^3 g \tag{4.169}$$

The value of the dimensionless function g is

$$g = \int \exp\left\{\frac{z\varepsilon}{kT}\left[-\frac{l(y)}{\alpha^4} + \frac{2m(y)}{\alpha^2}\right]\right\} y^{1/2} \, dy \tag{4.170}$$

from which, by substituting into Equation (4.168):

$$h = \Phi(T) \exp\left[-w(0)/kT\right] 2\pi a^3 g \tag{4.171}$$

In this expression, $w(0)/kT$ and g can be found by determining the dimensionless values $z\varepsilon/kT$ and α, making use of relationship (4.165), and with numerical integration of Equation (4.170).

By means of Equation (4.171) for h, the vapour pressure and formation energy of the hydrate and the values of the two other parameters can be calculated. Via this equation, we obtain the following value for the Langmuir constant:

$$C_{Ki} = \frac{1}{kT} 2\pi a_i^3 g_{Ki} \exp\left[-\frac{w_{Ki}(0)}{kT}\right] \tag{4.172}$$

From this and Equation (4.152), the vapour pressure is:

$$p_K = \frac{kT}{2\pi a_1^3 g_{K1}} \exp\left[\frac{w_{K1}(0)}{kT}\right] \frac{y_{K1}}{1-\sum_{J} y_{J1}} = \frac{kT}{2\pi a_2^3 g_{K2}} \exp\left[\frac{w_{K2}(0)}{kT}\right] \frac{y_{K2}}{1-\sum_{J} y_{J2}} \tag{4.173}$$

If kT/p_K is substituted by the molar volume of species K in the gas phase (V_K^G), we obtain:

$$\frac{y_{Ki}}{1-\sum_J y_{Ji}} = \frac{2\pi a_i^3 g_{Ki}}{V_K^G} \exp\left[-\frac{w_{Ki}(0)}{kT}\right] \tag{4.174}$$

where the first expression on the right-hand side is the ratio of the volume of molecule K in the free hydrate to the gas-phase molar volume, while the second expression corresponds essentially to the Boltzmann factor.

Making use of the previous considerations, the energy of the hydrate can be given by:

$$\frac{U-U^\beta}{kT^2} = N_Q \sum_{J,i} v_i y_i \left[\frac{\partial \ln \Phi_J(T)}{\partial T} + \frac{w_{Ji}(0)}{kT^2} + \frac{\partial \ln g_{Ji}}{\partial T}\right] \tag{4.175}$$

and

$$\frac{U-U^\beta}{N_Q} = \sum_{J,i} v_i y_i [w_{Ji}(0) + kT^2 \partial \ln g_{Ji}/\partial T] \tag{4.176}$$

The first term on the right-hand side of Equation (4.176) is the potential energy of a dissolved molecule situated at the centre of the cavity, while the second term is the "extra" potential energy of the dissolved molecule moving in the field $w(r)$, arising as a consequence of its motion.

In the opinion of Lennard-Jones and Devonshire, their cell theory can be well applied to the face-centred cubic lattice, and hence to the cavities in clathrates, but only if the interactions between hydrate-forming materials in adjacent cavities can be neglected. However, this is not the situation in the majority of cases, and more exact results can therefore be obtained by using the correction functions proposed by Wentorf and his co-workers [372].

On the basis of the publication by van der Waals and Platteeuw [272], by statistical thermodynamic calculation on some gas hydrates with H_I structures, the cavity radii a_1 and a_2 characterizing the two types of cavities in gas hydrates, and also the coordination numbers z_1 and z_2, were determined by means of X-ray analysis [98]. The used value of $\sigma_Q/2 = 0.125$ nm corresponds to the Pauling's van der Waals radius of covalently bound oxygen atoms. The value $\varepsilon_Q/k = 166.9$ K was chosen arbitrarily, taking as basis the measured and calculated vapour pressure of argon hydrate for a temperature of $0\,°C$. From these data, the values of the following relationships can be calculated:

$$\alpha_i = \frac{a_i^3}{\sqrt{2}(\sigma_K/2 + \sigma_Q/2)^3} \tag{4.177}$$

and:

$$\frac{z\varepsilon}{k} = z_i \left(\frac{\varepsilon_Q}{k}\right)^{1/2} \left(\frac{\varepsilon_K}{k}\right)^{1/2} \tag{4.178}$$

The values of the equilibrium vapour pressure (dissociation pressure) calculated by statistical mechanical means for the hydrates of several hydrate-forming components are given in Table 4.10. The agreement between the calculated and measured values is acceptable for the small and spherically symmetrical molecules. The calculated dissociation pressures of the hydrates of ethane and ethylene are much lower than the measured values. This is probably due to the facts that the rotation of these relatively large molecules in the cavities is hindered, and that the relationship used for the determination of the magnitude of the field of force is probably not applicable in the case of polyatomic molecules, particularly if the stability of the system is influenced by double-bonding too.

Table 4.10. Measured and calculated dissociation pressures of some gas hydrates [272]

Hydrate-forming substance	ε_K/k (K)	σ_K (nm)	Occupancy of cavities (Lennard-Jones–Devonshire)		Hydrate dissociation pressure (10^5 Pa)	
			y_1	y_2	measured	calculated by L-J-D method
Argon	119.5	0.3408	0.825	0.841	96.45	96.45
Krypton	166.7	0.3679	0.832	0.830	14.64	15.55
Xenon	225.3	0.4069	0.813	0.835	1.15	1.01
Methane	142.7	0.3810	0.818	0.836	26.26	19.19
CF_4	152.5	0.470	0.282	0.894	1.01	1.61
Ethane	243.0	0.3954	0.837	0.827	5.25	1.41
Ethylene	199.2	0.4523	0.523	0.879	5.49	0.50
Oxygen	117.5	0.353	0.821	0.839	—	63.63
Nitrogen	95.05	0.3698	0.810	0.845	—	90.90

The previous calculation methods, based on the general relationships of the Lennard-Jones 12–6 potentials, give acceptable results for the monoatomic gases and methane, but the results were not satisfactory for the non-spherically symmetrical molecules CO_2, N_2, O_2, C_2H_6 and C_2H_4. The cause of this may presumably be sought in the facts that possible distortions of the structures of the hydrate crystals were not taken into consideration by the authors, while the form and size, and hence the resulting hindered rotation of the enclosed molecule were neglected in the determination of the cavity potential.

In their later work, McKoy and Sinanoglu [103] tried to find a solution to this problem. They concluded that, in the application of the statistical theory, the internal partition functions of the gas molecules enclosed in the cavities can be regarded as identical to those relating to the free gas molecules, because the field strength developing in the cavity is too weak to have more than an insignificant influence on those energy levels which determine the partition functions.

According to van der Waals and Platteeuw [98], the chemical potential of the lattice molecules is independent of the mode of occupancy of the cavities, i.e. the dissolved molecules do not influence the properties of the host lattice. They established that the change in chemical potential accompanying clathrate formation corresponds to the chemical potential difference which exists between the common ice phase (α) and the unperturbed gas hydrate lattice (β).

McKoy and Sinanoglu [103] analytically determined the value of the change in the chemical potential for bromine hydrate, and drew further conclusions on the basis of this datum. They stated that the gas molecule resides with the greatest probability in the neighbourhood of the centre of the cavity, and its maximum departure from this position does not exceed 0.05–0.1 nm. In the cases of nitrogen and ethane, for example, the bond distances between the atoms lie in the range 0.10–0.15 nm, and if the centres of these molecules do not depart from the centre of the cavity by a distance greater than 0.1 nm, then the cavity wall cannot be nearer to the molecule than 0.25–0.30 nm at either end of the molecule; consequently, the enclosed molecule is not able to force any of the water molecules of the lattice from its equilibrium position.

Making use of the cell model of Lennard-Jones and Devonshire [374] in their calculations, McKoy and Sinanoglu [103] gave the following relationship for the cavity potential, including the dispersion and repulsion forces in the cavity, in the case of a small molecular displacement (for small r/a):

$$w(r) \equiv w(0)+(z\varepsilon_0/r^2)[22(r_0/a)^2-10(r_0/a)^6]r^2+Q(r^4) \qquad (4.179)$$

where r is the distance of the centre of the gas molecule from the centre of the cavity, a is the cavity radius, ε_0 and r_0 are intermolecular parameters, and z is the number of water molecules in the cavity wall. The value of $w(0)$ naturally expresses the potential relating to the centre of the cavity.

The value of the coefficient of the factor r^2 is $k/2$ in the relationship corresponding to the potential of a three-dimensional harmonic oscillator:

$$\tilde{w}(r) = w(0)+kr^2/2 \qquad (4.180)$$

The frequency of oscillation of the molecule is:

$$v_g = 2\pi^{-1}(k/m)^{1/2} \qquad (4.181)$$

where m denotes the mass of the molecule. The characteristic values of these frequencies are 10^4–10^5 s^{-1}. McKoy and Sinanoglu investigated the extent to which an oscillation with such a frequency can perturb the frequency spectrum of the empty water lattice, and (which follows from this) how the thermodynamic properties of the lattice change.

According to Blue [375], the vibrations of the water molecules in the ice lattice can be divided approximately into three correlated motion types:

(1) The translational vibration of the centres of gravity around the equilibrium position;

(2) The rotational oscillation of the rigid molecule;

(3) The internal vibration of the individual water molecules.

The concrete values and limiting values of these vibrations were determined by Ockman [376]. The lowest frequency in the frequency spectrum of the lattice is 2×10^{12}–2×10^{13} s^{-1}. If this is compared with the frequency of 10^4–10^5 s^{-1} for the gas molecules enclosed in the cavities, there is a high probability that the dissolved molecule does not exert any influence on the lattice structure; accordingly, the conclusions of the previous investigations, that the thermodynamic properties of the lattice are independent of the existence of the enclosed gas molecules, are considered acceptable.

The relative probability of residence of the molecules is well expressed by the Boltzmann factor: $\exp\left[-w(r)/kT\right]r^2$. When this is plotted as a function of the distance r (the distance from the centre of the cavity) on the basis of the data of McKoy and Sinanoglu [103], Figs. 4.8 and 4.9 can be obtained.

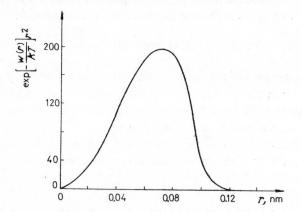

Fig. 4.8. Dependence of the Boltzmann probability factor
on the distance from the centre
of the cavity for N_2 hydrate [103]

It is evident from these figures that, at a distance $r>0.1$ nm, the probability of residence of the molecules is relatively low. This distance is small compared to the cavity diameter, and consequently it is not probable that the rotation of the gas molecules enclosed in the cavities would be strongly hindered; indeed, a certain degree of tumbling translational motion can be conceived, without resulting perturbation by the wall of the cavity.

In view of these data, McKoy and Sinanoglu refute the conclusion of van der Waals and Platteeuw [98] that the rotation of the non-spherically symmetrical gas molecules in the neighbourhood of the cavity wall is strongly hindered, since the molecules can virtually never reside close to the wall. With regard to the fore-

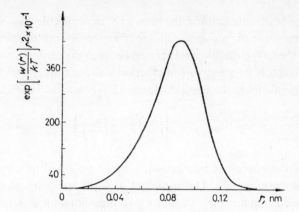

Fig. 4.9. Dependence of the Boltzmann probability factor
on the distance from the centre
of the cavity for C_2H_6 hydrate [103]

going considerations, the differences between the theoretical and experimental data of the previous authors presumably arise from the fact that the application of the Lennard-Jones 12–6 potentials to these hydrate systems was not completely appropriate.

If the force law, i.e. the Lennard-Jones 12–6 potential, Expression (4.164), is written in the form

$$\Phi(r) = \varepsilon_0 \left[\left(\frac{r_0}{r} \right)^{12} - 2 \left(\frac{r_0}{r} \right)^{6} \right] \tag{4.182}$$

the potential minimum ε_0 and the related r_0 are independent of the shape and size of the molecules taking part in the interaction. The dissociation pressures calculated from this potential expression for the spherical atoms Kr and Xe agree well with the experimental results, but for other molecules the expression gives substantially low values.

In the cases of the non-spherically symmetrical gas molecules (N_2, C_2H_6, C_2H_4, CO_2, etc.), the interaction between the cavity wall and the molecule cannot be centric, and therefore the value of the cell potential is influenced by the shape of the molecule. For the elimination of this during the calculations, McKoy and

Sinanoglu investigated whether the Kihara potential [377–382] gives better results than the Lennard-Jones 12–6 potential in the determination of the dissociation pressure. The Kihara potential includes the effect of the molecular dimensions on the interactions.

When the Kihara potential is applied to several gases, e.g. N_2, O_2, C_2H_6, C_2H_4 and CO_2, the second virial coefficient shows good agreement with the experimental results.

In the course of the calculations, the nucleus of the molecule was taken as centred between the atoms for the non-polar diatomic molecules, centred between the oxygen atoms for the CO_2 molecule, and centred between the carbon atoms for the hydrocarbons. Here too, the energy of interaction was given in a form corresponding to the equation involving the Lennard-Jones 12–6 potential, i.e. Equation (4.164):

$$\Phi(\varrho) = \varepsilon \left[\left(\frac{\varrho_m}{\varrho} \right)^{12} - 2 \left(\frac{\varrho_m}{\varrho} \right)^{6} \right] \tag{4.183}$$

where $\Phi(\varrho)$ is the energy of interaction, ε is the potential minimum, and ϱ_m is the position of the minimum. The argument of Φ (its independent variable), however, was taken in such a manner as to give the minimum distance (ϱ) between the nuclei of the molecules. The dissociation pressure data calculated or expressed on the basis of the relationship are to be found in Table 4.11.

Table 4.11. Dissociation pressure data measured experimentally and calculated by different methods for hydrates of several gases [103]

Gas	Hydrate dissociation pressure (10^5 Pa)				
	Measured	Calculated			
		L-J 12–6	Kihara	L-J 28–7	L-J 28–7
Ar	96.45	96.45	—	521.16	95.95
Kr	14.64	15.55	—	129.28	14.34
Xe	1.16	1.01	—	12.12	0.65
CH_4	262.60	19.19	13.13	151.50	16.16
			19.19		
N_2	141.4	90.9	116.15	535.3	72.72
O_2	101.0	63.63	121.2	388.85	76.76
CO_2	12.59	0.71	9.09	8.48	—
		1.70		20.20	
N_2O	10.1	0.6	8.28	9.09	—
		1.52		13.13	
C_2H_6	5.25	1.11	8.48	13.13	—
C_2H_4	5.49	0.5	1.30	3.03	—
			0.82		

Stupin [383] similarly used the Lennard-Jones 12–6 potential to determine how guest molecules with different sizes influence the properties of the water lattice in the case of hydrates of structural type H_{II}. At a D_2O content of 2%, he showed that the frequencies ν_{OD} of the stretching vibrations of hydrate-forming materials occupying only the large cavities increased in the sequence $SF_6.17H_2O$, $CClF_3$. $.17H_2O$, $CCl_2F_2.17H_2O$, $CBr_2F_2.17H_2O$. When the system also contained Ar, which can occupy the small cavities too, the frequencies ν_{OD} for $CBr_2F_2.xAr$. $.17H_2O$ and $SF_6.xAr.17H_2O$ became practically equal, independently of the sizes of the guest molecules.

It was demonstrated by Hamann and Lambert [380] that the Lennard-Jones 28–7 potential is likewise suitable for the description of the interactions of two quasi-spherically symmetrical molecules in the form

$$\Phi(r) = \frac{\varepsilon_0}{3\left[\left(\frac{r_0}{r}\right)^{28} - 4\left(\frac{r_0}{r}\right)^7\right]} \tag{4.184}$$

where ε_0 is similarly the potential minimum, and r_0 is the distance relating to this.

The concrete application of the Lennard-Jones 28–7 potential to gas hydrates proved to be difficult. When the various interactions are taken into consideration, the average interactions for the individual molecules can be determined with the aid of the Gegenbauer polynomials [381, 382].

The values of the spherically symmetrical cavity potentials calculated by means of various potential relationships can be compared, as may be seen in Figs. 4.10–4.13. The dissociation pressures calculated in this way are similarly given in Table 4.11. The final column of the table shows the data obtained by substituting the

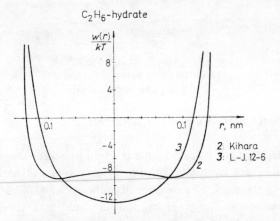

Fig. 4.10. Dependence of the spherically symmetrical cavity potential $\omega(r)$ on the radius for C_2H_6 hydrate [103]

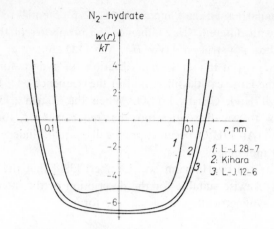

Fig. 4.11. Dependence of the spherically symmetrical cavity potential $\omega(r)$ on the radius for N_2 hydrate [103]

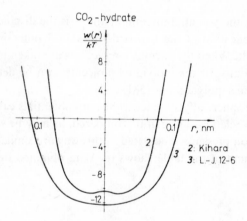

Fig. 4.12. Dependence of the spherically symmetrical cavity potential $\omega(r)$ on the radius for CO_2 hydrate [103]

constants of the Lennard-Jones 12–6 potentials into the Lennard-Jones 28–7 equation.

It can be stated that the calculated dissociation pressures are very sensitive to the intermolecular force constants; the differences can be well seen in Table 4.12. These data reflect the work of a number of authors [98, 377–380, 384, 385].

From a comparison of the graphical and tabulated data, therefore, it can be seen that the Lennard-Jones 12–6 potential can be applied satisfactorily for monoatomic gases and the practically spherically symmetrical methane. In the case of

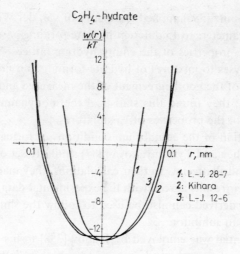

Fig. 4.13. Dependence of the spherically symmetrical cavity potential $\omega(r)$ on the radius for C_2H_4 hydrate [103]

rod-like molecules, better results may be obtained by means of the Kihara potential. For the gases studied, the Lennard-Jones 28–7 potential is the least suitable.

The parameters given by McKoy and Sinanoglu [103] may be conveniently used in the thermodynamic study of gas mixtures.

Table 4.12. Intermolecular force constants calculated in different ways [98,377–380,384,385]

Gas	L-J 12–6		Kihara			L-J 28–7	
	r_0 (nm)	ε_0/k (K)	Nuclear distance l, or radius a (nm)	φ_m (nm)	ε/k (K)	r_m (nm)	ε_m/k (K)
Ar	0.383	119.5	0.1094	0.347	124.0	0.336	240.0
Kr	0.413	166.7	—	—	—	0.348	340.0
Xe	0.457	225.3	—	—	—	0.388	470.0
CH_4	0.428	142.7	0.021	0.381	178.0	0.363	310.0
	—	—	0.032	0.335	205.0	—	—
O_2	0.388	118.0	0.110	0.314	153.0	0.342	240.0
N_2	0.415	95.1	—	—	—	0.364	190.0
CO_2	0.457	205.0	0.220	0.370	279.0	0.369	200.0
	0.504	189.0	0.329	0.272	400.0	—	—
C_2H_4	0.508	199.0	0.134	0.40	266.0	—	—
C_2H_6	0.444	243.0	0.154	0.259	609.0	—	—
N_2O	0.467	205.0	0.231	0.340	309.0	—	—
H_2O	0.280	167.0	—	—	—	0.270	338.0

Parrish and Prausnitz [386] applied the theory of van der Waals and Platteeuw with the Kihara parameters to 15 different hydrate-forming gases, and determined the thermodynamic properties of the empty hydrate lattices. They subsequently extended their analyses to mixtures of hydrate-forming and non-hydrate-forming gases. On the basis of the good agreement of the measured and calculated dissociation pressure data, they found this statistical thermodynamic method suitable for the description of the properties of gas mixtures.

In the determination of the equilibrium conditions of formation of the natural gas hydrates, Malishev and Tyushnyakova [387] similarly set out from the theory of van der Waals and Platteeuw. In their calculations, they made use of the Langmuir constants. Their results agreed with the experimental data [386] within 2–5%. The equations they derived can also be used to follow the shifts in the equilibria in the presence of an inhibitor.

The Kihara potential was employed by Ivanov [388] in his calculations of the Langmuir constants of the spherical, linear and sphero-cylindrical molecular gas hydrates. In his analysis of the theory of the localized sorption of gas hydrates [389], he took as basis the partition function of Bogolyubov, the integral equation of Arinshtein and the equation of the Langmuir isotherm.

The computer programme developed by Parrish and Prausnitz for the prediction of hydrate formation, utilizing the statistical mechanical theory of van der Waals and Platteeuw and the Kihara potential, was applied by Ng Heng-Joo and Robinson [390] to multicomponent systems containing isobutane; they also corrected it, thereby making it suitable for the determination of the three-phase and four-phase equilibrium points of ternary systems.

Dharmawardhana and his co-workers [391] similarly made use of the differences in enthalpy and chemical potential between the empty hydrate lattice and ice for an exact prediction of the conditions of dissociation of natural gas hydrates. The theory of Parrish and Prausnitz [386] was again taken as the basis for the determination of these factors.

A number of papers were recently published by Holder and his co-workers [392] with regard to the determination of the thermodynamic and molar properties of the hydrates of various gases. Via an investigation of the molecular properties of the noble gases, methane and water, and the interactions arising between them, they also provide an analytical compilation relating to the limitations of application of the van der Waals–Platteeuw and the Kihara potential functions.

Chapter 5

PROPERTIES AND CHARACTERISTIC DATA OF INDIVIDUAL GAS HYDRATE SYSTEMS

5.1. Systems with a single gas component

5.1.1. Nitrogen hydrate and oxygen hydrate

In the course of investigations connected with the gas hydrates, a number of authors (e. g. [1, 66]) have established that the air enclosed in the clathrate structure is enriched to some extent in oxygen. However, an independent hydrate could not be detected for either nitrogen or oxygen, and these gases were therefore classified as auxiliary gases ("Hilfsgase").

Nevertheless, the theoretical considerations by van der Waals and Platteeuw [98] led to the assumption that the hydrates of gaseous N_2 and O_2 might exist independently. van Cleeff and Diepen [159] reported two series of experiments in which they succeeded in producing nitrogen hydrate. In one experiment, they worked with nitrogen prepared by the thermal decomposition of sodium azide, and in the second series with factory-produced nitrogen with a purity of 99.999%. The results of the parallel studies showed good agreement.

Their investigations indicated that the lower quadruple point of the nitrogen hydrate system is to be found at a pressure of 1.408×10^7 Pa and a temperature of $-1.4 \,^\circ C$. It is characteristic of the stability that the pressure relating to $0 \,^\circ C$ is 1.621×10^7 Pa. The experimentally-obtained p vs. T data of the three-phase equilibrium curve are given in Table 5.1, and the curve itself in Fig. 5.1.

The equilibrium pressure of oxygen hydrate at $0 \,^\circ C$ was found to be $1.216 \times \times 10^7$ Pa [159], but the concrete experimental data were not published.

The equilibrium curve of nitrogen hydrate was also determined by Marshall and his co-workers [127]. As shown in Fig. 5.1, the data they measured in the comparable pressure range fit in very well with the results of van Cleeff and Diepen. The data determined by Marshall and his co-workers can be seen in Table 5.2.

In another publication [111], making use of the statistical mechanical relationships, the same authors determined the Lennard-Jones–Devonshire force constants, the fugacity–temperature equilibrium data below the ice-point, the chemical potentials (using the theory of solid solutions), and the heats of formation of nitrogen hydrate in the pressure interval 1.50–7.00×10^7 Pa (on the basis of the Clausius–Clapeyron equation).

Table 5.1. Equilibrium data on nitrogen hydrate [159]

p (10^5 Pa)	T (°C)	p (10^5 Pa)	T (°C)
hydrate–liquid–gas equilibrium			
109.1	−4.0	272.7	5.1
119.2	−3.1	278.7	5.1
131.2	−2.1	281.8	5.5
132.3	−2.2	297.9	6.0
137.3	−1.6	302.0	6.1
144.4	−1.1	338.3	7.1
152.5	−0.6	373.7	8.1
158.7	−0.3	384.8	8.5
158.7	−0.2	413.0	9.1
159.6	0.0	457.5	10.1
162.6	0.0	505.0	11.1
165.6	0.2	521.1	11.5
174.7	0.8	552.4	12.1
176.7	1.0	612.0	13.1
190.9	1.7	675.6	14.1
191.9	1.7	710.0	14.6
195.9	2.1	743.3	15.2
206.0	2.4	812.0	16.1
215.1	2.7	890.8	17.1
223.2	3.1	919.0	17.5
230.3	3.4	955.0	17.9
247.4	4.1		
ice–liquid–gas equilibrium			
139.4	−1.4	106.0	−1.0
126.2	−1.3	81.8	−0.7
126.2	−1.2	73.7	−0.6

For the nitrogen–water system, the following equation was found to hold by van Cleeff and Diepen [393] as regards the temperature-dependence of the hydrate vapour pressure in the neighbourhood of the quadruple point:

$$\log p = 14.1293 \pm 0.0006 - \frac{3257 \pm 23}{T} \tag{5.1}$$

while in the presence of ice:

$$\log p = 5.5598 \pm 0.0002 - \frac{927 \pm 17}{T} \tag{5.2}$$

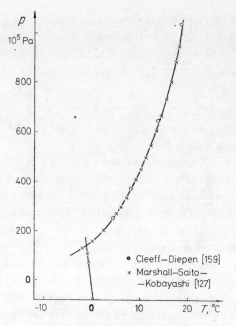

Fig. 5.1. Equilibrium diagram
of nitrogen hydrate [127, 159]

The corresponding data for the oxygen–water and oxygen–ice systems, respectively, can be calculated via the following equations:

$$\log p = 14.3082 \pm 0.0005 - \frac{3340 \pm 21}{T} \tag{5.3}$$

and

$$\log p = 5.1168 \pm 0.0006 - \frac{838 \pm 38}{T} \tag{5.4}$$

Table 5.2. Equilibrium data on nitrogen hydrate [127]

p (10^5 Pa)	T (°C)	p (10^5 Pa)	T (°C)
256.8	4.4	1743.7	24.3
378.2	8.0	1981.7	25.4
656.5	13.4	2140.4	26.3
1050.5	18.2	2262.4	27.1
1189.8	19.6	2764.1	29.1
1326.8	20.9	3272.4	31.2
1581.0	23.1	3388.0	32.0

The compositions of the hydrates were also determined in the knowledge of the heats of solution and the Clapeyron equation, taking into account corrections for the dissolution of gaseous nitrogen and oxygen in water.

The volume changes occurring during hydrate formation were determined from the relationship

$$\Delta V = V_{\text{hydrate}} - n V_{\text{ice}} - V_{\text{gas}} \tag{5.5}$$

The individual molar volume data can be calculated by utilizing the density and the compression (deviation) factor.

When the fundamental data available were used, it was found that at a pressure of 141.5 bar and a temperature of 0 °C the composition of nitrogen hydrate is $N_2 . (6.01 \pm 0.23) H_2O$, while at a pressure of 109.2 bar and a temperature of 0 °C that of oxygen hydrate is $O_2 . (6.06 \pm 0.21) H_2O$. The results reflect the fact that, near the quadruple point, the cavities are not fully occupied by gas molecules for either nitrogen hydrate or oxygen hydrate.

Bukhgalter and Dyegtyarev [394] investigated the conditions of formation of hydrates of natural gases containing nitrogen. For a natural gas with a nitrogen content of 63%, the experimentally determined hydrate formation temperatures agreed with those calculated by means of the equilibrium constants (with a maximum difference of 6%) in the pressure range 70–140 bar.

5.1.2. Hydrates of the noble gases

The hydrates of the noble gases were discussed as long ago as 1927 by Schroeder [1]. Argon hydrate was discovered by Villard [37] in 1923. Following this, de Forcrand [35, 36] systematically investigated the hydrate-forming properties of the noble gases. His findings can be summarized as follows: the hydrates of helium

Table 5.3. Equilibrium data on noble gas hydrates [35, 36]

Argon hydrate		Krypton hydrate		Xenon hydrate	
p (10^5 Pa)	T (°C)	p (10^5 Pa)	T (°C)	p (10^5 Pa)	T (°C)
1	−39.5	1	−24.8	1	−1.13
10	0.0	14.6	0.0	1.46	1.4
		48.0	12.5	17.2	23.5

and neon can exist only at very high pressure. According to his calculations, for example, helium hydrate, if it is formed at all, can be stable only above a pressure of several thousand bars, and in the neighbourhood of 0 °C the hydrate of neon cannot be formed below a pressure of 2.533×10^7 Pa.

The data to be seen in Table 5.3 on the hydrates of argon, krypton and xenon were published by de Forcrand [35, 36]. At the same time, by comparing the data on the noble gases, he concluded that at a temperature of 0 °C radon can form a hydrate only at a pressure lower than atmospheric.

Miroshnichenko [395] surveyed the thermodynamic and statistical physical properties of gas hydrates by means of a review of the properties of argon hydrate. He applied the theories to the possibility of hydrate formation by supercooled clouds and fogs containing noble gases.

The investigations of the conditions of hydrate formation by the noble gases played an important role in the determination of the physicochemical properties of the gas hydrates, for the interactions involved could be interpreted most easily

Table 5.4. Equilibrium data on argon hydrate
formation [124, 127]

p (10^5 Pa)	T (°C)	p (10^5 Pa)	T (°C)
109.6	1.16	1879.2	25.60
126.5	2.94	2121.8	26.70
135.3	3.72	2131.0	26.83
174.2	6.28	2426.7	27.7
327.3	11.60	2767.9	28.9
329.0	11.78	3151.7	29.7
513.6	15.26	3469.3	30.4
686.1	17.83	3864.2	31.0
839.6	19.33	3877.4	30.9
1445.7	23.42	3926.9	31.1

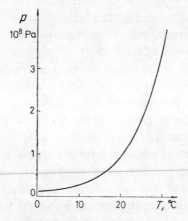

Fig. 5.2. Equilibrium diagram
of argon hydrate [124, 127]

in the cases of the data relating to these relatively small, practically spherically symmetrical monoatomic species. The principles and data of these calculations have been presented in Chapter 4.

With the development of the high-pressure measurement technique, more comprehensive investigations followed on the conditions of formation of argon hydrate. Marshall and his co-workers [124, 127] published detailed data on this system too; these are given here in Table 5.4 and Fig. 5.2. They also carried out thermodynamic and statistical mechanical calculations relating to argon hydrate [111].

With regard to the ease of availability of the various noble gases and the stabilities of their hydrates, the hydrates of krypton and xenon can be investigated the most readily. By taking into consideration the results of Schroeder [1] too, von Stackelberg and Müller [66] established the data relating to their stabilities.

In connection with the hydrates of the noble gases, Pauling and Hayward [190] made the interesting finding (already mentioned above) that xenon is an excellent anaesthetic agent. This can be explained by the fact that it forms clathrates with the water content of the blood; as a consequence, the ions and electrically charged groups too are localized, and hence conscious and reactive mental activity is inhibited. The structural scheme they published for xenon hydrate is shown in Fig. 5.3.

The dissociation pressure and other thermodynamic properties of xenon hydrates in the temperature interval 0–12 °C were also investigated by Ewing and Ionescu [396]. Their measurements relate too to the investigation of the possibilities of interactions in the haemoglobin–water–nitrogen–xenon system. In the course of their work, they studied the solubility of xenon in water and evaluated the Henry constant. They established that, in the temperature and pressure ranges 5–25 °C and $0–5\times10^5$ Pa, the solubility of xenon in water follows Henry's law until hydrate formation occurs. At a temperature of 5 °C, with the value $k = 1.1044\times10^7$ Pa mol^{-1}, their determination method gave a value of 2.52×10^5 Pa for the dissociation pressure of the hydrate.

Hydrate dissociation pressure values obtained on the basis of solubility examinations in the temperature interval in question are shown in Table 5.5. These results differ from those published by de Forcrand [35, 36]. The difference is presumably due to the different purities of the xenon samples. The xenon used by Ewing and Ionescu contained impurities in a quantity of 90 ppm (90 g t^{-1} = = 90 g/10^6g), as determined by mass-spectrometric measurements. Accordingly, the composition of the contamination was known exactly too, whereas at the time of the studies by de Forcrand it is not probable that this purity could be reached. Thus, the gas sample investigated by de Forcrand may also have contained such impurities as N_2, Ar, O_2 and CO_2, the stabilizing effects of which can be presumed.

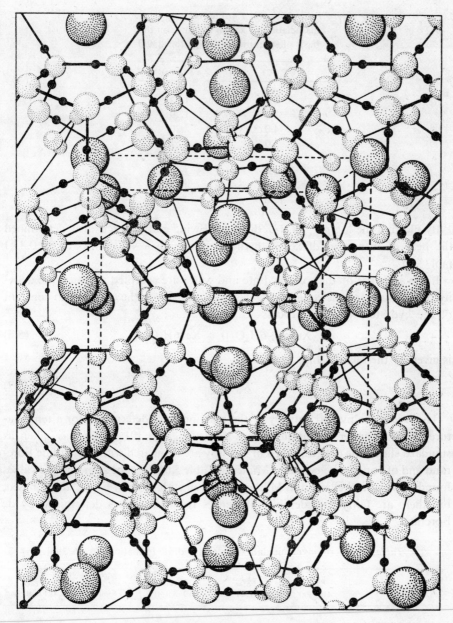

Fig. 5.3. Structure of xenon hydrate [190].

From *The Architecture of Molecules* by Linus Pauling and Roger Hayward, W. H. Freeman and Company. Copyright © 1964

Table 5.5. p vs. T values relating to the
dissociation of xenon hydrate, on the
basis of solubility investigations [396]

T (°C)	p (10^5 Pa)
0.0	1.53
2.0	1.79
4.0	2.20
5.0	2.51
6.0	2.83
8.0	3.42
10.0	4.17
12.0	4.96

By plotting the log p_{diss} vs. $1/T$ diagram, Ewing and Ionescu also determined
the heat of formation of xenon hydrate, obtaining a value of -64.9 ± 0.8 kJ
mol^{-1}.

If the activities of the solid hydrate and water are accepted as unity, the activity of gaseous xenon is equal to the gas pressure, and so the equilibrium constant
of the hydrate formation process

$$Xe(g) + 6H_2O(l) \rightleftharpoons Xe \cdot 6H_2O(s) \tag{5.6}$$

is given by the relationship

$$K_p = \frac{1}{p_{Xe}} \tag{5.7}$$

The free enthalpy and entropy of the reaction were determined for a temperature of 5 °C under conditions where the value $p_{Xe} = 101\,325$ Pa was taken as
standard state. The data obtained in this way are compared with those of de Forcrand and of von Stackelberg and Müller in Table 5.6. It can be seen that the values
found by von Stackelberg and Müller agree well with those of Ewing and Ionescu,
but there are fairly considerable differences from the values determined by de
Forcrand.

Table 5.6. Thermodynamic data on xenon hydrate formation [36, 66, 396]

Authors	$p_{diss}^{0\,°C}$ (10^5 Pa)	$T_{diss}^{1\,bar}$ (°C)	ΔH_{form} (kJ mol^{-1})	$\Delta G_{form}^{5\,°C}$ (kJ mol^{-1})	$\Delta S_{form}^{5\,°C}$ (kJ mol^{-1})
Ewing and Ionescu [396]	1.53	−3.6	64.8	2.09	−226.09
de Forcrand [36]	1.16	−1.13	76.5	1.80	−281.35
von Stackelberg and Müller [66]	1.51	−3.4	70.22	—	—

In the course of our own investigations, we too dealt with the conditions of formation and decomposition of krypton and xenon hydrates [397]. The resulting data are listed in Table 5.7. The compositions of the gas samples investigated were as follows: 98.5% Kr+1.2% Xe+0.3% impurities, and 98.6% Xe+1.3% Kr +0.1% impurities. The sequence relating to the quantities of the impurities was $Ar > N_2 > O_2 > CO_2$.

Table 5.7. Dissociation parameters of krypton and xenon hydrates [397]

T (°C)	Krypton hydrate p_{diss} (10^5 Pa)	Xenon hydrate p_{diss} (10^5 Pa)
−5.0	9.49	—
−2.0	12.21	1.11
0.0	14.54	1.51
2.0	18.28	1.92
4.0	22.62	2.22
6.0	27.97	2.73
8.0	35.14	3.43
10.0	41.60	4.44
12.0	48.88	5.65
14.0	56.76	7.07
16.0	66.25	8.89
18.0	76.45	10.80
20.0	87.26	12.82
22.0	99.68	15.25

The data in Table 5.7 are presented graphically in Fig. 5.4, which also features data above the upper quadruple point. A comparison of these data with those of Ewing and Ionescu for xenon hydrate demonstrates satisfactory agreement, but only up to a temperature of 10 °C. In our experiments at higher temperatures, we probably observed formation of a hydrate slightly more stable than to be expected on the basis of the solubility data.

In our investigations, the decomposition point of the hydrate was determined by establishing the related pressure–temperature pairs during the slow warming-up of the hydrate obtained by cooling of the gas–water system under appropriate conditions. Our visual observations during the experiments indicated that the exact determination of the dissociation parameters is influenced by many factors, for the process does not take place instantaneously. As the temperature of the hydrate system is slowly increased, visually well observable changes occur on the surface of the free water possibly present, in the interior of the liquid and in the hydrate phase alike. An example of such a change is that the curvature of the liquid surface, and consequently its surface tension, varies continuously in the temperature interval between the period before the decomposition of the crystal grains and the total breakdown of the hydrate film on the liquid surface, as a consequence of the pressure and volume changes accompanying the phenomenon, as well as of the structural rearrangement. Prior to the decomposition, the

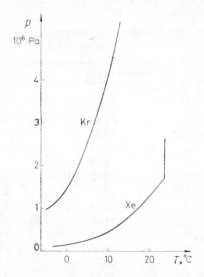

Fig. 5.4. Dissociation equilibria of
krypton hydrate and xenon hydrate [397]

previously not entirely transparent liquid phase clears, and the crystal grains undergo agglomeration; moreover, it can also occur that some of the crystals on the liquid surface fall to the bottom (if the tube was not completely filled with hydrate).

Directly before the perceptible decomposition, a slight decrease of the pressure can be observed, during which the surface of the crystals becomes more shiny, and the movement of the smaller crystals can also be experienced. In our interpretation, this can be explained by the loosening of the bonding forces in the clathrate structure. On further temperature increase, a few microbubbles appear; these scarcely leave the crystal surface, but remain suspended in the liquid. Subsequently, more and more gas bubbles can be seen in the environment of the crystal grains; they slowly become interconnected and (as a demonstration of the existence of a dynamic equilibrium) the resulting larger gas bubbles are surrounded by a thin hydrate film. At a given temperature this film disaggregates and the large gas bubbles are set free; however, they are not yet able to escape from the liquid, for at this temperature the surface of the liquid is still covered by a hydrate film. Nevertheless, as a result of the fact that the volume of the gas bubbles enclosed between or above the hydrate crystals continuously increases, partly because of the constant hydrate dissociation and partly because of the warming-up, the gas hydrates come under an overpressure, which delays the decomposition process. Intensive decomposition starts when the hydrate film on the water surface breaks down completely, and the gas accumulated in the liquid phase is able to escape into the gas space.

Depending on the type and amount of the hydrate, a longer or shorter warming-up time is necessary for dissociation of the total quantity of the gas hydrate. The rate of decomposition becomes perceptibly slower when the quantity of the solid phase falls below a limiting value, i.e. the ratio little gas/much hydrate corresponding to the beginning of the decomposition is reversed in the liquid.

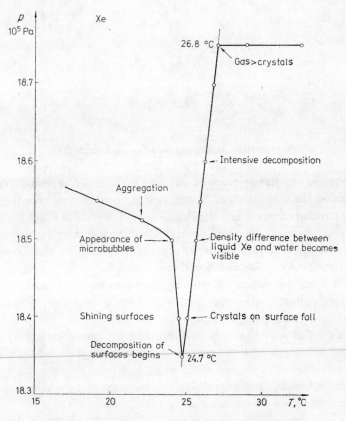

Fig. 5.5. Dissociation process of xenon hydrate [397]

From these observations too it may be seen that, depending upon the observation mode, the temperature of hydrate dissociation at a given pressure can be determined only within an interval of about 1–3 °C. However, even this value only relates to the conditions of the commencement of intensive decomposition: for the decomposition of a larger quantity of hydrate, heating to a temperature higher than this is necessary.

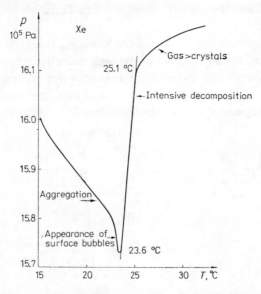

Fig. 5.6. Dissociation process of xenon hydrate [397]

In our experience, the beginning of intensive hydrate decomposition depends to some extent on the formation conditions as well, for these influence the compactness of the structure formed, and the directions and degrees to which the individual crystal grains aggregate on one another. These factors may be determining as regards the energy requirements of the preordering preceding the decomposition, and the duration of the dissociation process.

Figures 5.5 and 5.6 depict the working diagrams of the hydrate dissociation process in the vicinity of the upper quadruple point of xenon hydrate, in the form of plots of the p vs. T pairs accompanying the phenomenon [397].

From the NMR investigations by Garg and his co-workers [398] on xenon hydrate and other gas hydrates of type I, the relationships between the reorientation and diffusion properties of the water molecules bound in the hydrate are also known. It was found that both the reorientation and the diffusion processes depend upon the guest molecules, but the reorientation of the water molecules is always faster than the diffusion process.

5.1.3. Hydrates of the halogen elements

The investigations of the hydrates of the halogen elements are connected with the early discovery of the clathrate compounds themselves, as well as with the recognition of the fundamental hydrate structure properties. As described in previous chapters, the determination of the compositions of bromine and chlorine hydrates was of particular importance in the work of numerous authors. However, it is due to just this fact that a survey of the literature does not yield the full data

Table 5.8. Equilibrium data on chlorine hydrate [66,120, 399]

T (°C)	p (10^5 Pa)
0.0	0.33
5.0	0.58
9.6	1.00
15.0	1.77
20.0	2.91
25.0	4.60
28.7	6.06

for the overall hydrate vapour pressure curve. Using the part-data of a number of authors, primarily von Stackelberg and Müller [66], Iskenderov and Musayev [399] and Byk and Fomina [120], we have compiled the equilibrium data on chlorine hydrate, which are presented in Table 5.8 and Fig. 5.7.

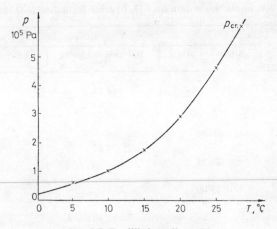

Fig. 5.7. Equilibrium diagram
of chlorine hydrate [66, 120, 399]

The phase diagrams of chlorine hydrate up to a pressure of 8.75 bar in the presence of water or 1–10% aqueous solutions of NaCl have also been given by Bozzo and his co-workers [400].

The main properties of bromine and iodine hydrates were referred to in Chapter 2, but the data relating to the vapour pressure curves are not available to us.

Glew and Hames [401] carried out hydrate experiments with BrCl produced in the presence of free halogen. Their results show that the hydrate formed in the neighbourhood of 5 °C is richer in bromine than the starting aqueous solution. They determined a value of $10\,450\pm860$ cal mol^{-1}=43.752 ± 3.600 kJ mol^{-1} for the fusion enthalpy of the hydrate of composition BrCl.(7.28±0.6)H$_2$O in equilibrium with the aqueous solution.

5.1.4. Carbon dioxide hydrate

The composition of this hydrate was determined in the very early years of gas hydrate research, and the concrete values of the lower and upper quadruple points of CO$_2$ hydrate were known even in the middle of last century. Information of this is to be found in the review by Schroeder [1]. These data were utilized by von Stackelberg and Müller [66], while Unruh and Katz [64] and Robinson and Mehta [156] published equilibrium p vs. T data on the hydrate; these are given here in Table 5.9 and Fig. 5.8.

Unruh and Katz [64] found in their experiments that CO$_2$ hydrate is produced very rapidly, in the form of fine flakes, and the liquid phase solidifies during a

Table 5.9. Equilibrium data on CO$_2$ hydrate formation [64, 156, 296]

T (°C)	p (10^5 Pa)		
	Unruh and Katz [64]	Robinson and Mehta [156]	Berecz and Balla-Achs [296]
0.77	—	14.14	12.72
2.03	—	15.95	15.35
3.00	—	17.97	17.88
4.05	20.93	—	20.50
5.77	—	24.82	26.76
6.00	26.51	—	27.57
7.56	—	32.10	33.13
7.72	33.09	—	34.64
8.72	37.82	—	39.09
8.90	—	39.39	40.10
9.89	46.16	—	65.65
10.10	—	45.81	90.90

few minutes even if the quantity of water present is far in excess of the stoichiometric proportion. The total pressure of the system was observed to decrease by
$3-5 \times 10^5$ Pa as a consequence of hydrate formation. The solid crystalline hydrate
grains formed were distributed throughout the entire liquid phase, but the quantity of crystals was much higher near the liquid surface than elsewhere. It can be
concluded from this phenomenon that the density of the hydrate is lower than

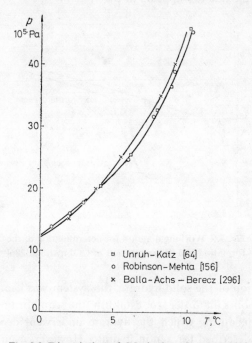

Fig. 5.8. Dissociation of CO_2 hydrate [64, 156, 296]

that of water, and since CO_2 has a relatively high solubility in water (as a result
of not only physical, but also chemical processes), the conditions and mechanism
of hydrate formation differ appreciably from those of other gases which dissolve
in water to only a very slight extent.

Figure 5.8 also includes our own experimental results, obtained using commer-
cial carbon dioxide containing nitrogen impurity [296]. Our experiments reinforced the empirical data reported by Unruh and Katz with regard to the pressure
decrease and the grain structure of the hydrate. To illustrate the pressure changes
accompanying formation of the hydrate, a working diagram of one of our experiments with carbon dioxide is shown in Fig. 5.9. From this figure it can be
seen that, as the gas–water system is cooled, the pressure decreases monotonously

until the thermodynamic conditions for hydrate nucleus formation are reached as a consequence of the slight supercooling of the system. From this point on, the more intensive decrease of pressure indicates the enclosure of gas in the clathrate structure. The state parameter changes accompanying the process of hydrate formation follow an identical course even if the equilibrium vapour pressure curve of the hydrate-forming constituent (CO_2) is exceeded during cooling, and thus liquid carbon dioxide too appears in the system.

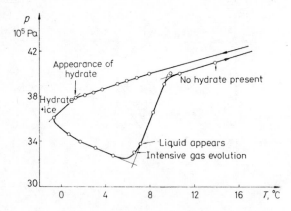

Fig. 5.9. Working diagram for determination of the formation and decomposition of CO_2 hydrate [296]

After the formation of the solid hydrate, the system was kept at constant pres sure for several hours in order to establish the equilibrium; slow warming up was then begun, during which the hydrate underwent decomposition in an analogous way as described for xenon hydrate. However, even when the experi ments were repeated many times and under various conditions, it did not prove possible to obtain the same state parameter values for the formation and the de composition of the hydrate. Under isobaric conditions, the temperature of forma tion was always 3–4 °C lower than the temperature of decomposition; indeed, depending on the quantity of hydrate formed, under isothermal conditions there was a difference of as much as $5–25 \times 10^5$ Pa between the formation and decom position pressures. It was observed that, as a result of low temperature in an appa ratus thermostated for equilibrium to be established, the quantity of hydrate in creased continuously, but the structural properties of the clathrate did not change; that is, within the standard deviation of the experimental errors, the vapour pres sure curve of CO_2 hydrate could be well approximated to, even if the slow heating of the system was begun at once, without prolonged thermostating. If the system was recooled before the crystalline hydrate had completely decomposed, in order that the still intact crystal grains should play the role of the inoculating crystals, and thus the hydrate separation and nucleus growth should start at a lower degree of supercooling, the difference between the formation and decomposition parameters did lessen, but total identity could not be attained. This is illustrated in Fig. 5.10, based on data obtained from four consecutive experiments.

Bozzo and his co-workers [400] determined the phase diagram of CO_2 hydrate in the presence of water or aqueous NaCl solutions up to a pressure of 4.5×10^6 Pa. The experimental data were used to determine the hydrate composition, its critical

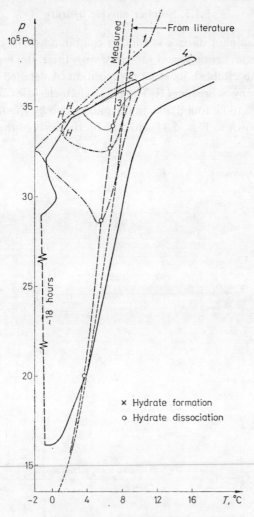

Fig. 5.10. Working diagram for determination
of the equilibrium of CO_2 hydrate [296]

decomposition conditions, and the value of the heat of formation, and special attention was paid to the connections between the salt effect and hydrate formation.

Some aspects of the thermophysical properties of CO_2 hydrate were reported by Groisman [402]. He carried out detailed investigations relating to the thermal conduction and thermal diffusion of the hydrate between -30 and $0\,°C$. In this temperature range it was found that the heat capacity of CO_2 hydrate, measured at constant volume, can be considered practically constant.

5.1.5. Sulphur dioxide hydrate

The hydrate of sulphur dioxide is another compound discovered at the beginning of the nineteenth century, and for many years thereafter much work was performed in efforts to establish its exact composition. A detailed account of this is given in the review by Schroeder [1]. The phase characteristics and a discussion of the p vs. T data are to be found also in the publication by Findlay and Campbell [403]; these are shown in Figs. 5.11 and 5.12 on the basis of the report of Parent

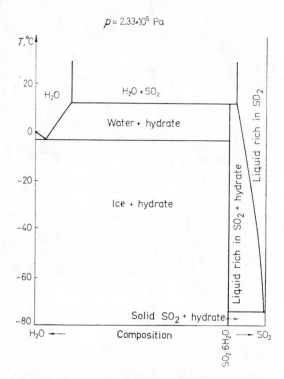

Fig. 5.11. Phase diagram of the SO_2–H_2O system [45, 403]

[45]. SO_2 hydrate decomposes readily into two liquid phases, and in this property it very closely resembles the hydrates of ethane and propane. Line BE in Fig. 5.12 indicates the ice–hydrate–aqueous SO_2 solution equilibrium, and line BC the ice–hydrate–vapour equilibrium. Curve BF describes the aqueous SO_2 solution–hydrate–vapour equilibrium, section FG gives the limit of the equilibrium between the two liquid phases and the vapour, and FH that between the

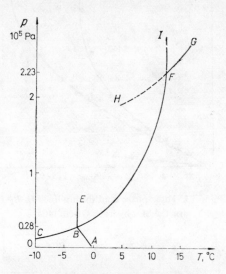

Fig. 5.12. p vs. T equilibrium diagram
for SO_2 hydrate [45]

liquid rich in SO_2, the hydrate and the vapour. FI is the melting curve of the hydrate, where liquid water, liquid SO_2 and the hydrate may be in equilibrium; and finally, along curve AB there is equilibrium between ice, the aqueous solution and the vapour.

In the range 1–400 MPa, van Berkum and Diepen [404] investigated the three-phase equilibrium: hydrate–liquid rich in SO_2–liquid rich in water. The three-phase curve has a maximum temperature (293.68 ± 0.1 K) at 3072 ± 20 bar. In their opinion the system exhibits a congruent melting curve from 2100 ± 10 bar up to higher pressures, a maximum temperature occurring at 294.05 ± 0.05 K and 3450 ± 20 bar. The critical mixing conditions could be only partly determined.

These authors also established the solubility of SO_2 in liquid water and its dependence on the temperature in the presence of a second liquid phase and the gas. The results of the measurements are shown in Fig. 5.13, where the earlier data of Tammann and Kriege [40] and Roozeboom [276] too are plotted. From the results of van Berkum and Diepen, the relatively simple polynomial expression

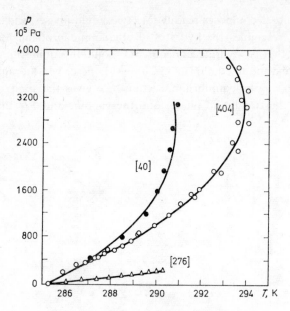

Fig. 5.13. Three-phase equilibrium lines $H-L_1-L_2$ for the SO_2 hydrate system [404]

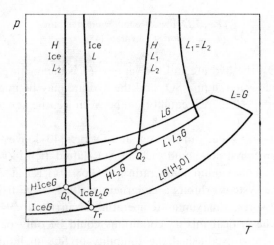

Fig. 5.14. Schematic p vs. T diagram for gas hydrate systems having two liquid phases. H: gas hydrate; L_1: liquid rich in hydrate former; L_2: liquid rich in water; Tr: triple point of water; Q_1 and Q_2: quadruple points [404]

for the curve is

$$T = 285.25 + 5.555 \times 10^{-3} p - 9.041 \times 10^{-7} p^2 \tag{5.8}$$

where T is measured in K and p in bar.

At pressures around that relating to the temperature maximum, the hydrate crystals are suspended in the surrounding liquid and are difficult to distinguish. At the temperature maximum, the mole fraction of SO_2 in the liquid phase rich in water (L_2) is 0.826 ± 0.001.

Figure 5.14 is a schematic outline of the p vs. T diagram for such gas hydrate systems.

5.1.6. Hydrogen sulphide hydrate

The first publications relating to the hydrogen sulphide hydrate system originate from de Forcrand [263, 405–407]; these were incorporated in the later review by Villard [201]. Very detailed data on the three-phase equilibrium of H_2S hydrates were given by Scheffer and Korvezee [408–410]. The positions of the quadruple points in the H_2S–H_2O system, and also the p vs. T equilibrium curves, were determined by Schreinemakers [411]. These data were checked and made more exact by Wright and Maas [412] at temperatures near 0 °C. The hydrate volume and the pressure relations were investigated in detail by Reamer and his co-workers [85], whose data were taken into account by Selleck and his co-workers [86] in their survey of the phase equilibrium properties of the H_2S–water system. Their experimental results extend to three main questions:

(1) the change of the equilibrium pressure as a function of temperature under isochoric conditions in both homogeneous and heterogeneous states;

(2) the change of the pressure as a function of the volume at given temperatures in the equilibrium system gas–liquid rich in water (the primary aim of these investigations was the determination of the bubble-point);

(3) the compositions of the individual phases in heterogeneous equilibria.

These experiments were carried out with H_2S–H_2O mixtures with 11 different compositions, and the results were compared with the data of Scheffer and Korvezee [408, 410] and Wright and Maas [412]. The experimental results agreed well with the data of Scheffer and Korvezee; the data on the three-phase equilibria were somewhat higher than those of Wright and Maas, but the differences were not significant. The p vs. T data found for various three-phase equilibria can be seen in Table 5.10 and Fig. 5.15.

In their review, Selleck and his co-workers [86] gave detailed information on the compositions of the various phases, and the equilibrium molar proportions.

Table 5.10. Equilibrium data on H$_2$S hydrate [86, 408, 410, 412]

Hydrate-liquid rich in water-gas H-L$_1$-G		Hydrate-ice-gas H-I-G		Hydrate-liquid rich in H$_2$S-gas H-L$_2$-G		Aqueous solution–liquid rich in H$_2$S-gas L$_1$-L$_2$-G		Hydrate-aqueous solution–liquid rich in H$_2$S H-L$_1$-L$_2$	
T (°C)	p (10^5 Pa)	T (°C)	p (10^5 Pa)	T (°C)	p (10^5 Pa)	T (°C)	p (10^5 Pa)	T (°C)	p (10^5 Pa)
0.39	0.94	−22.6	0.35	−14.0	7.07	29.5	22.92	29.5	22.92
4.44	1.61	−17.7	0.45	−12.2	7.47	32.2	24.34	29.6	35.35
10.0	2.83	−15.0	0.51	−6.7	8.88	37.7	27.57	29.9	70.70
12.0	3.53	−12.2	0.57	−1.1	10.50	43.3	31.10	30.0	80.29
15.5	5.15	−9.4	0.65	−0.9	10.60	48.9	34.84	30.2	106.05
18.7	7.07	−7.8	0.71	4.4	12.32	49.4	35.35	30.5	143.42
21.1	9.09	−6.7	0.74	9.5	14.14	54.4	39.09	30.8	176.75
22.5	10.60	−3.8	0.83	15.5	16.46	59.9	43.63	31.1	213.61
25.3	14.14	−1.11	0.93	18.4	17.67	65.5	48.48	31.4	247.40
26.6	16.36	−0.39	0.95	21.1	18.87	66.6	49.49	31.6	282.80
27.3	17.67			26.0	21.21	71.1	53.73	31.9	318.15
28.9	21.21			26.6	21.51	76.6	59.18	32.1	353.5
29.5	22.92			29.5	22.92	93.3	78.88	32.2	359.3
						98.9	88.88		
						99.9	91.91		

Hydrogen sulphide forms a relatively stable gas hydrate, and production of the hydrate is not difficult. Accordingly, it is understandable that its properties have been investigated in great detail, and that it can be considered one of the most well-known gas hydrates. Natural gases frequently contain hydrogen sulphide, and the hydrate of hydrogen sulphide is therefore of importance in a discussion of the hydrates of hydrocarbons. In 1954, Noaker and Katz [113] reported a detailed account of the conditions of formation of CH_4–H_2S hydrate, as functions of the composition, pressure and temperature, and gave the vapour–liquid equilibrium constants necessary for the calculations.

As already stated in Chapter 2, hydrogen sulphide is very prone to form double hydrates. The structural investigations by von Stackelberg and Frühbuss [68] led to the clarification of this problem.

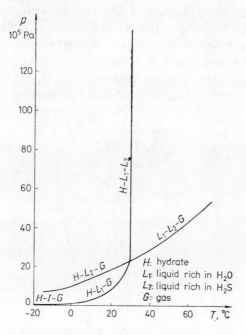

Fig. 5.15. p vs. *T* equilibrium diagram for H_2S
hydrate [86, 408, 410, 412]

Its great readiness to participate in hydrate formation meant that hydrogen sulphide was a useful tool in investigations of the effects of various anti-freeze materials, with the aim of changing the conditions of hydrate formation. This topic was treated in Chapter 3, in connection with the studies by Bond and Russell [82] (see Fig. 3.15).

5.1.7. Hydrates of pure hydrocarbons

5.1.7.1. Methane hydrate

The results of his experiments connected with methane hydrate, made in wide pressure and temperature ranges, were published first by Villard [30, 413] in 1888, and on this basis he stated that the highest temperature at which methane hydrate can exist is 21.5 °C. In 1940, Deaton and Frost [414, 415] reported new data on this system, giving experimental results relating to the three-phase (vapour–liquid rich in water–hydrate) equilibrium in the temperature range 0.5–12.8 °C. It was at this time that investigations of the hydrate-forming properties of hydrocarbons flourished. As one of the first results, Roberts and his co-workers [44] gave detailed information on the conditions of formation of methane and ethane hydrates. They determined the phase diagram of the methane–water system between the temperature limits −12.2 and +15.5 °C, and investigated the equilibrium conditions in the three-phase systems vapour–liquid–hydrate, vapour–ice–hydrate, and liquid–liquid–hydrate. The heat of formation of the hydrate was calculated from the equilibrium data, and the composition $CH_4.7H_2O$ was established. A short time later, further data were published by Roberts and his co-workers

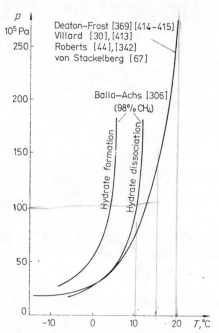

Fig. 5.16. p vs. T equilibrium diagram for CH_4 hydrate [30, 44, 67, 306, 342, 369, 413–415]

Table 5.11. Equilibrium data on CH₄ hydrate [44, 65, 127, 306, 342, 369, 413–415]

Villard [413] purity unknown		Deaton and Frost [414, 415] 99.9% CH₄		Roberts et al. [44, 342] 99.9% CH₄		Frost and Deaton [369] 99.7% CH₄		Berecz and Balla-Achs [306] 98.0% CH₄		Kobayashi and Katz [65] 99.9% CH₄		Marshall et al. [127] 99.9% CH₄	
T (°C)	p (10^5 Pa)	T (°C)	p (10^5 Pa)	T (°C)	p (10^5 Pa)	T (°C)	p (10^5 Pa)	T (°C)	p (10^5 Pa)	T (°C)	p (10^5 Pa)	T (°C)	p (10^5 Pa)
0.0	26.76	0.6	28.48	−14.1	16.82	−10.7	18.48	−6.0	16.16	22.4	348.55	17.2	164.02
1.1	30.30	1.1	29.89	0.0	27.17	−8.8	19.59	−2.7	20.20	22.7	361.98	22.0	308.96
5.5	47.47	2.2	33.43	7.8	60.20	−6.6	21.41	−0.5	25.25	27.8	664.58	22.6	347.64
8.5	64.13	2.8	35.35	13.4	110.24	−4.4	22.82	1.3	30.30	28.8	793.86	24.6	456.52
9.9	75.75	4.0	39.19	13.6	111.30	−2.2	24.64	2.6	35.35	37.7	2,828.0	26.7	586.3
10.8	83.83	6.0	49.08					3.8	40.40			28.1	674.07
14.3	124.73	7.2	55.04					5.0	45.45			33.2	1141.80
16.1	153.5	7.8	58.78					5.8	50.50			36.8	1573.28
17.3	180.3	8.4	62.42					7.5	60.60			39.2	1929.5
19.3	234.3	9.4	69.69					8.4	70.70			40.1	2125.6
20.3	267.6	11.0	83.63					9.2	80.80			40.6	2306.6
		12.8	100.7					9.8	90.90			41.5	2446.2
								10.3	101.0			43.2	2798.5
								10.7	111.1			44.7	3293.2
								11.0	121.2			46.0	3788.6
								11.6	151.5			46.4	4094.2

[342] on measurements relating to the phase diagram, and Frost and Deaton [369] then discussed the possibility of hydrate formation below 0 °C. Figure 5.16 and Table 5.11 were constructed from these data to illustrate the hydrate equilibrium conditions in the CH_4–H_2O system; these indicate that, depending on the temperature, the water–gas–hydrate equilibrium exists in the pressure interval 2.84–28.4×10^6 Pa, the possible temperature range being 0–21 °C.

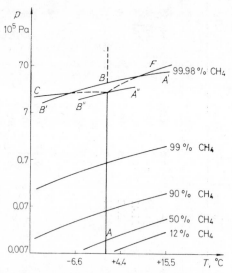

Fig. 5.17. Dew-points of $CH_4(g)$–$H_2O(l)$
mixtures [45]

Roberts and his co-workers paid attention to the determination of the dew-point, which is closely correlated with the moisture content of the gas, and is suitable as a course of informatory data on the conditions of hydrate formation at given values of p and T. The values of the dew-points of gaseous CH_4–water mixtures with various compositions can be seen in Fig. 5.17 (after Parent [45]).

Figure 5.16 and Table 5.11 give our own results, obtained in the determination of the dissociation parameters of methane hydrate (Berecz and Balla-Achs [306]). Our investigations were made with natural gas containing 98 % methane, the impurities consisting of nitrogen and higher hydrocarbons. As emerges from Fig. 5.16, up to a pressure of about 6.0×10^6 Pa, the conditions of dissociation of the hydrate do not differ appreciably from those for pure methane hydrate, even though the quantity of impurities present is fairly considerable. However, the difference becomes more and more significant as the pressure increases. As regards the extent and sign of the difference, the nature of the contaminating gases is the determining factor.

The development of the high-pressure measurement technique provided a possibility for an answer to be given to the question of whether methane hydrate can exist above 21.5 °C, as suggested by Villard [413]. This problem was investigated by Kobayashi and Katz [65] up to a pressure of 7.90×10^7 Pa. They established that methane hydrate formation can occur without any obstacle above the temperature limit of 21.5 °C given by Villard, but at higher pressures a supercooling of at least 4–5 °C is necessary for the formation of hydrate nuclei to begin. From their study of the phase properties of the CH_4–H_2O system, Kobayashi and Katz came to the conclusion that the critical phenomena arise only in the neighbourhood of 7×10^8 Pa, but if their theoretical considerations are taken into account, the p vs. T curve given by their experimental data can be extrapolated so that, at a pressure of 2.84×10^8 Pa, the dissociation temperature of the hydrate is 37.7 °C. Their data are presented in Table 5.11 and Fig. 5.18. This figure and table also contain the results of a later high-pressure study, made by Marshall and his co-workers [127]. These latter authors investigated the equilibrium conditions of methane hydrate in the temperature interval 18–47 °C and the pressure interval $1.50–40 \times 10^7$ Pa. Their experimentally-determined equilibrium values in the pressure range $1.5–7.0 \times 10^7$ Pa agree well with the data of Kobayashi and Katz [65] and Campbell and McLeod [416], and their p vs. T curve extrapolated to lower pressure also corresponds to the literature data. It can readily be seen in Fig. 5.18

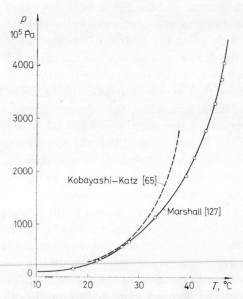

Fig. 5.18. Equilibrium diagram for CH_4 hydrate at high pressure [65, 127]. Reprinted by permission of the SPE-AIME

however, that, above a pressure of 7×10^7 Pa, the dissociation data measured by Marshall and his co-workers differ to an ever greater extent from those extrapolated previously by Kobayashi and Katz: the reason they put forward to explain this was that the hydrate structure changes as a consequence of the increasing pressure.

The equilibrium data were utilized to determine the composition of the hydrate too [127]; it was considered that in the fully closed constant-volume system they used, the amount (n_1) of the hydrate-forming gas at the beginning of hydrate dissociation can be calculated by means of the relationship

$$n_1 = \frac{p_1 V_1}{Z_1 R T_1} \tag{5.9}$$

On the total decomposition of the hydrate, the amount of the gas (n_2) is

$$n_2 = \frac{p_2 V_2}{Z_2 R T_2} + S n_{H_2O} \tag{5.10}$$

where V_1 and V_2 are the volumes of the gas phase, S is the solubility of the gas in water in mol/mol of water, and n_{H_2O} is the amount of water present. It can be assumed that $V_1 \neq V_2$, as the specific volumes of the hydrate and water are not exactly the same, but if only a small quantity of water is present the volume difference can be neglected, and the difference ($n_2 - n_1$) then corresponds to the quantity of gas bound by the structured water in the hydrate. Consequently, the ratio $n_{H_2O}/(n_2 - n_1)$ is in effect the hydration number, and its reciprocal gives information about the degree of occupancy of the cavities in the lattice structure of the hydrate. However, these calculations showed that the value corresponding to the theoretical hydration number could never be attained in the course of the experiments, even when the strongest supercooling was employed, and the total conversion of the quantity of water taken in the apparatus could not be achieved either. It was assumed that the probable cause of this was the fact that the residual free water molecules were wedged between the hydrate crystals formed, so that there was no possibility for the methane to build up a regular clathrate structure with these water molecules.

Glew [114] carried out detailed investigations of the solubility conditions of the methane–water system. Powell and Latimer [246] took into account the theories relating to the structure of water, and demonstrated that, if a substance interacting only weakly with water is dissolved in water, then the free volume of the solution decreases and the vibrational motion of the water molecules in the environment of the dissolved molecules becomes more limited too. In this case, the structure developing has a geometrically similar coordination to that established by von Stackelberg and Müller [66], Claussen [78, 79] and Pauling and Marsh [80] in

Table 5.12. Standard enthalpy and heat capacity values for the dissolution of methane in water [185, 245, 419–421]

T (°C)	Winkler [419], Morrison and Billet [420]		Claussen and Polglase [185] Culberson and McKetta [421]		From a combination of the former data		Eley [245]
	ΔH_2° (J mol⁻¹)	$\Delta C^\circ_{p_2}$ (J K⁻¹ mol⁻¹)	ΔH_2° (J mol⁻¹)	$\Delta C^\circ_{p_2}$ (J K⁻¹ mol⁻¹)	ΔH_2° (J mol⁻¹)	$\Delta C^\circ_{p_2}$ (J K⁻¹ mol⁻¹)	$\Delta C^\circ_{p_2}$ (J K⁻¹ mol⁻¹)
0	19 604.2	−257.49	18 772.4	−238.26	19 315.72	−253.72	−262.08
25	13 401.10	−249.96	13 058.3	−219.03	13 263.14	−229.90	−241.60
50	7 661.94	−222.79	7 824.96	−199.80	7 812.42	−206.07	−217.78
75	2 386.78	−211.51	3 072.3	−180.58	2 955.26	−182.25	−191.02
100	−2 424.4	−183.08	−1 199.55	−161.35	−1 304.16	−158.42	−160.93
125	—		−4 990.92	−142.12	−4 965.84	−134.59	−128.32
$\dfrac{\mathrm{d}\,\Delta C^\circ_{p_2}}{\mathrm{d}T}$ JK⁻² mol⁻¹		+0.752		+0.7524		+0.9614	0.752–0.9614

studies of the structures of gas hydrates. This solvation structure was proved independently by a number of authors, on the basis of reaction-kinetic [417, 418] and solubility [185, 369] investigations.

Investigating the Henry's law constant with regard to the solubility of methane in water, Winkler [419] considered the following equation:

$$CH_4(l_1) \rightleftharpoons CH_4(g) \qquad (5.11)$$

and found that between 0 and 50 °C the solubility of methane in water is never more than 0.22%. This was confirmed by Morrison and Billet [420]. Other investigations, however, from which those of Culberson and McKetta [421] and Claussen and Polglase [185] may be mentioned, indicated a methane solubility higher by 4.5% than the earlier value.

For the temperature-dependence of the solubility of methane in water, and equation of the type

$$\log S_2 = \frac{A}{T} + B \log T + CT + D \qquad (5.12)$$

can be used, which includes the fact that $\Delta C_{p_2}^{\circ}(l_1 \rightarrow g)$ is a linear function of temperature. The change in the molar heat of the dissolved gas was determined via the method of Eley [245], with the use of the following equation:

$$\Delta C_{p_2}^{\circ}(l_1 \rightarrow g) = R - C_c = R - \frac{d}{dT}\left[\frac{\alpha_1^{l_1}}{\beta_1^{l_1}} TV_2\right]^{l_1} \qquad (5.13)$$

from which it can be seen that the heat capacity of aqueous methane is composed of two parts: the heat capacity $C_{V_2}^{g}$ possessed by the methane molecule enclosed in the cavity, and the heat capacity C_c necessary for the volume of the cavity to remain V_2^{l} despite the thermal effects. In the equation, $\alpha_1^{l_1}$ and $\beta_1^{l_1}$ [422] are the proportionality factors corresponding to the expansion and compression of water, and $V_2^{l_1}$ is the molar volume of the aqueous methane. The standard enthalpy and heat capacity changes calculated between the temperature limits 0 and 125 °C are shown in Table 5.12.

The solubility of methane in water in high-pressure systems was determined by Culberson and McKetta [421] by thermodynamic means, and the resulting values of the molar volume of aqueous methane, $V_2^{l_1}$, can be seen in Table 5.13. The molar volume is a linear function of T; its value at 0 °C is $V_2^{l_1} = 34.37 \pm 0.54$ cm^3, while its coefficient of thermal expansion is $1180 \pm 190 \times 10^{-6}$ K^{-1}. Table 5.13 also presents the molar volume data $\Phi_2^{l_1}$ on aqueous methane obtained by Krichevsky and his co-workers [423, 424] and Masterson [425] by dilatometry or densitometry. These latter are somewhat higher values than the data calculated thermodynamically, which presumably arises from the fact that, even for

Table 5.13. The molar volume of aqueous methane [421, 423–425]

T (°C)	Culberson and McKetta [421] $V_2^{l_1}$ (cm³ mol⁻¹)	Krichevsky et al. [423, 424] $\Phi_2^{l_1}$ (cm³ mol⁻¹)	Masterson [425] $\Phi_2^{l_1}$ (cm³ mol⁻¹)
0	34.4±0.5	36±0.5	—
16.8	—	—	(33.2)
23.0	—	—	36.3
25.0	35.6±0.4	37±0.5	—
29.1	—	—	38.0
35.1	—	—	38.2
37.8	35.4±0.4	—	—
50.0	—	38±0.5	—
71.1	37.4±0.8	—	—
104.4	40.3±0.9	—	—
137.8	39.4±0.5	—	—

dilute solutions, calculations were made with the $V_1^{ol_1}$ value for water, and not with its partial molar value, $V_1^{l_1}$; it was neglected that the partial molar volume of water actually increases linearly with the molar ratio of the dissolved gas and water in the solution, according to the equation

$$V_1^{l_1} = V_1^{ol_1} + (\Phi_2^{l_1} - V_2^{l_1})\frac{n_2}{n_1} \tag{5.14}$$

As mentioned previously, the equilibrium of the methane hydrate–water–gas system

$$CH_4 . nH_2O(s) \rightleftharpoons nH_2O(l_1) + CH_4(g) \tag{5.15}$$

has been investigated by many authors (Villard, Roberts and his co-workers, Deaton and Frost, Campbell and McLeod, and Kobayashi and Katz). It was concluded by Glew [114], however, that the thermodynamic functions characterizing the equilibrium on the basis of this equation are not exact enough. For example, if the hydrate phase is denoted by h, and taking $\Delta C_p(h \rightarrow l_1 g) = 48.2$ cal K^{-1} mol^{-1} = 201.8 J K^{-1} mol^{-1} for the standard enthalpy change, the data of Villard [413] give a $\Delta H^\circ(h \rightarrow l_1 g)$ value of 12.896 kcal mol^{-1} = 53.993 kJ mol^{-1}, those of Roberts and his co-workers [44, 342] give 12.818 kcal mol^{-1} = 53.666 kJ mol^{-1}, those of Deaton and Frost [414, 415] give 13.093 kcal mol^{-1} = 54.818 kJ mol^{-1}, those of Campbell and McLeod [416] give 12.420 kcal mol^{-1} = 52.00 kJ mol^{-1}, and those of Kobayashi and Katz [65] give 12.830 ±0.140 kcal mol^{-1} = 53.716±0.586 kJ mol^{-1} at 0 °C. In spite of the differences, these values conform well with the heat of formation value of 13–15 kcal mol^{-1} = 55–63 kJ mol^{-1} characteristic of hydrates of type H_1 (see Section 1.4).

If methane hydrate decomposes into two liquid phases:

$$CH_4 \cdot nH_2O(s) \rightleftharpoons nH_2O(l_1) + CH_4(l_1) \qquad (5.16)$$

the reaction corresponds to the difference between Equations (5.15) and (5.11), and thus the thermodynamic function can be written as the difference between the thermodynamic functions relating to these two equations:

$$\Delta X^\circ(h \rightarrow l_1) = \Delta X^\circ(h \rightarrow l_1 g) - \Delta X^\circ(l_1 \rightarrow g) \qquad (5.17)$$

In the knowledge of the temperature-dependence of the value of the equilibrium constant, for reaction (5.16) with the transformation $(h \rightarrow l)$ the standard enthalpy change (ΔH°) at $0\,°C$ is 34.449 ± 544 kJ mol^{-1}.

Another possible phase transformation for methane hydrate is the dissociation process

$$CH_4 \cdot nH_2O(s) \rightleftharpoons nH_2O(s_1) + CH_4(g) \qquad (5.18)$$

By taking the equilibrium constant determined according to the method of Glew [202], together with $\Delta C_p\,(h \rightarrow s_1 g) = -8.36$ J K^{-1} mol^{-1}, a value of $\Delta H^\circ(h \rightarrow s_1 g) = 19.062 \pm 427$ kJ mol^{-1} is obtained for process (5.18).

From the above listing of the standard enthalpies of the possible reactions it can be seen that the enthalpy changes for processes (5.11) and (5.18) are equal within the limiting error. It follows from this that the interaction energy of the methane is practically the same in the hydration layer in the solution as in the coordination layer of the crystalline hydrate.

For processes (5.15) and (5.18), the difference between the reaction heats is 34.654 kJ mol^{-1}. This corresponds to the reaction heat of the transformation

$$nH_2O(s_1) \rightleftharpoons nH_2O(l_1) \qquad (5.19)$$

which is the same as the heat of sorption necessary for the melting of n moles of ice bound in the hydrate. From the known heat of melting of ice:

$$\Delta H^\circ = 34\,654 \text{ J mol}^{-1} = n \times 6010.99 \text{ J mol}^{-1} \qquad (5.20)$$

and hence the number of water molecules per methane molecule, as given by calorimetric measurements, is

$$n = \frac{34\,654}{6010.99} = 5.765 \qquad (5.21)$$

This corresponds to the general composition given for the hydrate structure H_1.

In this way, the thermochemical data for methane hydrate at $0\,°C$ indicate a degree of occupancy of the hydrate lattice of 99.7%, in comparison with the value of 83% calculated earlier on a theoretical basis by van der Waals and Platteeuw [98] and Glew [256].

If the calculations based on the solubility of methane are similarly carried out for other hydrate-forming gases, the enthalpy changes for the processes of dissolution and of the hydrate formation equilibrium, referring to 0 °C, can likewise be established. The results of such calculations are to be seen in Table 5.14. There is

Table 5.14. Enthalpy changes for gas dissolution and hydrate formation [158]

Gas	Coordination number	$\Delta H_2^\circ(l_1 \rightarrow g)$ (J mol^{-1})	$\Delta H^\circ(h \rightarrow s, g)$ (J mol^{-1})
CH$_4$	20	19 315.8	19 021.5
H$_2$S	20	21 485.2	19 031.5
C$_2$H$_6$	24	23 240.9	24 453.0
Cl$_2$	24	25 832.4	27 170.0
SO$_2$	24	31 015.6	32 186.0
CH$_3$Br	24	30 890.2	33 858.0
Br$_2$	24	36 575.0	34 694.0
C$_3$H$_8$	28	28 674.8	26 334.0
CH$_3$I	28	35 530.0	30 514.0
C$_2$H$_5$Cl	28	35 112.0	36 366.0

of necessity a correlation between the data enumerated here, but the difference is in all cases greater than for methane; the reason is that, for the more polarized solutes with dipole moments, the calorimetric relations can be followed only with greater error.

In the interpretation of the foregoing data, it would perhaps be appropriate to take into account too the finding of Sloan and his co-workers [426] that, in

Table 5.15. The density of methane hydrate, determined by two different methods [158]

p (10^5 Pa)	T (°C)	ϱ_1 (g cm^{-3})	ϱ_2 (g cm^{-3})
26.26	0	0.897	0.804
158.5	17	0.915	0.907
436.3	25	0.925	0.911
1515.0	37	0.956	0.914

the pressure range 70–100 bar, gaseous methane is actually much drier than would be expected from a graphical determination of the moisture content of the gas, and it is certain that this moisture content is not identical with that of dry natural gases containing predominantly methane. This fact definitely has an effect on the composition of the hydrate.

The equilibrium properties of methane and ethane hydrates at pressures lower than atmospheric were determined by Falabella and Vanpee [427] up to 148 K. The suitable mixing of the gas and ice led to the immediate formation of the hydrate. By extrapolating the experimental data to a pressure of 1.01×10^5 Pa, they obtained an equilibrium temperature of 193.2 K for methane hydrate; compared to the value of 244 K in the literature, this does not show such a good agreement as

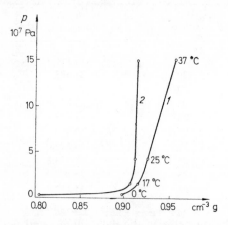

Fig. 5.19. Density of CH_4 hydrate, as determined by different methods [158]

for their experimental data relating to lower temperatures. In the data obtained by extrapolation for ethane hydrate, there is a difference of more than 16 °C for atmospheric conditions.

References can also be found to the determination of the density of methane hydrate. Makogon [158] reports data on density determinations with two different methods; these are to be seen in Table 5.15 and Fig. 5.19.

5.1.7.2. Ethane hydrate

As regards the clarification of the equilibrium conditions for ethane hydrate, primarily the results of Deaton and Frost [414, 428, 429] and Roberts and his co-workers [44, 342] are the most reliable and the most detailed, although the purities of the gases they used were not satisfactory either for comparative investigations.

In 1937, Deaton and Frost worked with a gas mixture with the composition 93.6% $C_2H_6 + 3.8\%$ $C_2H_4 + 2.6\%$ CH_4; in 1938, the composition of the gas they used was 97.1% $C_2H_6 + 2.1\%$ $C_3H_8 + 0.8\%$ CH_4. With this latter gas, they observed

Table 5.16. Equilibrium data on ethane hydrate [44, 342, 369, 414, 428, 429]

Deaton and Frost [428, 429]		Deaton and Frost [414]		Roberts et al. [44, 342]		Frost and Deaton [369]	
T (°C)	p (10^5 Pa)	T (°C)	p (10^5 Pa)	T (°C)	p (10^5 Pa)	T (°C)	p (10^5 Pa)
0.6	5.18	1.7	5.95	−12.3	3.01	−9.61	3.40
3.3	7.45	2.8	6.82	−3.9	4.52	−6.66	3.67
6.0	10.27	4.4	8.38	0.25	5.58	−3.89	4.15
8.8	14.97	5.6	9.59	2.27	6.85	−1.11	4.52
11.6	22.01	6.0	10.36	4.40	8.98		
13.9	30.39	6.6	11.15	6.0	10,75		
		7.0	12.00	6.5	11.59		
		7.8	12.93	8.0	13.50		
		8.4	13.84	9.6	16.82		
		8.9	15.20	11.27	21.91		
		9.4	16.05	12.60	26.01		
		10.0	17.39	13.80	31.31		
		11.0	20.45	14.90	70.13		
		12.2	23.73	15.05	50.70		
		13.4	28.11	15.20	70.63		

a higher hydrate stability; this may be attributable to the presence of propane. They later published equilibrium data obtained from experiments made with a gas of the same composition. They also worked with a gas mixture with an ethane content of 98.3%, in the temperature range below 0 °C [369]. Roberts and his co-workers carried out work with a gas with an ethane content of 99.5% and, besides the various three-phase hydrate equilibria, determined the dew-point diagram of the ethane–water system. These groups of authors were in complete agreement in stating that ethane hydrate can exist only at temperatures below

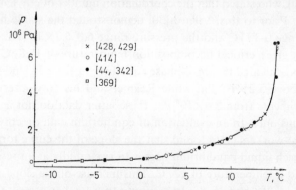

Fig. 5.20. p vs. *T* equilibrium diagram for C_2H_6 hydrate
[44, 342, 369, 414, 428, 429]

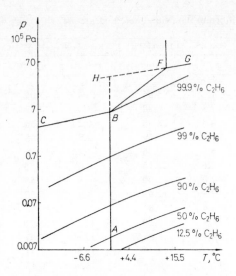

Fig. 5.21. Dew-points of C_2H_6–H_2O mixtures
[342]. Reprinted by permission of the SPE-AIME

15 °C. The characteristic quadruple point data on the hydrate are reported by von Stackelberg and Müller [66].

The equilibrium data on ethane hydrate are summarized in Table 5.16 and Fig. 5.20, and the dew-point relations of ethane–water mixtures in Fig. 5.21.

5.1.7.3. Ethylene hydrate

The results of the early investigations on ethylene hydrate were first reviewed by Claussen [79], who stated that the coordination number of ethylene with regard to water is 7.4. Prior to this, Villard [30] demonstrated the stable hydrate in the temperature range 0–17 °C and the pressure range 6.0–6.7×10⁵ Pa. On extrapolation, his data give a critical decomposition temperature of 18.8 °C. Diepen and Scheffer [430] investigated the three-phase equilibria in the ethylene–water system at pressures above 5.2×10⁶ Pa, while Reamer and his co-workers [89] carried out their experiments from 3.6×10⁶ Pa; these latter data do not agree very well with the previous ones. In an evaluation of equilibrium data on ethylene hydrate, Snell and his co-workers [116] found that the slope of the curve for the hydrate–vapour–water-rich liquid equilibrium begins to rise rapidly at a pressure of about 4.9×10⁶ Pa and a temperature of 17 °C. This increase of the slope attains a maximum at 7.8×10⁶ Pa and 18 °C. In response to a further pressure increase, the slope begins gradually to decrease. In their calculations with the aim of the prediction of the equilibrium conditions, the data of Otto and Robinson [115] and Carson and Katz [46] were used too. It was concluded that the p–T dependence of the ethylene hydrate formation equilibrium exhibits two inflections at values above 4.96×10^6 Pa and 7.8×10^6 Pa. A change in the crystal structure is known to be accompanied by a change in the slope of the vapour pressure curve, and thus it was assumed that the change in slope of the vapour pressure curve of ethylene hydrate is connected with a change in the number of water molecules associated with the hydrocarbon molecule, in this way the crystal structure of the hydrate also varying as a function of the pressure.

Table 5.17. Equilibrium data on ethylene hydrate [89]

Isobaric conditions		Isothermal conditions	
p (10^5 Pa)	T (°C)	p (10^5 Pa)	T (°C)
10.15	1.5	5.75	−1.17
11.98	5.2	6.46	0.0
12.72	5.2	7.07	1.4
14.34	5.5	8.99	3.9
18.18	5.1	12.12	5.2
19.88	7.3	15.95	6.5
21.41	6.5	18.28	8.0
22.72	8.5	21.41	8.8
24.6	9.6	23.91	10.2
25.65	9.7	28.38	11.4
27.27	10.5	31.20	12.6
30.30	10.2		
32.11	11.5		
33.53	12.0		
36.26	13.0		

The p vs. T data of Reamer and his co-workers [89] for ethylene hydrate are shown in Table 5.17 and Fig. 5.22. Table 5.17 gives side by side the results of the equilibrium investigations carried out under isothermal and isobaric conditions. The data reveal that the establishment of the equilibrium is strongly influenced by many factors, e.g. the duration of the investigation, the water quantity present, etc.

The data of Snell and his co-workers [116] and Diepen and Scheffer [430] are utilized in Fig. 5.23 to extend the results of the equilibrium investigations to higher pressures.

Table 5.18 gives the high-pressure equilibrium conditions relating to ethylene hydrate for the hydrate–liquid ethylene–gas system, based on the publication by van Cleeff and Diepen [431]. Here too, the experimental data suggest that the

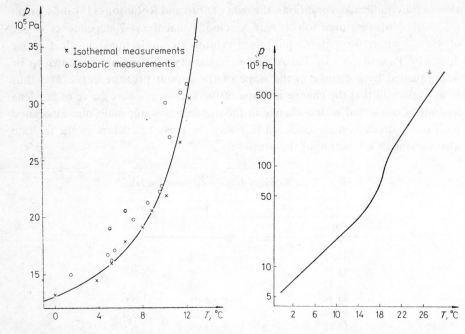

Fig. 5.22. p vs. T equilibrium diagram for ethylene hydrate [89]. Reprinted by permission of the SPE-AIME

Fig. 5.23. p vs. T equilibrium diagram for ethylene hydrate at high pressure [116, 430]

discontinuity observed above 18 °C must occur because of structural changes. In order to interpret the break-point, the temperature-dependence of the compression factor of ethylene was investigated; this value too displays a sudden fall near 18 °C, from which it follows that plots of dp/dT and $d \log p/dT$ against

Table 5.18. Equilibrium data on ethylene hydrate [431]

T (°C)	p (10^5 Pa)	T (°C)	p (10^5 Pa)
17.1	46.5	22.1	248.0
17.9	55.0	23.1	312.0
18.1	57.9	24.1	377.0
18.3	61.9	25.1	443.0
18.5	67.7	26.3	532.0
18.7	76.9	27.3	608.0
18.9	84.0	28.1	670.0
19.5	106.0	29.0	745.0
20.1	137.0	30.2	844.0
21.1	191.0	31.9	1015.0

T give a maximum here. In the case of ethylene, therefore, deviation from the ideal gas law is very considerable in this temperature–pressure range.

The solubility of ethylene in water was determined at high pressures [431], and on the basis of the experimental results a more accurate value of the compression (deviation) factor was proposed.

In their investigation of the process of formation of ethylene hydrate, Morlat and his co-workers [432] were primarily concerned with the hydrate equilibrium conditions of systems for the desalination of water with ethylene. By determining the partial pressure of the water vapour, they came to the conclusion that in the course of hydrate formation the small cavities in the water lattice are initially only temporarily occupied by ethylene molecules; however, this stabilizes the lattice so that, with the permanent occupation of the small cavities, the system attains the composition $C_2H_4 . 7.67H_2O$.

By means of a chromatographic analysis of the composition of mixed ethane and ethylene hydrates, Koshelev and his co-workers [368] proved that the hydrate is always richer in ethane, in spite of the fact that this cannot be concluded from their partition coefficients.

5.1.7.4. Propane hydrate

As regards a description of the equilibrium conditions for propane hydrate, the works of Deaton and Frost [360, 369, 415] are outstanding. These are well supplemented by the investigations of Wilcox and his co-workers [357] and Miller and Strong [50]. Later, Reamer and his co-workers [89] and Robinson and Mehta [156] dealt with equilibrium studies on propane hydrate.

The above authors worked with gaseous propane with purities of 99.5–99.9%. With regard to the size of the propane molecule, this compound forms a liquid hydrate, the crystal structure of which is well known from the work of Claussen [78, 79] and von Stackelberg and his co-workers [66–68].

The two-phase and three-phase equilibrium data are presented in Table 5.19 and Fig. 5.24.

We have already dealt in Section 4.1 with the investigations by Miller and Strong [50] relating to the determination of the composition of propane hydrate and with their results, and Equation (4.41), describing the relationship between the fugacity, the activity and the hydration number, was also given there. On the basis of the calculations of Miller and Strong [50], the temperature-dependence of the value of f_2 is well expressed by

$$\log f_2 = 26.253 - 6.7992 \times 10^3 \, T^{-1} \tag{5.22}$$

Data on the fugacity of propane for various p vs. T pairs were published by Sage and his co-workers [433]; these can be extrapolated to the temperature

Table 5.19. Equilibrium data on propane hydrate [50, 89, 156, 357, 360, 369, 415]

Deaton and Frost [360]		Deaton and Frost [415]		Frost and Deaton [369]		Wilcox et al. [357]		Miller and Strong [50]		Reamer et al. [89]		Robinson and Mehta [156]	
p (10^5 Pa)	T (°C)	p (10^5 Pa)	T (°C)	p (10^5 Pa)	T (°C)	p (10^5 Pa)	T (°C)	p (10^5 Pa)	T (°C)	p (10^5 Pa)	T (°C)	p (10^5 Pa)	T (°C)
1.84	0.55	1.87	0.55	1.02	−11.9	8.20	5.72	1.70	0.05	2.46	1.17	2.12	1.12
2.40	1.66	2.38	1.66	1.18	−9.0	13.28	5.38	1.77	0.25	3.12	2.55	2.47	1.67
2.76	2.22	2.77	2.22	1.35	−5.75	18.02	5.38	1.77	0.35	4.24	4.05	3.39	3.23
3.11	2.77	3.08	2.77	1.38	−5.55	20.85	5.94	1.92	0.60	7.00	5.44	4.66	4.70
3.96	3.89	3.95	3.89	1.52	−3.3	29.79	6.00	1.94	0.70	15.15	5.50	5.55	5.78
				1.73	−0.9	43.55	5.60	2.26	1.40	20.98	5.61		
				1.76	−0.3	62.72	5.72	3.00	2.70				
				1.86	0.22			3.25	3.05				
				1.97	0.44			3.53	3.50				
				1.98	0.61			3.75	3.70				
				2.10	0.72			3.99	3.95				
				2.98	2.38			4.02	4.00				
				4.11	3.78			4.70	4.70				
								4.83	4.90				
								4.90	4.95				
								5.61	5.60				
								5.80	5.70				

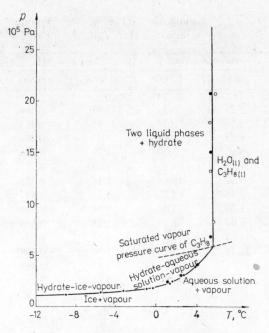

Fig. 5.24. Equilibrium diagram for C_3H_8 hydrate
[50, 89, 156, 357, 360, 369, 415]. Adapted by
permission of the *Am. Chem. Soc.*

limits 0–5.7 °C characteristic of propane hydrate. Any aqueous solution can be used for the determination of the composition of the hydrate, so long as the freezing-point depression does not exceed 4 °C. In the knowledge of the gas fugacity relating to the p vs. T data on the hydrate, determined under various conditions, there is a possibility for the establishment of the composition of the hydrate.

5.1.7.5. Propylene hydrate

Investigations into propylene hydrate have become important only in the past decade. Previously, Reamer and his co-workers [434] had already established that the stability of propylene hydrate is much lower than that of propane hydrate, despite the fact that their physical properties are very similar. More detailed studies on propylene hydrate were made by Clarke and his co-workers [117], using propylene with a purity of 99.4%, between the temperature limits −13 °C and +6 °C. The relationship between the total pressure of the investigated system and the equilibrium temperature of the hydrate was given by the following equation:

$$\log p = 4.30\ 595 \pm 0.00\ 006 - (968.4 \pm 0.8)T^{-1} \qquad (5.23)$$

By taking into account the vapour pressure of water, the following equation can

Table 5.20. Equilibrium data on propylene hydrate
formation [117, 434]

Reamer et al. [434]		Clarke et al. [117]	
p (10^5 Pa)	T (°C)	p (10^5 Pa)	T (°C)
5.29	0.5	4.61	−0.161
5.56	0.78	4.67	−0.113
6.25	1.16	4.72	−0.064
6.97	5.00	4.79	0.006
7.13	4.2	5.04	0.217
9.12	14.1	5.16	0.325
9.39	15.2	5.31	0.442
11.20	23.0	5.38	0.493
6.94	1.39	5.45	0.558
14.68	1.61	5.55	0.594
21.79	1.78	5.76	0.798
		5.91	0.897
		5.95	0.943

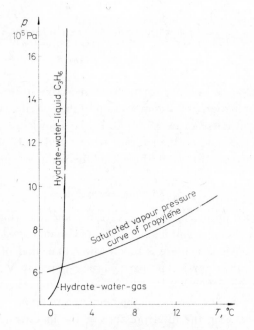

Fig. 5.25. p vs. T equilibrium diagram for C_3H_6
hydrate [117, 434]. Adapted by permissions of
SPE–AIME and the National Research Council
of Canada

be written for the fugacity of propylene:

$$\log f = 24.8309 \pm 0.0001 - (6610 \pm 20)T^{-1} \qquad (5.24)$$

The heat of decomposition of the hydrate is $126\,944 \pm 377$ J mol^{-1}.

It can be concluded from the data that propylene forms a hydrate with the composition $C_3H_6 . 17H_2O$. The lower quadruple point was found at $-0.134\,°C$, and the upper quadruple point at $0.958\,°C$. This shows quite clearly that the hydrate can be formed only in the presence of liquid phase rich in hydrocarbon. Similar results are reported by Otto and Robinson [115].

Data relating to the formation of propylene hydrate are presented in Table 5.20 and Fig. 5.25.

5.1.7.6. Butane hydrates

With regard to the formation of n-butane hydrate, the only publication at our disposal originates from Wilcox and his co-workers [357], who remark that the production and isolation of the hydrate are extremely difficult. They succeeded in preparing relatively pure n-butane hydrate on one occasion. The crystal melted at $1.0\,°C$.

Reliable data are not available on the hydrate of isobutane either, although reference to this compound can be found in the data tabulated by Brown [435].

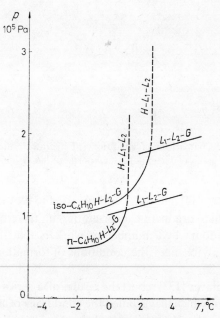

Fig. 5.26. p vs. T equilibrium diagram for
C_4H_{10} hydrate [45]

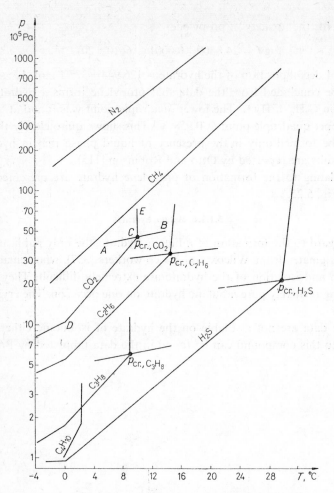

Fig. 5.27. Comparison of conditions of hydrate formation for
several gases of technological importance [46, 65, 361, 521]

Deaton and Frost [414, 415] and other authors tried to plot the phase diagrams of
these hydrates, but there are many uncertainties in their data. However, it has
been proved beyond doubt by a number of workers that the presence of either
butane very appreciably changes the conditions of formation and decomposition
of methane hydrate.

Musayev and Korotayev [133] report the results of a series of laboratory experiments, in the course of which they studied the conditions of hydrate formation for
the systems liquid C_4H_{10}–H_2O, liquid C_4H_{10}–N_2–H_2O, liquid C_4H_{10}–O_2–H_2O,
and gaseous C_4H_{10}–N_2–H_2O, using technical butane with a purity of 94.9%, to

an upper pressure limit of 5×10^6 Pa. They concluded that butane hydrate cannot be formed under any conditions in these systems; only ice separates out at $-2.8\,°C$. As a result of these experiments, they stated that the use of an anti-freeze agent is not necessary in the transportation of liquid technical butane by pipeline.

The few literature data available have been utilized on the basis of the review by Parent [45], and the phase diagrams of the hydrates of n-butane and isobutane can be seen in Fig. 5.26.

Data on the conditions of formation of isobutane hydrate, and on the mixed hydrates of methane and isobutane are reported by Wu Bing-Jing and Robinson [436].

The stabilities of the hydrates of the gases discussed above, and the p and T values characteristic of hydrate formation, can readily be compared by means of Fig. 5.27.

5.1.8. Freon hydrates

In connection with the study of the conditions of formation of the hydrates of freons, which play an extremely important role in the refrigerating industry, two publications must be emphasized: those of Chinworth and Katz [63] and Wittstruck [166]. Both groups determined the equilibrium phase data relating to three frequently employed freons, F12, F11 and F22 (with halon numbers 122, 113 and 121, respectively); in addition, data were given by Chinworth and Katz on methyl chloride, and by Wittstruck on freon F13B1 (with halon number 1301). The common measurements by the two groups led to fairly different results. From their publications it is not possible to draw conclusions as to the purities of the substances used, but it can be supposed that Wittstruck worked with purer material. The hydrate phase equilibrium data for these cooling media are shown in Tables 5.21 and 5.22, and in Figs. 5.28–5.32.

From the various findings, it is worthwhile mentioning that hydrate formation was the easiest from $CHClF_2$, a sudden slight pressure decrease being sufficient for crystal precipitation in many cases. Formation of the hydrate of $CBrF_3$ was the most difficult, and consequently this compound decomposes the most easily, so much so that in many cases only two p vs. T pairs could be recorded during the decomposition. The hydrates of CCl_2F_2 and $CHClF_2$ are fairly stable.

Wittstruck also tried to produce the hydrate of freon F114 ($C_2Cl_2F_4$, with halon number 242), but between the temperature limits $-2\,°C$ and $+10\,°C$ his experiments met with no success. He justified this by the fact that the molar volume of freon F114 at $0\,°C$ is 111.5 cm³/mol, and the molecule is therefore too large to be able to form a hydrate.

By means of X-ray diffraction investigations, the clathrate structures of the freon hydrates were checked and it was demonstrated that the hydrates of freons

Table 5.21. Equilibrium data on hydrate formation by several freons [63, 166]

Freon 11 = CCl₃F Boiling-point = 23.8 °C Hydrate: CCl₃F.16.6H₂O				Freon 12 = CCl₂F₂ Boiling-point = −29.8 °C Hydrate: CCl₂F₂.15.6H₂O				Freon 22 = CHClF₂ Boiling-point = −40.8 °C Hydrate: CHClF₂.12.6H₂O			
Wittstruck [166]		Chinworth and Katz [63]		Wittstruck [166]		Chinworth and Katz [63]		Wittstruck [166]		Chinworth and Katz [63]	
p (10^5 Pa)	T (°C)	p (10^5 Pa)	T (°C)	p (10^5 Pa)	T (°C)	p (10^5 Pa)	T (°C)	p (10^5 Pa)	T (°C)	p (10^5 Pa)	T (°C)
0.63	8.3	0.70	4.16	4.43	11.9	3.70	4.22	7.93	16.3	7.77	7.22
0.46	7.2	0.93	8.88	4.18	11.6	3.86	5.33	7.26	15.8	8.23	10.00
0.43	6.8	1.08	13.33	2.93	10.9	3.91	6.33	6.48	15.0	9.91	18.05
0.36	6.1	1.10	15.54	2.83	9.8	4.13	7.50	5.45	13.6	10.67	21.10
0.32	5.7	1.25	21.1	2.43	9.1	4.27	9.30	4.67	12.4	11.40	23.90
0.28	5.3	1.42	26.6	1.94	8.2	4.75	12.90	4.15	11.5	15.35	17.16
0.17	3.5	1.72	32.2	1.78	7.7	5.58	18.40	3.73	10.6	22.97	17.14
0.12	1.8	2.06	37.1	1.58	7.3	8.79	10.00	3.05	9.1	4.27	0.83
0.07	−0.5	3.13	8.5	1.46	6.9	11.55	11.11	2.76	8.4	4.62	2.78
0.07	−2.7	0.58	1.11	1.26	6.2	29.00	10.55	2.24	6.9	4.83	3.33
0.05	−6.6	0.62	2.77	1.07	5.4	1.06	0.89	1.84	5.4	6.48	8.72
		0.64	4.44	0.91	4.4	1.20	1.00	1.49	3.7	6.92	11.65
		0.68	5.00	0.75	3.7	1.30	1.66	1.28	2.6	7.71	13.22
		0.73	6.17	0.66	3.1	1.73	2.66	1.19	2.0	8.16	15.27
		0.75	6.66	0.61	2.8	2.99	9.16	1.01	0.7	9.19	17.00
		0.80	7.83	0.54	2.0	3.96	7.50	0.89	−0.3	9.64	17.22
		0.89	8.27	0.44	0.8	4.36	9.90	0.85	−2.2	9.80	17.50
				0.35	−2.2	4.39	9.90	0.85	−3.1		
				0.27	−6.8			0.74	−5.5		

Table 5.22. Equilibrium data on hydrate formation
by freons [63, 166]

CH$_3$Cl hydrate [63]		CBrF$_3$.15.6H$_2$O [166]	
p (10^5 Pa)	T (°C)	p (10^5 Pa)	T (°C)
3.81	9.0	4.27	7.2
4.12	12.2	3.47	6.4
4.52	15.3	2.42	4.5
5.05	18.9	1.66	3.0
5.67	23.0	1.32	1.7
6.01	25.1	1.07	0.8
11.55	21.3	0.85	−0.9
22.02	20.9	0.79	−2.5
1.80	11.6	0.69	−5.0
2.32	13.9	0.65	−7.0
3.13	16.1		
2.24	16.9		
4.02	18.6		
5.37	21.1		

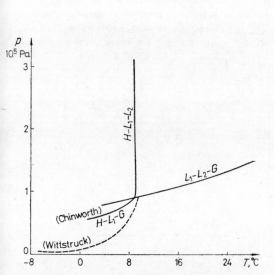

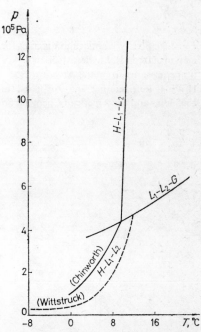

Fig. 5.28. p vs. T equilibrium diagram for formation of freon F11 hydrate [63, 166]. Adapted by permissions from the American Society of Heating, Refrigerating and Air-Conditioning Engineers and the American Chemical Society

Fig. 5.29. p vs. T equilibrium diagram for formation of freon F12 hydrate [63, 166]. Adapted by permissions from the American Society of Heating, Refrigerating and Air-Conditioning Engineers and the American Chemical Society

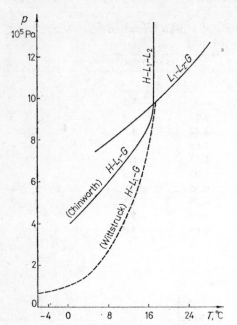

Fig. 5.30. p vs. T equilibrium diagram for formation of freon F22 hydrate [63, 166]. Adapted by permissions from the American Society of Heating, Refrigerating and Air-Conditioning Engineers and the American Chemical Society

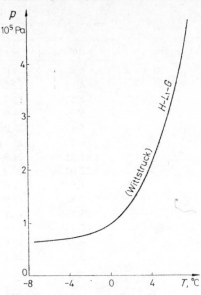

Fig. 5.31. p vs. T equilibrium diagram for formation of freon F13B1 hydrate [166]. Adapted by permissions from the American Society of Heating, Refrigerating and Air-Conditioning Engineers and the American Chemical Society

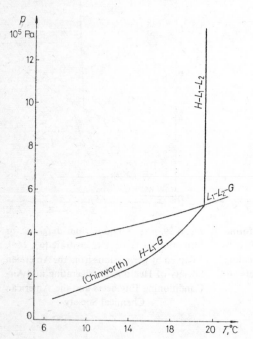

Fig. 5.32. p vs. T equilibrium diagram for formation of CH_3Cl hydrate [63]. Adapted by permissions from the American Society of Heating, Refrigerating and Air-Conditioning Engineers and the American Chemical Society

F11, F12 and F13B1 crystallize in the H_{II} structural type, while the hydrate of freon F22 is characterized by a crystal structure of type H_I, but the cavities are only partly occupied by the freon, since the molecules cannot fit into the small cavities.

5.2. Hydrate systems containing more than one gas component

The hydrates of the various gases may have structures corresponding to the fundamental types H_I or H_{II}. A mixture containing more than one gas constituent can similarly form a hydrate structure of one kind, but it is possible that the hydrates of the individual constituents are characterized by different structures.

If a system containing only one type of hydrate is formed, the phenomenon corresponding to complete mixing of the solid solutions is manifested, and thus a mixed hydrate is formed. The CH_4–H_2S–H_2O system is a good example of mixed hydrates of type H_I, forming a continuous series. The complete mixing of hydrates with H_{II} structures was demonstrated by von Stackelberg and Meinhold [67] for the C_2H_5Cl–CH_3CHCl_2–H_2O and C_2H_5Cl–$CHCl_3$–H_2O systems, but the same feature is displayed by the hydrates formed in the propane–propylene–water system too.

If two binary systems form hydrates of different types, then partial mixing of the solid solutions can be detected in the ternary system. Such a situation arises with the methane–propane–water and the hydrogen sulphide–propane–water systems.

Experimental results relating to some mixed hydrates will be treated below.

5.2.1. Methane–carbon dioxide hydrate

Detailed information on this hydrate system can be found in the publication by Unruh and Katz [64]. They stated that the process of hydrate formation in a CO_2–CH_4 mixture is much slower than with pure gaseous CO_2. The data obtained during their investigation of the conditions of hydrate formation for the ternary system are given in Table 5.23. The solubility of pure CO_2 in water was taken into account in the value of the calculated CO_2 content.

Figure 5.33, based on the data of Unruh and Katz, illustrates the conditions of hydrate formation from ternary mixtures with various compositions, as a function of the composition. The methane used in the experiments had a purity of 99.0%, and the carbon dioxide a purity of 99.5%. In their evaluation of the ex-

Table 5.23. Hydrate formation by CH_4–CO_2 gas mixtures [64]

Gas sample	T (°C)	p (10^5 Pa)	Actual CO_2 content of gas investigated (mole %)	Calculated CO_2 content of gas (mole %)	Gas volume (cm³)	Volume of liquid phase (cm³)
1	3.83	29.13	53.9	34	47	47
	5.72	35.49		30	38	56
2	5.77	35.19	54.6	36	47	47
	7.77	43.48		32	38	56
	9.72	53.02		28	28	66
	11.55	64.29		23	20	74
3	2.33	20.43	77.7	60	42	52
	6.00	31.60		44	20	74
4	3.22	32.80	27.4	12.5	42	52
	5.22	40.51		8.5	28	66
	7.83	52.34		7.0	21	73
	10.61	40.40		5.5	15	79
5	6.39	30.75	82.4	71	47	47
	9.00	43.83		63	28	66
	10.61	54.08		52	15	79
	12.33	70.70		41	8.5	85.5
	12.50	71.76		41	8.5	85.5

perimental data, Unruh and Katz also gave the isotherms of the ternary system as a function of the composition.

In Section 3.2.1, we have described the results of our own experiments (Berecz and Balla-Achs [296]) on the hydrate-forming properties of the CH_4–CO_2

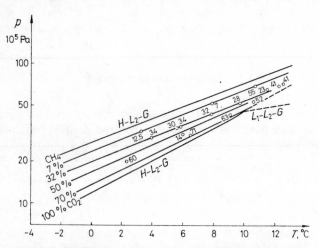

Fig. 5.33. Equilibrium diagram for CH_4–CO_2 hydrate [64].
Reprinted by permission of the SPE-AIME

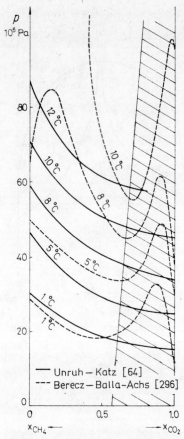

Fig. 5.34. Equilibrium isotherms for CH_4–CO_2 hydrate [64,296]. Reprinted by permission of the SPE-AIME

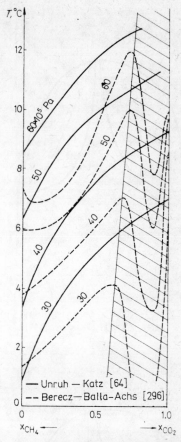

Fig. 5.35. Equilibrium isobars for CH_4–CO_2 hydrate [64,296]. Reprinted by permission of the SPE-AIME

system. The data presented in Table 3.4 are utilized in Figs. 5.34 and 5.35 to compare the isotherms and isobars obtained at the various compositions in the experiments of Unruh and Katz and of Berecz and Balla-Achs. Full agreement between the results of the two experiments was naturally not to be expected, because of the different purities of the gases used. This is particularly so in the case of the gaseous methane: Berecz and Balla-Achs employed methane with a purity of 98%, and Unruh and Katz methane with a purity of 99.5%. The stabilizing effect observed with the less pure methane can therefore be explained as being due to the greater quantity of higher hydrocarbons present. There is an essential difference between the results of these two experiments: the results of Berecz and Balla-Achs draw attention to the considerable change in the condi-

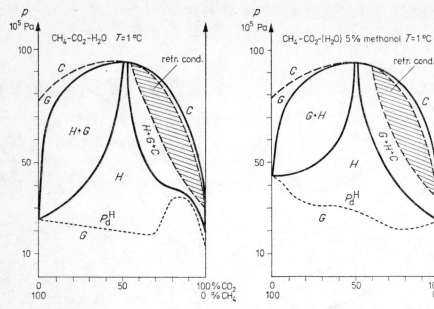

Fig. 5.36. Isothermal phase equilibria in the CH_4–CO_2–H_2O system [437]

Fig. 5.37. Isothermal phase equilibria in the CH_4–CO_2–5% aqueous methanol system [437]

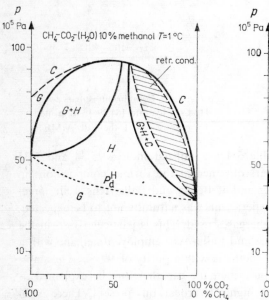

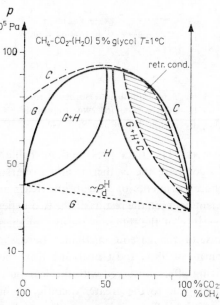

Fig. 5.38. Isothermal phase equilibria in the CH_4–CO_2–10% aqueous methanol system [437]

Fig. 5.39. Isothermal phase equilibria in the CH_4–CO_2–5% aqueous glycol system [437]

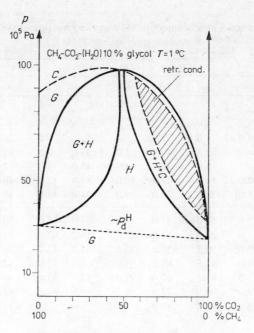

Fig. 5.40. Isothermal phase equilibria in the
CH_4–CO_2–10% aqueous glycol system [437]

tions of hydrate formation or decomposition as a consequence of the possibility of retrograde condensation in the ternary system and the appearance of the condensed phase.

In later work, Berecz and Balla-Achs [437] investigated in detail the methane and carbon dioxide contents of hydrates formed from gas mixtures with different composition, as well as the effects of the inhibitors methanol and ethylene glycol on the phase composition. The results indicated that the conditions of hydrate formation undergo a fundamental change as a result of retrograde condensation in mixtures containing more than 50% carbon dioxide; the condensed phase destabilizes the hydrates. Phase equilibrium diagrams constructed from measurements carried out at $+1\ ^\circ C$ are shown in Figs. 5.36–5.40, and the data of these diagrams are summarized in Table 5.24.

In the figures, G denotes the gas phase, H the hydrate phase and C the condensate. The p_d^H curves show the decomposition pressures of the hydrates, measured experimentally at $+1\ ^\circ C$.

The compositions of hydrates formed under heterogeneous conditions are influenced considerably by the rate of cooling and the extent of supercooling. The composition of the first hydrate nuclei originating during the cooling still coincides with that of the gas phase. On further cooling (with regard to the lever

Table 5.24. Equilibrium CO_2 contents (in %) of gas and hydrate phases in the CH_4-CO_2 hydrate system at $+1$ °C [437]

p (10^5 Pa)	H_2O		5% CH_3OH		10% CH_3OH		5% glycol		10% glycol	
	gas	hydrate	gas	hydrate	gas	hydrate	gas	hydrate	gas	hydrate
30	1.0	14.0	—	—	—	—	—	—	—	—
40	1.5	29.0	—	—	—	—	—	—	0.5	11.0
50	2.5	39.0	1.0	27.0	—	—	1.5	28.0	1.5	28.0
60	4.0	44.0	3.5	40.5	1.0	34.0	4.0	42.0	2.5	39.0
70	6.0	47.5	5.0	45.5	4.0	45.0	8.0	47.0	5.5	44.0
80	13.0	50.0	11.0	48.0	10.0	51.0	15.0	49.0	12.0	46.0
90	30.0	51.0	28.0	50.0	29.0	54.0	30.0	49.0	27.0	48.0
90	65.0	55.0	67.0	51.0	71.0	59.0	67.0	53.0	67.0	54.0
80	80.0	55.0	83.0	53.0	83.0	59.5	82.0	54.0	82.0	56.0
70	89.0	57.0	90.5	56.0	90.5	61.0	89.0	56.0	89.5	60.0
60	94.0	59.0	95.5	60.5	95.0	65.0	95.0	62.0	95.0	65.0
50	96.0	63.0	97.5	67.0	97.0	72.0	98.0	69.0	97.0	73.0
40	98.0	73.0	99.0	75.0	99.0	85.0	99.7	82.0	99.0	84.0
35	99.0	87.0	99.8	87.0	—	—	—	—	99.8	97.0

rule characteristic for the phase diagrams), formation of the phase richer in CO_2 also begins. During the change in the composition of the crystalline phase, the temperature of hydrate formation changes too, which leads in turn to a variation in the composition of the gas phase. The composition of the crystal nuclei remains unchanged in this process, but that of the next layers deposited on these will not be the same, and hence the crystal will develop a layered structure. When small crystals are precipitated, the homogenization of the crystals can occur by means of diffusion, mainly in the vicinity of the decomposition point of the hydrate. If the crystal nuclei are large, this process requires a very long time. The diffusion itself is also a slow process, and is hindered still further by a relatively stable crystal structure, where a rearrangement of the molecules enclosed in the cavities should occur. If the cooling cycle is retarded, the size of the crystal nuclei is increased, but the time necessary for homogenization by means of diffusion is also longer. On fast cooling, the hydrate lattice soon becomes blocked in the many small crystals, and fewer gas molecules can be incorporated into the cavities. All this has an effect on the accuracy of the experimental data too. The scatter in our experimental results was particularly large if a supercooling of more than 5 °C was applied, because the hydrate crystals formed during the supercooling and grown at higher temperatures could not undergo homogenization in the rewarming to the required temperature. The hydrates rich in carbon dioxide and containing a small proportion of methane have very loose structures, especially in the presence of inhibitors. The phase equilibrium studies here therefore exhibited a much lower scatter than in the case of methane hydrates containing a little carbon dioxide, because the homogenization by diffusion was less hindered. In the two heterogeneous phase equilibrium regions in the figures, there is either a negligibly small difference between the compositions of the gas and hydrate phases, or, when a longer time is available for the establishment of equilibrium, the compositions of the gas and the hydrate phases are shifted in accordance with the possible heterogeneous regions. This interval becomes narrower as the pressure increases, and at a pressure of $8.5-9.0 \times 10^6$ Pa the system approaches the structural state characteristic of the eutectic mixture system.

It cannot be neglected in these experiments that, for both chemical and physical reasons, CO_2 dissolves to a greater extent in water than in alcoholic solutions. Its solubility increases with the decrease of the temperature, but the liquid phase appearing as a result of the condensation forms a more unstable hydrate than that of methane. At higher pressures, this fact too plays a part in the hydrate of methane being more stable than that of carbon dioxide in alcoholic solutions.

The regions in the phase diagrams characteristic of the composition of methane hydrates containing carbon dioxide are the broadest in the pressure range $4-6 \times \times 10^6$ Pa. As the temperature is raised, this interval shifts in the direction of higher pressures. The ratio of the CO_2 contents measured in the hydrate and in the gas depends on the initial CO_2 content of the gas mixture and also on the temperature of hydrate formation. For example, Table 5.25 shows the ratio of the CO_2 contents of the hydrate and the gas phase at 6×10^6 Pa in a methane–carbon dioxide mixture with a CO_2 content of 96%, as functions of the inhibitor concentration and the temperature.

Investigation of the composition data for the heterogeneous regions reveals that this pressure of 6×10^6 Pa is that at which the enrichment of carbon dioxide in

Table 5.25. Variation in the ratio of the CO_2 contents of the hydrate and gas phases as functions of the temperature and inhibitor concentration [437]

(Pressure $= 6 \times 10^6$ Pa; CO_2-CH_4 gas mixture with a CO_2 content of 96%.)

Liquid medium	1 °C	3 °C	5 °C
	$CO_{2\,(hydrate)}/CO_{2\,(gas)}$		
Water	11.0	20.2	24.7
5% methanol	11.4	21.0	26.0
10% methanol	34.0	4.6	—
5% glycol	10.5	17.5	16.3
10% glycol	15.6	19.3	21.0

15

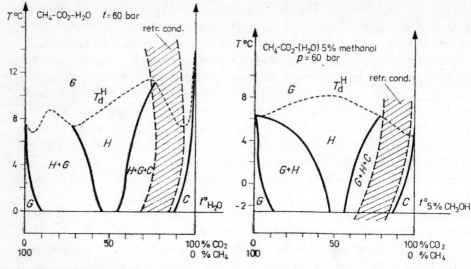

Fig. 5.41. Isobaric phase equilibria in the CH₄–CO₂–H₂O system [437]

Fig. 5.42. Isobaric phase equilibria in the CH₄–CO₂–5% aqueous methanol system [437]

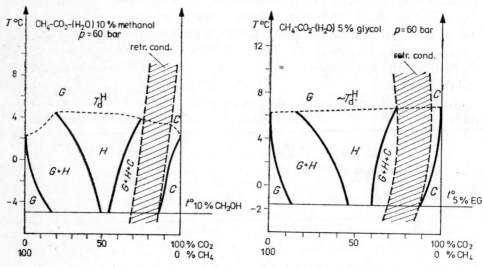

Fig. 5.43. Isobaric phase equilibria in the CH₄–CO₂–10% aqueous methanol system [437]

Fig. 5.44. Isobaric phase equilibria in the CH₄–CO₂–5% aqueous glycol system [437]

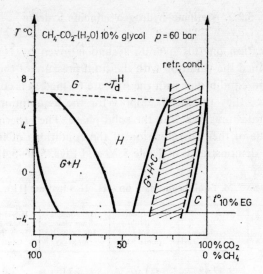

Fig. 5.45. Isobaric phase equilibria in the
CH_4–CO_2–10% aqueous glycol system [437]

the hydrate is the highest. This fact is of significance from the point of view of selection of the optimum pressure conditions in the separation of the CO_2 content from natural gases.

The tabulated data point to two further important phenomena:

(1) The medium with a methanol content of 5% does not inhibit, but rather stabilizes the hydrate (see Fig. 3.23).

(2) The inhibitory effects of alcoholic solutions with different concentrations follow the sequence of the freezing-points of the aqueous solutions (except when the methanol content $\leqslant 5\%$).

The inhibitory effect of the alcohols is manifested primarily in the phenomenon that the hydrates formed have looser structures as the alcohol concentration is increased, and even when a prolonged period was available for the stabilization they did not become more compact than in the presence of water without inhibitor. A similar effect was produced by the condensed phase appearing in gas mixtures with higher carbon dioxide contents.

The above features are demonstrated more illustratively by the isobaric phase equilibrium diagrams in Figs. 5.41–5.45, based on values measured at 6×10^6 Pa.

5.2.2. Methane–hydrogen sulphide hydrate

Detailed information on this hydrate system is given by Noaker and Katz [113], who determined the variation with the total pressure of the ratio H_2S/CH_4 for the gas phase in equilibrium with the hydrate formed at a constant temperature. On this basis, they produced support for the assumption that complete mixing of the hydrates must occur in the solid phase. The experimental data obtained in the course of their investigation of the conditions of formation of the mixed hydrate are demonstrated in Table 5.26 and Figs. 5.46–5.48.

Table 5.26. Equilibrium data on CH_4–H_2S hydrate [113]

T (°C)	p (10^5 Pa)	H_2S content of initial gas (mole %)	H_2S content of gas phase (measured) (mole %)	H_2S content of gas phase (calculated) (mole %)
15.5	49.49	11.6	8.23	7.0
11.1	26.46		9.51	9.0
9.1	31.20	7.1		6.3
13.9	49.08			6.3
17.0	69.69			7.0
6.1	22.62		5.73	6.5
17.0	65.34		6.6	7.0
5.6	28.98	3.11		3.0
9.7	43.83			3.1
14.4	68.17		2.92	2.95
3.3	20.80		3.78	3.9
5.3	33.22	1.09		1.0
9.2	47.36			1.04
11.7	68.57		1.11	1.06
14.4	21.51	22.2		22.0
22.2	51.91		19.8	21.0
6.7	10.60		21.4	22.0
8.3	21.21	12.5		9.5
14.1	36.76			11.0
18.9	61.50		11.5	11.5

Hydrate formation in the CH_4–H_2S–H_2O system containing more than 27% hydrogen sulphide was investigated in the quaternary equilibrium system by Khoroshilov and Bukhgalter [438]. From their experimental results, they put forward an empirical relationship, according to which the equilibrium pressure of the four-phase system (in bars) can be described in terms of the φ_{H_2S} volume fraction of hydrogen sulphide, as follows:

$$p = 2137(\varphi_{H_2S} \cdot 100)^{-1.005} \tag{5.25}$$

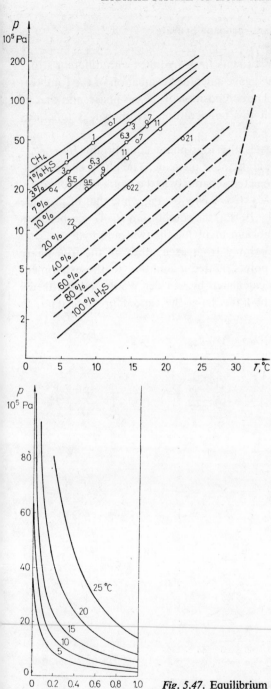

Fig. 5.46. Equilibrium diagram for CH₄–H₂S hydrate [113]. Reprinted by permission of the SPE-AIME

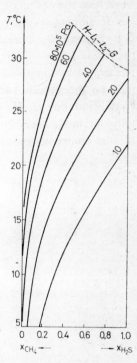

Fig. 5.48. Equilibrium isobars for CH₄–H₂S hydrate [113]. Reprinted by permission of the SPE-AIME

Fig. 5.47. Equilibrium isotherms for CH₄–H₂S hydrate [113]. Reprinted by permission of the SPE-AIME

5.2.3. Methane–propane hydrate

In this hydrate system, the gas constituents form hydrates with different structures. van der Waals and Platteeuw[98] gave a detailed evaluation of the composition of the hydrate formed at $-3\,°C$. The composition of the gas phase, and that of the hydrate in equilibrium with it at this temperature, are shown as functions of the pressure in Fig. 5.49.

In this figure, line AB demonstrates the four-phase equilibrium $H_I–H_{II}–G–$ice, where the gas phase consists of almost pure methane. The phase characteristics developing around point A are depicted separately in the figure. The dissociation pressure of methane hydrate at $-3\,°C$ is 2.34×10^6 Pa, while that of propane hydrate is 1.50×10^5 Pa. Above 2.34×10^6 Pa, the equilibrium $H_{II}–G–$ice becomes metastable as compared to the equilibrium involving the two pure hydrates, and the two three-phase surfaces are separated from each other by the four-phase equilibrium line $H_I–H_{II}–G–$ice. The points plotted in addition to the experimentally-obtained curves are the results calculated by van der Waals and Platteeuw for the compositions of the mixed hydrates from the vapour pressure data, in

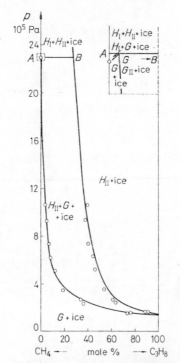

Fig. 5.49. Equilibrium isotherm for
CH_4–C_3H_8 hydrate at $-3\,°$ C [98]

Table 5.27. Degrees of occupancy of hydrate cavities in the CH_4–C_3H_8 hydrate system at $-3°$ C [98]

p (10⁵ Pa)	Proportion of small cavities occupied by CH_4 y_{M_1}	Occupancy of larger cavities		Remarks
		by methane y_{M_2}	by propane y_{P_2}	
36.16	0.8836	0.8522	0	metastable CH_4 hydrate
20.20	0.8088	0.1795	0.7646	
10.10	0.6762	0.0309	0.9496	stable
5.05	0.4952	0.0060	0.9860	"homogeneous"
3.03	0.3317	0.0017	0.9937	solutions
2.02	0.1739	0.0005	0.9965	
1.58	0	0	0.99795	C_3H_8 hydrate

accordance with the Lennard-Jones–Devonshire theory. The composition data calculated for a temperature of $-3\,°C$ are listed as a function of the pressure in Table 5.27.

This system was likewise investigated by Carson and Katz [46], who stated that the four-phase equilibrium can be described by means of a continuous curve in the p vs. T diagram. It may be presumed that they worked in only a section of the

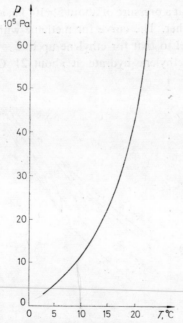

Fig. 5.50. Equilibrium of CH_4–C_3H_8 hydrate, on the basis of [46]. Reprinted by permission of the SPE-AIME

four-phase line H_{II}–L_1–L_2–G. Their results are depicted, together with the four-phase equilibrium data, in Fig. 5.50.

In the course of their practical application of the theory of gas hydrates, Byk and Fomina [439] dealt not only with the properties of the individual hydrates of methane and propane, but also with those formed from mixtures of these gases. By taking into consideration the Gibbs phase rule, they found that it is possible for the two hydrates with their different structures to exist side by side in a gaseous mixture of methane and propane under isobaric conditions.

Malyshev and his co-workers [440] also reported a study of the common occurrence of hydrates of types H_I and H_{II}, similarly in a propane–methane mixture.

5.2.4. Methane–ethylene hydrate

This hydrate system was investigated in great detail by Snell and his co-workers [116], who drew attention to many interesting phenomena. They established that, below a pressure of 5×10^6 Pa, the hydrate equilibrium curves for the vapour phase containing more than 70 mole % ethylene cannot be distinguished from those of pure ethylene hydrate.

As may be seen in Fig. 5.23 too, the equilibrium curve for pure ethylene hydrate has a break-point at a pressure of about 5×10^6 Pa, and its slope subsequently becomes somewhat higher. The curve for methane with an ethylene content of 42 mole % runs parallel to that for ethylene up to 5×10^6 Pa and intersects the equilibrium curve for ethylene hydrate at about $21\,°C$ and 10^7 Pa. At higher

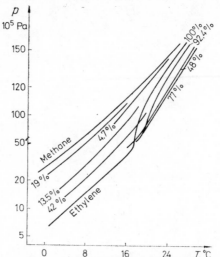

Fig. 5.51. p vs. T equilibrium diagram for
CH$_4$–C$_2$H$_4$ hydrate [116]

pressures it was observed that the mixture containing 47.7 mole % ethylene has a special composition, for the hydrate is stabilized between 48 and 84 mole %. Between 84 and 92 mole %, however, its stability is lower than that of the mixture with an ethylene content of 47.7 mole %, and the curve approaches that for pure ethylene, i.e. the slope of the vapour pressure curve initially decreases with increasing ethylene concentration, but then rises. This means at the same time that, at pressures higher than 5×10^6 Pa, the temperature of hydrate formation passes through a maximum at all pressure values; consequently, for a given composition at any pressure there is a maximum temperature above which hydrate cannot be formed.

This interesting change in the conditions of hydrate formation is illustrated in Fig. 5.51, and the effect of the composition on hydrate formation in Fig. 5.52.

The investigation of the conditions of hydrate formation provided a possibility for the determination of the equilibrium ratio of solid and gaseous ethylene in the methane–ethylene–water system. This can be seen in Fig. 5.53, where the isobars are of gradually diminishing slope until K attains a value of 0.8. At this value, the slope begins to rise sharply. The curves clearly show the temperature maxima formed above a pressure of 7×10^6 Pa for $K > 0.9$ too, demonstrating the partic-

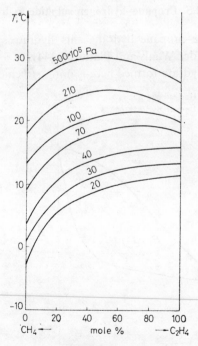

Fig. 5.52. Equilibrium isobars for
CH_4–C_2H_4 hydrate [116]

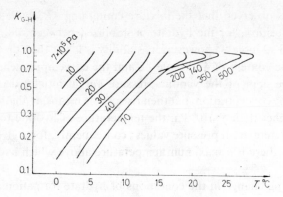

Fig. 5.53. Solid–vapour equilibrium constant
of CH_4–C_2H_4 hydrate [116]

ular case when the temperature of formation of the mixed hydrate is higher than those for the two pure constituents. A similar phenomenon is encountered in the propane–hydrogen sulphide–water system.

5.2.5. Propane–hydrogen sulphide hydrate

Similarly to methane–propane hydrate, this hydrate system too was investigated in detail by van der Waals and Platteeuw [441] at a temperature of $-3\,°C$. They found that the hydrate formed here is not a stoichiometric double hydrate

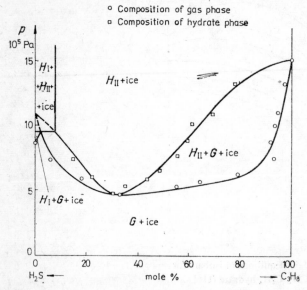

Fig. 5.54. Equilibrium diagram for
C_3H_8–H_2S hydrate at $-3\,°C$ [441]

either, but a mixed hydrate with varying composition, as can be seen in Fig. 5.54. Although the resulting hydrate structure resembles that of methane–propane hydrate, an azeotropic system is formed here. If the hydrogen sulphide–propane ratio corresponds to the azeotropic proportions 3:1, the decomposition pressure is constant at −3 °C, and this pressure is lower than the decomposition pressures of the hydrates of the two pure constituents at this temperature.

5.2.6. Propane–nitrogen hydrate

The initial conditions of hydrate formation in the propane–nitrogen–water system were determined experimentally by Ng Heng-Joo and his co-workers [442] in the L_1–H–G, L_1–L_2–H and L_1–L_2–G equilibrium regions (where L_1 is the liquid phase rich in water, L_2 is the liquid phase rich in propane, and G is the vapour phase). The measurements were carried out in the temperature range 275–293 K and between the pressure limits 0.3 and 17.0 MPa. The propane content of the gas phase varied between 0.94 and 75 mole % in the L_1–H–G region, and the propane content of the liquid phase between 83.1 and 99.0 mole % in the L_1–L_2–H region. The propane content of the gas phase lay between the limits 18.1 and 71.1 mole % in the four-phase equilibrium region.

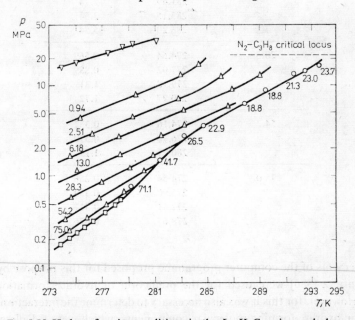

Fig. 5.55. Hydrate-forming conditions in the L_1–H–G region and along the L_1–L_2–H–G locus for the N_2–C_3H_8–H_2O system [386]

Table 5.28. Initial conditions of hydrate formation
by various $C_3H_8-N_2$ mixtures [386]

Composition of gas phase (mole % C_3H_8)	T (K)	p (MPa)
0.94	275.33	4.59
	279.55	8.16
	283.00	13.68
	284.28	18.09
2.51	276.29	3.03
	279.28	4.51
	282.74	7.35
	287.10	13.64
6.18	274.51	1.72
	278.26	2.85
	283.05	5.50
	286.95	9.47
	289.17	13.71
13.00	275.10	1.10
	278.41	1.72
	281.49	2.74
	283.15	3.54
	286.20	5.54
28.3	274.64	0.569
	277.02	0.889
	279.18	1.31
	280.77	1.72
54.2	274.16	0.332
	276.76	0.570
	280.27	1.19
75.0	274.54	0.256
	275.90	0.359
	277.41	0.517
	278.67	0.676

The elements of the computer programme proposed for this purpose by Parrish and Prausnitz [386] were used for the prediction of hydrate formation in this system. However, for this it was also necessary to determine the interaction parameters of the hydrate-forming molecular components, which were calculated on the basis of statistical thermodynamic relations for nitrogen and propane.

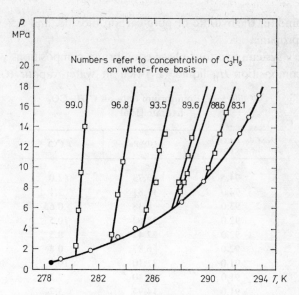

Fig. 5.56. Hydrate-forming conditions in the L_1–L_2–H region
for the N_2–C_3H_8–H_2O system [386]

In the course of the experiments, the initial conditions of hydrate formation
were determined for seven gas mixtures with different compositions. The results
are shown in Table 5.28.

The results of Parrish and Prausnitz from investigations of the L_1–H–G and
L_1–L_2–H equilibrium regions are demonstrated in Figs. 5.55 and 5.56.

The experimental results were compared with those given by computer, and
applied for the prediction of hydrate formation. On this basis, the absolute error
of the determination was established as 5.7% in the L_1–H–G equilibrium region.
This fact demonstrated the applicability of the method, since without the use of
interaction parameters this error was much greater, at 25.3%.

5.2.7. Propane–carbon dioxide hydrate

The conditions of hydrate formation in the propane–carbon dioxide–water
system were considered in detail by Robinson and Mehta [156] in a wide concentration range. This gas system too contains constituents which form hydrates with
different types of structures.

The same authors [443] found that the maximum temperature of hydrate formation in this three-component system is 14.2 °C, which is higher than the values

of 10.15 °C found for the hydrate of pure carbon dioxide, and 5.5 °C for the hydrate of pure propane.

In the ternary system, two four-phase equilibrium compositions were detected: these were a composition H_{II}–liquid (L_1) rich in water–vapour (G)–liquid pro-

Table 5.29. Equilibrium data on C_3H_8–CO_2 hydrate [156]

Mole % CO_2 in the gas phase	p (10^5 Pa)	T (°C)
94.5	43.73	12.0
94.0	11.81	3.1
93.0	8.38	0.66
92.0	32.62	10.5
92.0	22.42	8.5
92.0	6.87	0.8
91.0	31.10	10.3
91.0	10.40	7.2
91.0	14.94	5.7
90.0	12.82	5.2
87.0	16.16	7.8
87.0	5.25	0.7
86.0	12.32	6.3
85.0	8.48	2.2
85.0	7.77	3.3
84.0	34.64	13.0
79.0	3.64	0.8
77.0	7.27	4.9
76.0	6.66	4.2
75.0	19.69	10.6
75.0	9.99	7.0
74.0	13.33	8.7
58.0	16.96	10.5
57.8	4.24	2.5
52.5	7.07	5.4
52.0	10.90	7.9
40.0	8.08	6.5
40.0	3.33	1.7
39.0	4.44	3.2
37.0	7.68	6.4
35.0	5.95	5.1
28.0	3.13	2.0
18.0	5.05	5.9
17.0	5.15	5.0
17.0	6.56	5.9
16.0	4.84	4.6
14.0	3.43	2.8

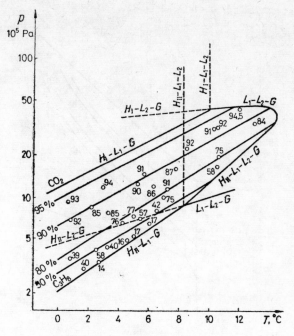

Fig. 5.57. p vs. T equilibrium diagram for
$C_3H_8-CO_2$ hydrate [156]

pane (L_2) rich in CO_2, and an $H_I-L_1-G-L_2$ equilibrium composition. The experimental procedure used was not suitable for the separation of the two hydrates with different structures, but it could be concluded that the univariant equilibrium $H_{II}-L_1-G-L_2$ developed only above 95 mole % CO_2. The three-phase equilibrium data are presented in Table 5.29. The equilibrium p vs. T data pairs in Fig. 5.57 were compiled from these data, with consideration of the composition of the gas mixture.

Craig [444] too dealt in detail with the four-phase equilibrium investigation of the propane–carbon dioxide–water system.

5.2.8. Methane–propylene hydrate

The conditions of formation of methane–propylene hydrate are given by Otto and Robinson [115]. They found that the system is very sensitive to the quantity of propylene, so much so that at a temperature of 10 °C, for example, the formation pressure of methane hydrate is lowered by about 3.5×10^6 Pa by a propylene content of 1.4%. In the presence of more than 25 mole % propylene, a four-

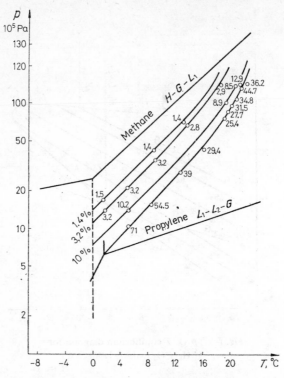

Fig. 5.58. p vs. T equilibrium diagram for
CH$_4$–C$_3$H$_6$ hydrate [115]

phase equilibrium is formed at all temperatures. In an investigation of the change in the conditions of hydrate formation with the composition, the composition of the vapour phase was determined by mass-spectrometry. It was stated that the highest pressure at which hydrate can be formed is 9.73×10^6 Pa. This pressure is associated with a temperature of 20.72 °C and a propylene content of 34.8 mole %. At this point, a critical phenomenon appeared, the liquid rich in hydrocarbon and the vapour were transformed into one phase, and with this the four-phase equilibrium ceased. At a pressure of 9.79×10^6 Pa, that is slightly above the critical one, the system lost its colour, but as the pressure or temperature was lowered, the colour changed first to yellowish, and then to brownish-red. If the pressure was kept at 9.79×10^6 Pa, the maximum colour intensity was reached at 9.6 °C.

The experimental data on the equilibrium conditions of hydrate formation are illustrated in Fig. 5.58, and their change with the composition in Fig. 5.59.

It can be seen from Fig. 5.58 that, as the pressure rises, the quantity of the liquid phase rich in hydrocarbon increases too, and the vapour phase becomes progressively richer in methane. Above 7×10^6 Pa, the methane concentration decreases again. The maximum is situated at a methane content of 74.3 mole %.

The schematic phase diagram of the methane–propylene–water system at a pressure of 3.19×10^6 Pa [115] is shown in Fig. 5.60. The four-phase equilibrium temperature at this pressure is 13.7 °C. The phases H–G–L_1–L_2 are in equilibrium here, besides which G–L_1–L_2, H–G–L_1 and H–L_1–L_2 can exist in three-phase equilibrium. In the knowledge of the state parameters of the three-phase equilibrium loci, the temperature corresponding to the triple point demonstrates schematically what equilibrium relations are possible in the three-phase regions at the critical temperature relating to the given pressure of 3.19×10^6 Pa, or at temperatures higher or lower than this. This is shown in Fig. 5.61, in which part (a) is the triangle diagram of the equilibrium G–L_1–L_2 between the temperatures 13.7 and 65.5 °C. This three-phase equilibrium is limited by the three two-phase equilibrium areas adjacent to it, and by the monophase areas characteristic of the three pure constituents. In part (b) of the figure, the four-phase equilibrium too

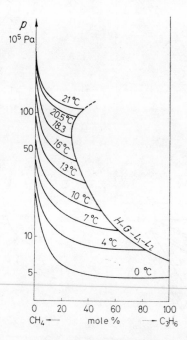

Fig. 5.59. Equilibrium isotherms
for CH_4–C_3H_6 hydrate [115]

16

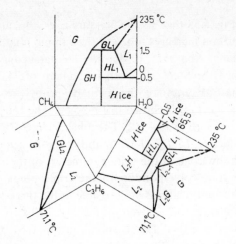

Fig. 5.60. Schematic phase diagram of CH_4–C_3H_6 hydrate at 3.19×10^6 Pa [115]

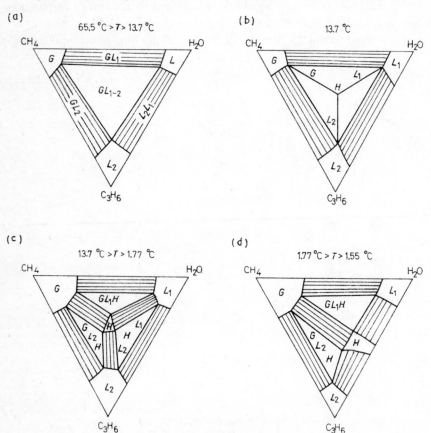

Fig. 5.61. Phase equilibria of CH_4–C_3H_6 hydrate [115]

appears, at a temperature of 13.7 °C. In parts (c) and (d), the possibilities of appearance of the hydrate phase at lower temperatures are illustrated within the two-phase and three-phase areas, respectively.

5.2.9. Propane–propylene hydrate

The conditions of hydrate formation in the propane–propylene–water system were investigated by Reamer and his co-workers [89] at a constant value of the molar ratio of the two gas constituents:

$$C = \frac{x_{propane}}{x_{propane} + x_{propylene}} = \frac{y_{propane}}{y_{propane} + y_{propylene}} \tag{5.26}$$

where y denotes the mole fraction in the vapour phase, and x that in the liquid phase. The hydrates produced were transparent, and their melting-points lay between the decomposition temperatures of the hydrates of the two pure constituents. The p vs. T data relating to hydrate decomposition for two mixtures of the ternary system with different compositions are listed in Table 5.30 and depicted in Fig. 5.62. It is evident from the figure that the pressure and temperature of the beginning of hydrate decomposition are functions of the composition parameter C. This dependence is apparent for both the equilibria $H-L_1-L_2$ and $H-L_1-G$.

Starting from the assumption that there is no mutual solubility between the water and liquid hydrocarbon phases, and that the solubility is ideal in the gas phase, the expected pressure of hydrate formation was calculated and the actual

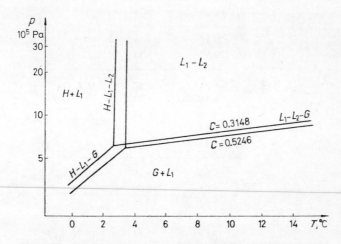

Fig. 5.62. p vs. T equilibrium diagram for C_3H_8–C_3H_6 hydrate [89].
Reprinted by permission of the SPE-AIME

Table 5.30. Equilibrium data on C_3H_8–C_3H_6
hydrate [89]

Composition parameter C	p (10^5 Pa)	T (°C)	Phases present
0.5246	3.08	0.55	H–L_1–G
	3.62	1.28	
	3.79	1.17	
	4.08	1.44	
	4.43	2.00	
	4.62	1.83	
	4.69	1.83	
	5.12	3.00	
	5.14	2.78	
	5.17	2.94	
	5.26	4.00	L_1–L_2–G
	6.38	5.16	
	6.58	7.50	
	7.45	12.39	
	8.52	15.27	
	9.77	21.77	
	6.94	3.50	H–L_1–L_2
	8.25	3.50	
	8.96	3.50	
	9.73	3.55	
	12.95	3.55	
	18.69	3.78	
	19.58	3.78	
	20.37	3.78	
	20.97	3.83	
0.3148	4.04	0.67	H–L_1–G
	4.56	1.33	
	5.10	1.89	
	5.29	2.00	
	5.47	2.05	
	6.64	5.39	L_1–L_2–G
	8.01	12.00	
	9.32	17.70	
	7.59	2.55	H–L_1–L_2
	13.78	2.78	
	19.67	2.83	

value was measured at the temperatures 4.5 and 21 °C. As can be seen in Table 5.31, the measured and calculated values showed good agreement.

Table 5.31. Equilibrium data on C_3H_8–C_3H_6 hydrate [89]

Composition parameter C	$T=4.5$ °C		$T=21.1$ °C	
	p (10^5 Pa)			
	measured	calculated	measured	calculated
0.3148	6.50	6.50	10.29	10.30
0.3711	6.46	6.43	10.19	10.21
0.5246	6.24	6.25	9.91	9.94
0.7589	5.96	5.95	9.45	9.46

The temperature vs. composition diagram of the hydrate shown in Fig. 5.63 for a pressure of 2.84×10^6 Pa was plotted from data obtained by the direct analysis of phases in equilibrium with one another.

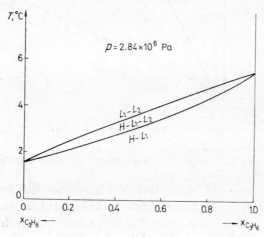

Fig. 5.63. Phase equilibria of C_3H_8–C_3H_6 hydrate at 2.84×10^6 Pa [89]. Reprinted by permission of the SPE-AIME

5.2.10. Chloroform–hydrogen sulphide hydrate

Equilibrium conditions more complicated than those in the previously-mentioned hydrate systems are encountered in the chloroform–hydrogen sulphide–water system, studied by von Stackelberg and Frühbuss [68]. The binary hydrates were well characterized in the experiments of Selleck and his co-workers [86]

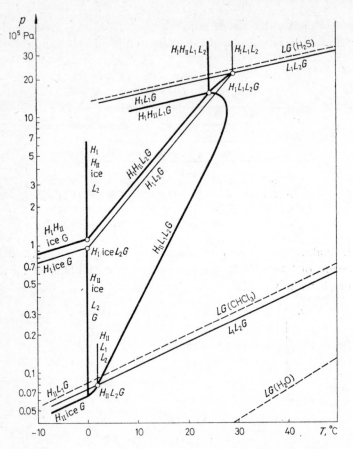

Fig. 5.64. p vs. *T* equilibrium diagram for CHCl₃–H₂S hydrate [68]

and of von Stackelberg and Müller [66]. If it is assumed that hydrogen sulphide and chloroform do not undergo separation in the liquid phase, then in this system L_1 means a non-aqueous liquid phase, the composition of which varies continuously as a function of the quantities of the pure constituents of the mixture.

Figure 5.64 depicts the equilibrium *p* vs. *T* diagram for this system. The equilibria of the monocomponent systems are indicated by dashed lines, those of the binary systems by thin lines, and those of the ternary systems by thicker lines. As regards these latter, the experimental data in the lower part of the line H_{II}–L_1–L_2–G are the most exact; these were determined by von Stackelberg and Frühbuss. The data of van der Waals and Platteeuw [98] too contributed to the complete construction of this figure.

In the course of the experiments it was found that, if chloroform is added to the H₂S–H₂O system in which the three phases H_1–G–ice are in equilibrium, the total pressure increases until the mixed hydrate with H_{II} structure is formed; the four-phase line H_I–H_{II}–G–ice must therefore lie above the line H_I–G–ice. The situation is the same above the melting-point of ice as regards the relative positions of the lines for the equilibria H_I–L_1–G and H_I–H_{II}–G–L_2.

The vapour pressures of water and chloroform are much lower than that of hydrogen sulphide at the same temperature. Accordingly, at the quintuple point, formed where the lines for the equilibria H_I–H_{II}–L_1–G and H_I–H_{II}–L_2–G intersect, the pressure and temperature are lower than at the binary H_I–L_1–L_2–G quadruple point. At the quintuple point, the system is in all probability metastable, and its position can therefore be determined only approximately.

A temperature maximum too appears in the four-phase line H_{II}–L_1–L_2–G. This means a limiting temperature, above which a hydrate with H_{II} structure cannot be formed.

5.2.11. Carbon tetrachloride–hydrogen sulphide hydrate

Similarly as in the case of chloroform, a complex hydrate structure is observed in this carbon tetrachloride system too. The situation is further complicated by the fact that the CCl_4–H_2O system itself does not yield a hydrate. For the study of this ternary system, information is again provided by the publications of von Stackelberg and Frühbuss [68] and van der Waals and Platteeuw [98]. Figure 5.65 shows the equilibrium conditions.

Particularly as a result of the investigations on these two systems, van der Waals and Platteeuw [98] query the conclusion of von Stackelberg and his co-workers that a distinction can be made between double hydrates with stoichiometric composition and non-stoichiometric mixed hydrates, and they also question whether the existence of double hydrates can be confirmed at all. As regards the ternary systems listed above, a number of the components are able to form hydrates with the H_{II} structure and should therefore be able to form a double hydrate with H_2S, but nevertheless the formation of a solid solution could be demonstrated.

5.2.12. Difluoroethane–hydrogen sulphide hydrate

Particular hydrate-formation conditions can be observed in the CH_3CHF_2–H_2S–H_2O system. Here, both pure components form a hydrate with H_I structure, but in spite of this a hydrate with H_{II} structure was detected by von Stackelberg and Jahns [70] in the ternary system. According to van der Waals and Platteeuw [98], this can only be interpreted in that CH_3CHF_2 actually forms a H_{II} structure,

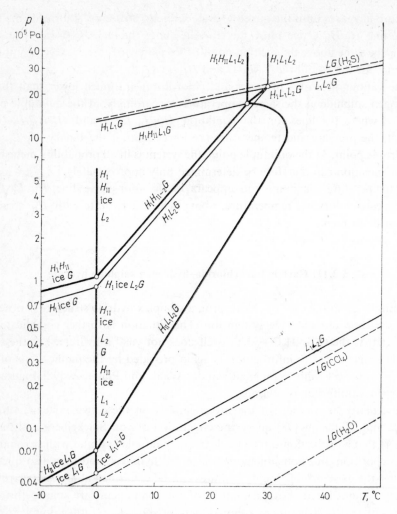

Fig. 5.65. p vs. *T* equilibrium diagram for CCl₄-H₂S hydrate [68, 98]

but the dissociation pressure of the hydrate is near that possible at the same temperature for the H_I structure.

The phase diagram determined experimentally for the CH_3CHF_2-H_2S-H_2O system can be seen in Fig. 5.66. It may be assumed that the primary role in the occurrence of the structural change is played by the fact that the hydrogen sulphide is incorporated into the many small cavities in the H_{II} structure, while the difluoroethane is not.

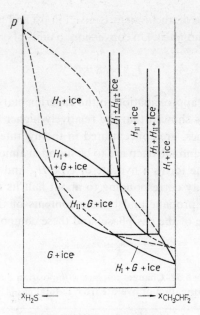

Fig. 5.66. Phase equilibria
of CH_3CHF_2–H_2S hydrate at
temperatures below 0 °C [98]

5.3. Hydrates of natural gases

The hydrates of natural gases are solid compounds containing a number of gas components. The melting-point of the mixed hydrate is determined primarily by the composition of the hydrate, and consequently by the pressure, but it is also influenced substantially by the secondary factors (flow velocity, pressure pulsation), which alter the conditions of hydrate formation to some extent.

In the industrial processing of natural gases, the water content present gives rise to much trouble from the aspects of quantitative measurement and transportation, but the problems multiply if the pressure and temperature conditions of the production or transportation permit the possibility of hydrate formation too.

The attention of industrial specialists was first drawn to the conditions of natural gas hydrate formation by Hammerschmidt [41]. On the basis of the first series of experiments, he proposed the following equation for the determination of the temperatures of decomposition of the hydrates of natural gases:

$$T_{dec.} = 8.9 p^{0.285} \tag{5.27}$$

where the total pressure of the system is given in psi and the decomposition temperature in degrees Fahrenheit. On conversion to units of bar and °C, the equation assumes the form

$$T_{dec.} = 8.9p^{0.0787} \qquad (5.28)$$

This gives a very good approximation to the experimental results.

The first experiments showed that the relatively higher hydrocarbons, such as propane and the butanes, are concentrated in the hydrate. It may be seen from Table 5.32 that ethane and nitrogen are to be found in almost unchanged amounts in the hydrate. Methane forms a hydrate of type H_I, and it is incorporated into the hydrate in a quantity corresponding to about half its starting concentration. At the same time, the propane and butane contents of the hydrate are several times higher than those of the gas phase, and these compounds form hydrates of type H_{II}.

Table 5.32. Changes in phase composition during formation of hydrate from natural gas [41]

Components	Gas composition (mole %)	Hydrate composition (mole %)	Distribution coefficient
CH_4	82.50	56.95	0.7
C_2H_6	5.99	5.66	1.0
C_3H_8	3.26	24.97	8.0
iso-C_4H_{10}	0.30	4.69	16.0
n-C_4H_{10}	0.49	0.83	2.0
C_{5+}	0.07	0.00	0.0
CO_2	0.20	0.44	2.2
N_2	7.19	6.46	0.9

Evaluation of the data in Table 5.32 suggests that a hydrate structure is formed here in which 62.5% of the cavities are occupied by small molecules, 30.5% of them by large molecules, and the remaining 7.0% by such similarly small molecules as CO_2 and N_2. In the presence of these latter constituents, the process of hydrate formation is much more intensive.

In the publication by Simanek and Pick [445] it is emphasized that the crystal lattice in the mixed hydrates is more compactly filled than in the hydrates formed by the individual components; they justified this on the grounds that the natural gas mixtures participate more intensively in hydrate formation, and the resulting hydrate structure is more stable than those of the hydrates of system containing only one gas component.

The enrichment of the C_3 hydrocarbons in the hydrate was observed by other authors too. For instance, the data in Table 5.33 were published by Kedzierski and Chowaniec [143].

Table 5.33. Changes in phase composition during formation of hydrates from natural gas [143]

Components	Gas composition (mole %)	Hydrate composition (mole %)	Molar ratio (hydrate/gas)
CH_4	96.37	76.96	0.79
C_2H_6	0.61	0.49	0.80
C_3H_8	0.33	11.54	34.97
$n\text{-}C_4H_{10}$	0.08	3.73	47.25
$iso\text{-}C_4H_{10}$	0.11	7.28	66.18
N_2	2.5	—	—

The effect of carbon dioxide in intensifying hydrate formation is presumably based on the fact that the water vapour content of natural gases increases with increasing CO_2 concentration. This increase is particularly considerable at higher pressures; assuming ideal behaviour, this cannot be calculated by means of the mixture rule from the data relating to the pure components (Török and his co-

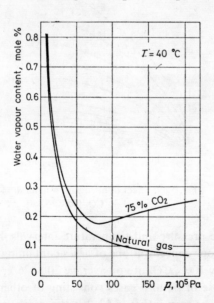

Fig. 5.67. Dependence of the moisture content of natural gas on the pressure [446]

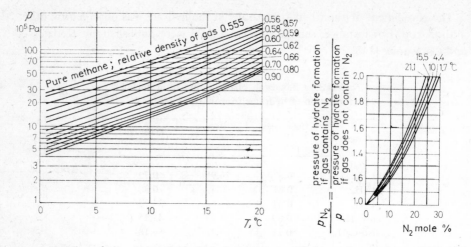

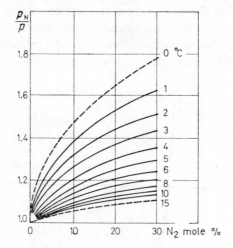

Fig. 5.68. Effect of nitrogen on the hydrate formation equilibrium of natural gas [357, 447] Adapted from the book D. L. Katz, *Handbook of Natural Gas Engineering* (p. 213, Fig. 5.4) by permissions of the McGraw-Hill Book Company and the American Chemical Society

Fig. 5.69. Effect of nitrogen on natural gas rich in CO_2 [306]

workers [446]). In the pressure and temperature intervals 4–20×10^6 Pa and 20–$100\,°C$, the water vapour concentrations of gases containing 25% CO_2, 50% CO_2, 75% CO_2 and virtually 100% CO_2 are on average 20%, 36%, 53% and 82% higher, respectively, than those of natural gases consisting of otherwise practically pure hydrocarbons. It may be seen in Fig. 5.67 that the moisture content of natural gas containing 75% CO_2 passes through a minimum as a function of pressure.

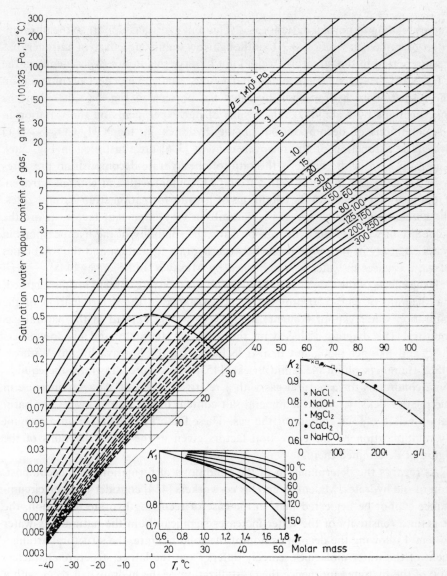

Fig. 5.70. Diagram for determination of the saturation water vapour content of natural gas with a relative density of 0.6 [449, 450]. Reprinted by permission of the SPE-AIME

This minimum is the sharper, the higher the CO_2 content and the lower the temperature.

However, in complete contrast with this, as the nitrogen content increases the water vapour content decreases. In the estimation of the conditions of hydrate formation, therefore, an error will definitely arise from the use of dew-point curves relating to normal natural gases. This lends further support to the phenomenon described in Section 3.2.1: although the stabilizing effect of nitrogen has been reported in the literature, we did not observe this in our experiments on the CH_4–N_2–H_2O system (Berecz and Balla-Achs [306]). The effect of nitrogen on the formation of hydrocarbon hydrates can be seen in Fig. 5.68, based on the investigations by Katz [447]. The results of our experiments on the variation in the conditions of decomposition of the hydrates in the CH_4–CO_2–N_2–H_2O system as a function of the nitrogen content [306] are illustrated in Fig. 5.69. The ratio p_N/p in the figure is the ratio of the hydrate decomposition pressures for CH_4–CO_2 gas mixtures with a fixed CO_2 content with and without nitrogen. From the point of view of the conditions of hydrate formation from natural gases, the moisture content of the gas is naturally a very important factor, as are the solubility of water in the hydrocarbons, and the solubility of the hydrocarbons in water. A study of the correlations between hydrate formation and the solubilities of natural gases in water was published by Musayev and Saniyev [448].

If the effects of the CO_2 and N_2 contents on the moisture content are disregarded, the diagram of Fig. 5.70 provides great assistance in the determination of the saturation water vapour content; this is very widely applied in the natural gas industry. The diagram is based on the work of McCarthy and his co-workers [449] and Skinner [450], but it also includes data of Kobayashi and Withrow [451], Hammerschmidt [452] and Brickel [453]. The figure illustrates the equilibrium conditions for a natural gas with a relative density of 0.6. An increase in the relative density, or in the mineral salt content of the water, lowers the saturation water vapour content of the gas. These factors can be taken into account by multiplication by the correction factors given in the insert diagrams of the figure or determined there.

As regards the determination of the conditions of formation and decomposition of the hydrates, Malenkov and his co-workers [454] consider that the circumstance cannot be neglected that, in the case of recurring hydrate formation, the structural sensitivity of the water increases significantly in the natural gas–water system. Following the decomposition of the solid hydrate, the structural arrangement of the water molecules scarcely differs from the lattice structure characteristic of the hydrate, and hence the crystallization of the hydrate can recur with a lower energy demand.

Deaton and Frost [361] investigated the saturation water vapour contents of three natural gases with different compositions, of air and of helium; their results for a temperature of 15 °C are shown in Fig. 5.71. Table 5.34 provides information on the compositions of the investigated gases.

In the thermodynamic investigation of problems arising during the production of natural gases, little attention was earlier paid to the equilibrium study of gas systems containing polar solvents, such as inhibitors, especially in the range of low temperatures and high pressures. Nevertheless, it is of great importance from this aspect that methanol, for instance, is very effective in practice for the removal of the moisture, and possibly CO_2 and H_2S contents of natural gases. In the presence of these latter gas components, the solubility of methanol in the natural

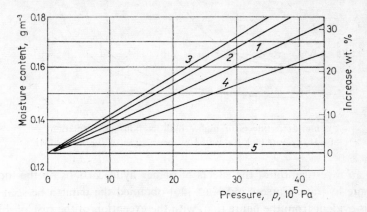

Fig. 5.71. Saturation water vapour contents of various gases at 15 °C [361]

gas increases, and this has an influence on the conditions of hydrate formation too. Experiments and calculations on the solubility of methanol in compressed natural gases were carried out by Won [455]. With regard to the various individual gas systems, the studies of Hemmaplardh and King [456] are of importance.

Table 5.34. Saturation water vapour contents of various gases at 15 °C [361]

Components	1	2	3	4	5
	mole %				
Air	—	—	—	100	—
Carbon dioxide	0.3	0.8	0.6	—	—
Nitrogen	9.4	25.0	1.0	—	2.0
Helium	—	—	—	—	98.0
Methane	79.4	68.4	94.36	—	—
Ethane	5.9	3.7	2.64	—	—
Propane	3.3	1.9	0.96	—	—
Butane	1.7	1.2	0.44	—	—

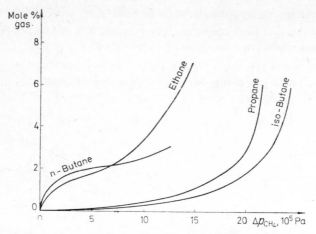

Fig. 5.72. Effects of higher hydrocarbons on formation
of CH_4 hydrate [46]. Reprinted by permission of the SPE-AIME

From an investigation of the effects of higher hydrocarbons on the formation
of methane hydrate, Carson and Katz [46] obtained the data to be seen in Fig.
5.72. It is evident from the figure that, with the exception of the case of *n*-butane,
the effect of the hydrocarbons in elevating the temperature of formation of methane hydrate increases with the increase in length of the carbon chain. From this,
it might possibly be assumed that this feature holds for pentane too. However,
since the concentration of pentane in the vapour space never exceeded 1.2–1.5
mole % for these gases, there is a great probability that, even at higher pressures,
its relative concentration was too low for it to form a mixed hydrate with methane.

Byk and Fomina [457] and Malyshev and Tyushnyakova [387] dealt with the
possibilities of application of equations of state, in connection with the conditions
of formation in the ice–hydrate–gas and water–hydrate–condensed gas systems.

Carson and Katz [46] examined the possibility of calculation of the equilibria
involving the hydrates of the natural gases. They found that the equilibrium constants for the formation of the hydrates of ethane, propane and isobutane can be
calculated from the data relating to their binary mixtures with methane. They
determined these K vs. T (°C) value pairs experimentally. From their investigations of the conditions of formation of the hydrates of pure CO_2 and of pure H_2S
(which occur as impurities in natural gases), they came to the conclusion that the
equilibrium constant for formation of the hydrate of CO_2 lies about halfway
between those for ethane and methane, while the tendency of hydrogen sulphide
to undergo hydrate formation is similar to that of isobutane. However, it turned
out that these considerations can be applied only in the case of gases with low
CO_2 and H_2S contents. Since the vapour–solid equilibrium constant gives the

mole fraction ratio of the hydrocarbon content of the vapour phase and the hydrocarbon bound in the hydrate, as regards the natural gas the sum of the mole fractions of the hydrocarbons in the solid hydrate can be simply calculated by means of the expression $\sum' \dfrac{y_{vapour}}{K}$. As the gas pressure is proportional to the gas composition, the pressure of hydrate formation at a given temperature can be calculated. Such an evaluation from the equilibrium constant data is illustrated in Table 5.35 for a gas system at a temperature of 10 °C.

Table 5.35. Calculation of hydrate composition from equilibrium constants [46]

Gas components	y	K (at 2.1×10^6 Pa)	y/K	K (at 2.45×10^6 Pa)	y/K
Methane	0.784	2.04	0.384	1.90	0.4120
Ethane	0.060	0.79	0.076	0.63	0.0953
Propane	0.036	0.113	0.318	0.090	0.4000
Isobutane	0.005	0.0725	0.069	0.060	0.0834
n-Butane	0.019	0.79	0.024	0.630	0.0302
Nitrogen	0.094		0.000		0.0000
Carbon dioxide	0.002	1.41	0.0014	1.26	0.0016
	1.000		0.8724		1.0225

Since the equality $\sum' \dfrac{y}{K} = 1$ must hold under the conditions of hydrate formation, hydrate formation for the given gas can be expected at a pressure of 2.43×10^6 Pa on the basis of the tabulated data. The values of the equilibrium constants used in the calculations can be seen in Figs. 4.1–4.6.

In the calculation of the equilibrium pressure of hydrate formation, therefore, the procedure is as follows: at given temperatures, two different pressure values are taken, at which the equilibrium constants of formation of the hydrates of the constituents are determined, and the exact pressure values are decided by means of extrapolation or interpolation. The data obtained in this manner, however, give only the expected value of the beginning of hydrate formation (without supercooling), since from the aspect of hydrate formation the vapour pressure of the liquid water is determining, while the conditions of hydrate decomposition are governed by the vapour pressure of the water vapour above the hydrate. The hydrate decomposes only if the vapour pressure of water is less at a given temperature than the vapour pressure of the water above the hydrate. The experimentally-obtained data of Makogon and Sarkisyants [157], shown in Figs. 5.73 and 5.74, illustrate

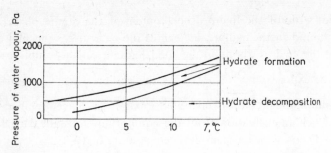

Fig. 5.73. Dependence of hydrate formation and decomposition on moisture content [157]

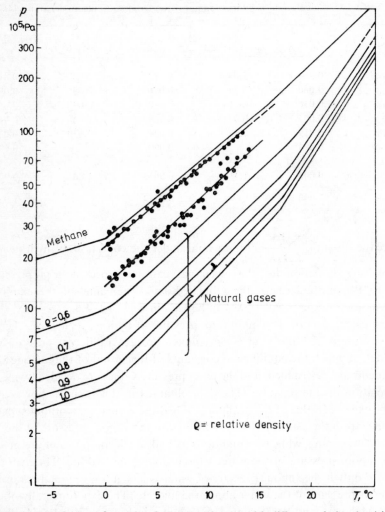

Fig. 5.74. Data on hydrate formation from natural gases with different relative densities [157]

the difference between the conditions of formation and decomposition of the hydrate in the case of a natural gas. The gas used in the experiments contained 82.7 mole % CH_4, 3.4 mole % C_2H_6, 1.5 mole % C_3H_8, 0.3 mole % C_4H_{10}, 7.0 mole % H_2, 5.0 mole % N_2 and 0.1 mole % CO_2.

After comparing the hydrate equilibrium data on different natural gases, Khoroshilov and his co-workers [458] published an empirical relationship for the determination of the conditions of hydrate formation:

$$\ln z = A - BT \qquad (5.29)$$

where $\ln z$ was defined in Equation (3.16), while the constants can be determined by the method described in connection with the Langmuir adsorption equations. The values of constants A and B were reported for hydrogen sulphide-free natural gas at pressures above and below 7 MPa, and also for gases containing hydrogen sulphide (Table 5.36).

Table 5.36. Values of constants A and B in Equation (5.29) [458]

Constants	Without H₂S content		For gases containing H₂S
	($p<7$ MPa)	($p>7$ MPa)	
A	3.515 570 5	8.975 110	5.406 96
B	0.014 336 0	0.033 039	0.021 33

On the basis of their data, a computer programme was prepared for the establishment of the conditions of hydrate formation.

The determination of the equilibrium conditions of formation and decomposition of the hydrates of natural gases was also dealt with by Malenkov [459] and Backhurst and Harker [460], and a detailed survey of this theme can be found in the dissertation of Falabella [461] too.

If the hydrocarbon gases contain condensate or oil, the conditions of hydrate formation change. The stability of the hydrate generally decreases, as explained in the previous chapters. For a natural gas containing 96 mole % methane and 2.63 mole % ethane, the effect of the gas–precipitate ratio on the equilibrium conditions of hydrate formation are demonstrated in Figs. 5.75 and 5.76, on the basis of the experiments of Dzhavadov and Tribus [462].

The conditions of formation and decomposition of the hydrates of liquid hydrocarbons were investigated intensively by Musayev and Chernikhin [128]. A comparison of the liquid hydrocarbons with the compositions given in Table 5.37 shows that they form hydrates at a lower temperature and a higher pressure

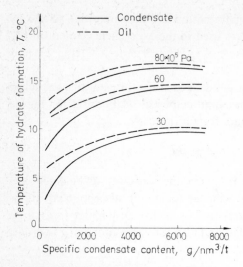

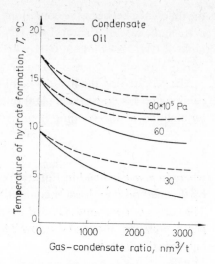

Fig. 5.75. Effect of gas–condensate ratio on hydrate formation from natural gas [462]

Fig. 5.76. Effect of specific condensate content of natural gas on hydrate formation [462]

Table 5.37. Gas compositions relating to hydrate formation demonstrated in Fig. 5.77 [128]

Components	A	B
	(mole %)	
Nitrogen	0.16	—
Methane	—	0.26
Ethane	7.10	5.00
Propane	90.84	92.70
n-Butane	0.33	1.77
Isobutane	1.47	0.27
Pentane	0.10	—

than do the gases, but on enrichment of the heavy components the stability of the hydrate in the investigated system increases. This is illustrated in Fig. 5.77.

In an investigation of the influence of liquid hydrocarbons on the conditions of hydrate formation, the properties connected with the retrograde condensation of the hydrocarbon systems must not be neglected. Reference to this can be found in the paper of Katz and Bergman [463].

The studies by Ng Heng-Joo and his co-workers [464–466] provide detailed information with regard to the prediction of hydrate formation in three-phase water–liquid hydrocarbon–hydrate systems.

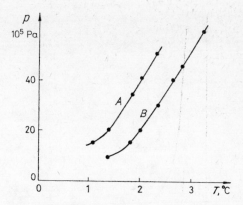

Fig. 5.77. Dependence of hydrate stability
on quantity of condensed phase [128]

Hydrate formation by natural gases at high pressure was studied by McLeod and Campbell [112]; up to the 1950's, the pressure limit of 2.84×10^7 Pa was generally not exceeded. Besides the parameters $\log p = f(T)$ characteristic of the hydrate formation, this research work attempted almost without exception to establish relationships involving the gas density for the phenomena relating to hydrate-separation, and it was for essentially this purpose that the vapour–solid equilibrium data were used. The only series of high-pressure experiments preceding the studies by McLeod and Campbell was performed by Kobayashi and Katz [65], but even they worked merely with pure methane up to a pressure limit of $7.97 \times \times 10^7$ Pa.

Subsequently, many authors tried to adapt to the field of high pressures by extrapolating the equilibrium data measured at lower pressures. However, the resulting data involved large errors, and doubt can therefore be cast on them; for example, as may be seen in the $\log p$ vs. T curves of Fig. 5.27, the slopes of these curves generally change at higher pressures.

From systematic investigations carried out up to a pressure of 7×10^7 Pa, it emerged that the hydrate equilibrium curves for the natural gases generally intersect that for methane at pressures exceeding the values to be found in the literature. This phenomenon was explained in terms of the pressure-sensitivity of the hydrates. It was also established that, in hydrate formation by natural gases, the effects of the butanes are insignificant at such high pressures; if anything, they rather exert only an inhibitory effect, and do not influence the pressure-sensitivity of the crystal lattice.

These researches prove that calculations made with the vapour–solid equilibrium constants give satisfactory results so long as the equilibrium curves for the hydrates of the natural gases run parallel with that for methane hydrate.

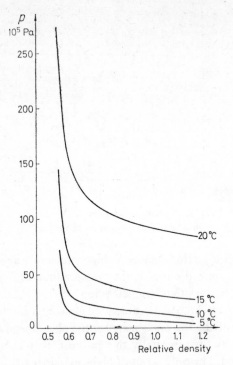

Fig. 5.78. Isotherms of hydrate formation from natural gases with different relative densities [47, 64]. Reprinted by permission of the SPE-AIME

For the natural gases, the limiting value was generally found at a pressure of about 2.8×10^7 Pa.

For the determination of the equilibrium conditions relating to pressures in excess of 2.8×10^7 Pa, a semi-empirical relationship, obtained via the Clausius–Clapeyron equation, was proposed by McLeod and Campbell [112]:

$$C = mT^2 \qquad (5.30)$$

The interpretation of this equation was discussed in detail in Section 4.1.

In the course of the production and transportation of natural gases, expansion processes take place of necessity. The cooling accompanying the expansion can lead to hydrate formation in the case of natural gases containing water. The natural gases employed in the practice of the gas industry have relative densities in the range 0.6–1.0, depending on the composition. This means that, with log p plotted against T, their hydrate formation equilibria generally lie in the area between the equilibrium curves for methane and ethane hydrates. Katz [47] experimentally determined the conditions of hydrate formation for natural gases with various specific weights, and he compared the measured data with the values determined from the equilibrium constants. These results led to the isotherms in Fig. 5.78 for natural gases with various densities. The characteristic data on the natural gases investigated are listed in Table 5.38. These data reflect the results

Table 5.38. Composition vs. density data for natural gases investigated [47]

Calculated relative gas density	0.603	0.704	0.803	0.906	1.023
Composition	mole fraction				
Methane	0.9267	0.8605	0.7350	0.6198	0.5471
Ethane	0.0529	0.0606	0.1340	0.1777	0.1745
Propane	0.0138	0.0339	0.0690	0.1118	0.1330
Isobutane	0.00182	0.0084	0.0080	0.0150	0.0210
n-Butane	0.00338	0.0136	0.0240	0.0414	0.0640
Pentane and higher homologues	0.0014	0.0230	0.0300	0.0343	0.0604
Pressure of hydrate formation, in 10^5 Pa, at 10 °C:					
calculated	32.66	22.83	17.95	14.35	13.50
measured	31.81	22.48	18.45	15.55	13.15

of earlier work too [357, 467–470]. With the aid of Fig. 5.78, the diagram in Fig. 5.79 was plotted up to an upper pressure limit of 2.80×10^7 Pa. These curves give (at the corresponding pressure) the minimum temperatures to which the natural gases with given density can be cooled without the danger of hydrate formation during their expansion. In the construction of the enthalpy lines valid for free expansion, the data of Brown [48] were taken into account at the various gas densities. Figure 5.80 refers to the expansion of a natural gas with a relative density of 0.6, expanding from a starting condition with a pressure of 1.4175×10^7 Pa and a temperature of 43.3 °C (A), or a pressure of 1.3476×10^7 Pa and a temperature of 48.8 °C (B). The point of intersection of a curve with the hydrate formation curve reveals the limiting situation, when formation of the hydrate of the given gas begins.

In the knowledge of the entropy and enthalpy data, the initial and final pressure and temperature value pairs can be plotted for natural gases with various densities; these indicate the possibility of expansion without the danger of hydrate formation. Such a diagram for a natural gas with a relative density of 0.6 is shown in Fig. 5.81. At higher pressures the curves turn back: in the course of the expansion

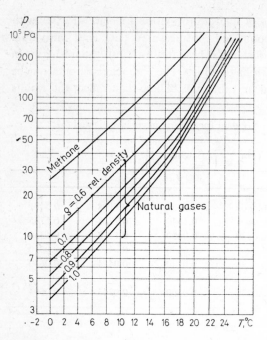

Fig. 5.79. Limiting curves of cooling without danger of hydrate formation during expansion for natural gases with different relative densities [47]. Reprinted by permission of the SPE-AIME

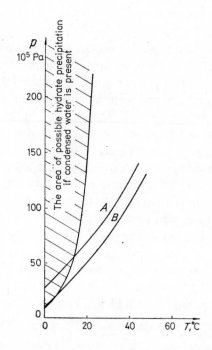

Fig. 5.80. Determination of conditions of hydrate formation for natural gas with a relative density of 0.6 on expansion from initial temperatures of 43.3 °C (*A*) and 48.8 °C (*B*) [47]. Reprinted by permission of the SPE-AIME

of natural gases starting from a pressure in the interval 3.50–4.25×10^7 Pa a temperature rise can be observed, and it is only when the pressure limit is passed that the temperature falls. In this pressure range, the compression factor of a natural gas decreases as a function of the temperature. From Fig. 5.81 the following conclusions can be read off:

(1) A natural gas with a starting pressure of 1.4175×10^7 Pa, a temperature of $38\,°C$ and a relative density of 0.6 can be expanded without the danger of hydrate formation up to about 7.4×10^6 Pa.

(2) If the previous gas has an initial temperature of $60\,°C$, then no hydrate formation occurs during the expansion, not even until atmospheric pressure is reached.

(3) For a gas with a relative density of 0.6 to expand from a pressure of $1.065 \times \times 10^7$ Pa to 3.55×10^6 Pa without the danger of hydrate formation, the starting temperature must be at least $40\,°C$.

(4) If the starting parameters of a natural gas with a relative density of 0.6 are 7.093×10^7 Pa and $60\,°C$, its temperature at 3.04×10^7 Pa too will be $60\,°C$,

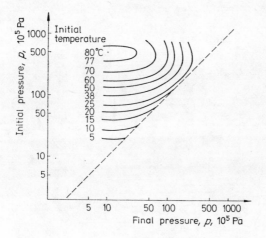

Fig. 5.81. Initial and final pressure–temperature data on expansion without hydrate formation for natural gas with a relative density of 0.6 [47]. Reprinted by permission of the SPE–AIME

and for this gas in the presence of water the hydrate will appear at a pressure of 8.6×10^6 Pa.

From similar diagrams for natural gases with various densities, Katz [47] found the method suitable for the determination of the conditions of free expansion of natural gases with any desired density by means of interpolation, but he drew attention to the fact that these curves can be accepted only for natural gases with compositions and densities similar to those listed in Table 5.38. Of course, these considerations are only valid for the case when the natural gas contains enough moisture for the possibility of hydrate formation to exist.

The theoretical and practical questions concerning the conditions of hydrate formation from the gases or gas mixtures discussed in Chapter 5, the structure formed, the inhibitors that may be used and their mechanisms of action in the case of a necessary change in the formation conditions, etc. can be interpreted on the basis of the considerations in the previous chapters.

Chapter 6

SOME TECHNOLOGICAL ASPECTS OF THE ROLE AND APPLICATION OF GAS HYDRATES

In the previous chapters, mention has already been made of the role of gas hydrates in some technological processes: e.g. in Chapter 1 in connection with the literature survey, in Chapter 3 with regard to the stability and the possibilities and modes of influencing this, and particularly in Section 5.3. The effects of gas hydrates are frequently so extensive that, especially in certain technological phases of the gas industry, such as the production and transportation, the appearance of gas hydrates causes very serious technological problems, and is regarded as a phenomenon that is undesired and to be avoided. It is not surprising, therefore, if the choice of subject in a great proportion of the publications connected with the hydrates of the constituents of natural gases (excepting the explicitly theoretical papers) involves the pressing need for the scientific clarification of such technological questions and the establishment of suitable parameters for the technological processes.

During the past twenty years, a particularly great influence has been exerted on the development of the technological role and application of gas hydrates, and on the further intensive investigations of the conditions connected with this, by the oil and gas production begun in the northern regions of the Earth (Siberia, Alaska, Canada); here, the temperature and pressure conditions in the depths are such as to give a possibility for the accumulation of the gas deposits from the beginning in the form of gas hydrates, and also for the occurrence of the undesired formation of gas hydrates during the exploitation. Very much attention has been paid to this question by Soviet researchers, whose work is well summarized by some short monographs [158, 471], a technical review [472] and a paper [473], but another significant publication on this subject is that of Katz [474].

In the northern part of Yakutia there is a very extensive permafrost area. In the zones where the thermal gradient is small, the natural gases can be found in the form of a solid hydrate in a layer of considerable thickness. The quantitative conditions of these gas hydrate deposits are determined by the thermodynamic features of the medium. The hydrates formed in such porous media have been subjected to investigation in numerous institutes of the Academy of Sciences of the Soviet Union and in many university laboratories.

According to Makogon and his co-workers [475], the hydrate zone in this area can be found in the upper layer of the sedimentary rock. Safir [476] stated that in the gas field of Mezoyak, for example, the hypsometric level of the gas–water contact is between -779 and -811 m. Test borings indicated that the hydrate formation temperature in this area is between 8.4 and 12.5 °C. The composition of the gas bound in hydrate form is 98.2–98.5% CH_4, 0.034–0.12% C_2H_6, 0.0015–0.006% C_3H_8, 0.00021–0.0019% C_4 hydrocarbons, 0.00027–0.0008% C_5 hydrocarbons, 0.3–0.5% CO_2 and 0.21–0.22% N_2.

It was similarly found by the same group [477] that in the districts of Krasnoyarsk and Tyumen in Siberia the gas deposits are embedded in a sandstone + limestone layer of about the same depth and age as the natural gas deposits in Uzbekistan; however, whereas the formation temperature range in Siberia is 9–31 °C, in Uzbekistan it is 37–59 °C. As the lower temperature favours the separation of drier fractions with higher methane contents, the Siberian gases are dry, in contrast with the gases in Uzbekistan, which have higher hydrocarbon contents. The hydrates of the hydrocarbons are absorbed in the sedimentary rock, and as it were seal off the gas pockets. In Northern Siberia, about 47% of the gas reserves are accumulated in the form of hydrate in such closed pockets.

The investigations of Khoroshilov and his co-workers [478] revealed that the sediment in Orenburg is characterized by an anomalously low formation temperature; this must presumably be attributed to the high thermal conductivity of the rocks covering the field. The gas composition of these reserves is as follows: 81.6–87% C_1, 3–5% C_2, 1.3–1.8% C_3, 0.23–0.37% iso-C_4, 0.45–0.74% n-C_4, 0.5–0.6% CO_2, 4–7.5% N_2 and 1.3–3.0% H_2S. The danger of hydrate formation is considerable during the recovery of this gas. The equilibrium temperature of hydration is influenced substantially by the hydrogen sulphide content of the gas. This was taken into consideration, and an empirical relationship was derived between the temperature (T) of hydrate formation, the gas pressure (p) in bar and the hydrogen sulphide content (c) in vol.%:

$$T = 14.7 \log p + 1.07c - 15.7 \qquad (6.1)$$

Information about the sites of hydrate formation in the permafrost layer (Sheshukov [479]) or in the thick sand layer serving as its foundation (Bily and Dick [480]) can be concluded from the low secondary γ-activity, the absence of negative anomalies in the specific conductivity curves, the low hypsometric level of the sediment, and the fact that there is no gas inflow to the positions of lower pressure. From their model experiments on the detection of the presence of sediments containing gas hydrates, Stoll and his co-workers [481] consider the observation of an anomalously high velocity of sound as conclusive evidence. On the opening of the wells in these areas, the well temperature is lower than in the production period. This fact is explained by Cherskiy and his co-workers [482] in that when the front zone is opened the mud chokes the gas to some extent; this results in a temperature decrease, which favours hydrate formation. This possibility is enhanced by the circumstance that in the gas–liquid flow that develops there is a very great turbulence. Protection against hydrate formation in the well head is provided by the addition of warmed mud or inhibitors.

The primary factors influencing hydrate formation are the overburden pressure, the temperature conditions, the salt content of the brine, and the energy of the capillary forces acting in the porous medium. The probability of occurrence of gas hydrates increases at high pressures and low temperatures. Reservoirs have been found in this porous medium in which 1 m^3 gas hydrate corresponded to 200 m^3 gas.

These factors are taken into consideration by Trofimuk and his co-workers [483, 484], who give a detailed account of how the depths of the hydrate zones vary at the poles, in the tropics and in the temperate or subtropical parts of the Earth. For example, the methane formed by the biological conversion of organic materials is transformed to hydrate at a depth of about 250 m in the neighbourhood of the poles, in contrast with a depth of at least 650 m in the tropics.

Tucholke and his co-workers [485] report that horizontal gas–water, gas–oil and oil–water interfaces can be detected in marine sediments from the deviations in the seismic reflection data. From their distribution map of the horizontal anomalies, they also determined the geological structures of areas under the sea, and the positions of the phase boundaries between gas and hydrate in the sediments.

The investigations of Makogon and his co-workers [486] revealed that a gas hydrate deposit can be found at a depth of only 120 m under the Arctic Ocean; this layer is mixed with other marine deposits, making it impermeable for the gas. Makogon [487] likewise published a diagram (Fig. 6.1) illustrating how the initial conditions of hydrate formation of various individual gases and of natural gas with a relative density of 0.6 depend on the depth of the ocean, as well as on the temperature and pressure.

Yefremova and Zhizhchenko [488] observed that the deposits of the Black Sea and the Caspian Sea contain large quantities of methane and carbon dioxide. This gas layer can be found 2–3 m under the sea-bottom. In core samples taken from a depth of about 2000 m, the existence of a hydrate layer with a microcrystalline structure was observed about 6.5 m below the bed of the Black Sea. The gases liberated as a consequence of the sudden pressure fall caused this structure to disintegrate, for the evolution of gaseous methane from its hydrate is accompanied by a volume increase of about 180 times. The gases released largely diffuse from the deposit towards the water, and accumulate at a depth of about 2–3 m under the sea-bottom. If the gases contain hydrogen sulphide and higher hydrocarbons too, it is found that the hydrates occur in separate strata, the hydrate of methane being situated about 500 m deeper, in a stratum with a temperature around 5 °C.

In the oceanic depressions, a pressure of 3×10^7 Pa and a maximum temperature of $+5$ °C are not rare at a certain depth. Under such conditions, the gases

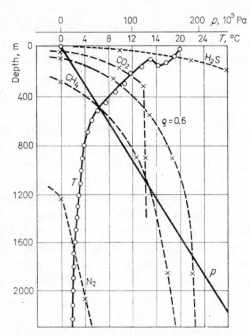

Fig. 6.1. Dependence of formation of hydrates of various gases on the depth of the ocean, the pressure and the temperature [487]

solidify during their migration or diffusion. The quantity of natural gases existing in the solid hydrate form in the Soviet Union is estimated to be about 10^{14} m³. This gas can be recovered and utilized only if the gas incorporated in the hydrate can be freed under the formation conditions. The equipment necessary for the exploitation and pipeline transportation would then not differ appreciably from that used in normal gas wells.

The decomposition of gas hydrates under the formation conditions can be achieved by means of a number of methods. One of the solutions is to lower the formation pressure to a value such that the gas hydrate must decompose at the formation temperature. In another suitable solution, the hydrate deposit must be warmed up (in part or fully) so that it reaches the decomposition temperature of the hydrate at the given formation pressure. A third possibility is to introduce into the hydrate phase catalysts promoting the decomposition of the hydrate, preferably under the pay zone.

The decrease of the formation pressure would in theory be the simplest solution. However, if the temperature is too low and there are many heavy components in the natural gas, then a considerable pressure decrease must be ensured. Even so, the method is only successful if the layer has a good permeability, for otherwise the flow rate from the production well will be low.

During the partial or total warming of the gas hydrate layer, the gas evolved must be drawn off at once, so that the formation pressure does not rise too significantly, as a dangerous gas blow-out can occur if the pressure is allowed to build up. The deposits can be warmed by pumping in water, steam or hot gas, or by introducing materials under the pay zone which produce heat on reacting with the brine. The injection of hot thermal water under the pay zone appears to be the most economical procedure, but of course this can only be conceivable if there is a sufficient quantity of such thermal water in the vicinity.

Makogon and Khalikov [489] give a method of calculating the extent of hydrate decomposition and the resulting gas production on the thermal treatment of the deposits in the depths. For this calculation one must determine the number of gas wells necessary, lying at a finite distance from one another, starting from the effectiveness of the thermal treatment.

The Siberian permafrost layer is a swampy taiga of low temperature, covered with deep snow. The frozen surface layer is several hundred metres thick, so that the construction of gas wells and a pipeline system involves extraordinarily great efforts. In addition, the danger exists that the gas coming to the surface will be reconverted to hydrate in the well-head or in the pipeline, or that the bore hole will collapse in the course of the exploitation, because of the melting occurring in the neighbourhood of the operating well. As a result of these extreme difficulties, and the factors possibly disturbing the natural conditions irreversibly, the recovery of the gas from the solid pay zone is possible only if the work is planned, organized and carried out with very great care.

With regard to the gas hydrates present in the porous rocks in the Northern Siberian gas fields, Verigin and his co-workers [490] prepared a one-dimensional linear programme relating to their decomposition as a result of a decrease in pressure; this programme can be used to determine the fundamental characteristics of the process. In this way they obtained information on the movement of the hydrate surface referred to a stratum with unit cross-section, on the expected yield of gas suddenly released in the decomposition, and on the distribution of the pressure in the gas.

The methods of checking on gas hydrate formation during the opening-up and operating of gas wells in the northern territories were summarized in the book by Dyegtyarev and Bukhgalter [491], published in 1976.

In an earlier work, Bukhgalter [492] gave information about the methods for the control and inhibition of hydrates in the gas produced from the lower pay zone at Kaluga, used as a peak reservoir.

Recognition of the characteristic features of gas hydrates clearly demonstrated the potentiality of gas storage in porous media, and possibly in the form of gas hydrates, and a number of publications and proposals have appeared in this respect [50, 158, 350].

As the problems relating to gas hydrates centre almost exclusively on the natural gas industry in Hungary too, we shall present below a brief picture concentrating mainly on the technological role and possibilities of application of the gas hydrates in connection with the production and transportation of natural gas, and much less mention will be made of the other areas, e.g. the methods of water demineralization (sweet-water production), chemical technological methods connected with the gas hydrates, such as the possibilities of the separation of gases by means of gas hydrates, the possibilities for the increase of gas pressure without the use of a compressor, the possibilities of gas drying, the elimination of fogs and clouds, the prevention of blow-outs in coal mines by means of gas hydrate formation, etc.

6.1. Role and application of gas hydrates in the natural gas industry

Practical problems connected with the gas hydrates arise only in the natural gas industry, primarily in the production and transportation of the gas, in either the gaseous or the liquid state. At the same time, the application of the gas hydrate method is becoming of ever greater importance as regards gas storage in exhausted pay zones, the preliminary stages of gas purification, and the separation of gaseous and liquid components.

The hydrate features of the produced and transported natural gases are substantially influenced by their inert content; this means their content of nitrogen, carbon dioxide and hydrogen sulphide, present in a quantity of 5–10 mole % or even more. The occurrence of natural gases with an inert content is frequent in every part of the world, as emerges from the review by Török [493]. In Hungary, there are natural gases with an inert content in 200 production sites of 90 gas field sources.

6.1.1. Gas hydrates and natural gas production

The questions posing problems in natural gas production that are connected with the gas hydrates can be divided into two basic groups. One of these groups involves the following questions:

Can gas hydrate be formed in the porous rock structure of the deep-lying pay zone, and if so, under what conditions?

Are the conditions of hydrate formation generally influenced by the porous medium, and if so, to what extent and in what direction?

The questions in the other group:

What are the conditions for the formation of gas hydrate plugs in the tubing? In what way can the formation of these be eliminated?

Let us examine these questions in the above order.

6.1.1.1. The role of porous medium in gas hydrate formation

The question of the formation of gas hydrates in deep porous rocks has been dealt with in the greatest detail by the researchers of the Gubkin University for Oil Chemistry and the Gas Industry in Moscow. The work carried out there, or in collaboration with this university, is summarized in the three monographs written by Makogon and his co-workers [157, 158, 472], and in that by Dyegtyarev and his co-workers [471]. The former monographs mainly treat the elaboration of the theoretical and experimental bases indispensably necessary for the direction of the technological processes, whereas the latter has a much more marked technological aspect, dealing in detail with the practical means of solving the problems effectively.

The investigations by Makogon and his co-workers were particularly interesting, as they were the first researchers to carry out successful studies and calculations on the possibility of the existence, formation and accumulation of the gas hydrates in the deep-lying porous rocks. Makogon and Skhalyakho [472, 494] performed their experiments up to a pressure of 2.7×10^7 Pa, with various porous media consisting chiefly of quartz sand, in vessels with a volume of 300–980 cm^3 that satisfactorily ensured the constancy of the temperature and pressure. The quartz sand was taken from pay zones the thermodynamic parameters of which corresponded to the possibilities of hydrate formation, and which were subsequently appropriately saturated with gas and water. The process was controlled by the variation of the pressure and the electric resistance, but the gas composition above the hydrate and the temperature jumps occurring in the course of the hydrate formation process were also studied, and the process was even followed visually too.

Results obtained in this manner are shown in Figs. 6.2 and 6.3. Figure 6.2 illustrates the p vs. T curves resulting at low initial pressures for the hydrate formation process: for the hydrate of a gas with a relative initial density of 0.6012 in the porous medium (curve 1); for the same gas hydrate in the case of a free gas–water contact, i.e. not for the porous medium (curve 2); for the methane hydrate–water system in the case of a non-porous medium (curve 3); and for the methane hydrate–water system in the porous medium (curve 4). It is clear from the figure that the curves referring to the cases in the porous and the non-porous media do not coincide.

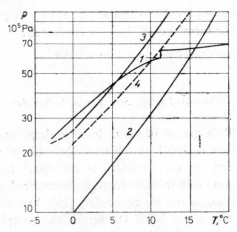

Fig. 6.2. Dependence of hydrate formation in porous and non-porous media on the pressure and temperature at low initial pressures [472]

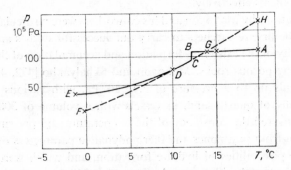

Fig. 6.3. Equilibrium conditions of formation of methane hydrates in cases of a free water–gas contact (curve *FH*) and a porous medium (curve *BCDE*) [472]

The same can be seen in Fig. 6.3 too, where curve FH shows the conditions in the non-porous medium, and curve BCDE those in the porous medium for the methane hydrate–water system. Point B indicates the initial pressure and temperature values for hydrate formation in the porous medium. Hydrate is formed in the porous medium in accordance with curve CD when $p_h < p_g$, i.e. the water vapour pressure above the hydrate (p_h) is less than the vapour pressure of pure water in the porous medium (p_g). Curve DE denotes hydrate formation in the porous medium when $p_h > p_g$.

Otherwise, Fig. 6.3 shows well that the appearance of the crystallization nuclei causes a decrease in the influence of the porous medium on the conditions of hydrate formation, or the influence can even be eliminated completely, for these conditions are then identical with those corresponding to the free gas–water interface (see section CD, where the two curves coincide). Consequently, in the presence of crystallization nuclei, the effect of the porous medium on the conditions of hydrate formation no longer exists if $p_h < p_g$, while if $p_h > p_g$ this effect on the hydrate formation process increases in proportion to the difference between the two pressures.

Makogon and his co-workers [472, 494] similarly demonstrated that the effect of the porous medium on the initial conditions of hydrate formation is the greater, the smaller the mean effective capillary radius, while it is the weaker, the higher the pressure and the larger the relative molecular mass of the gas.

In real gas pay zones, the distribution of the pores is clearly not uniform; there are always local sections and zones in which the capillary radii are greater by even one–two orders of magnitude than the mean value relating to the overall system. These sections will be the initial centres of gas hydrate formation in the porous medium, and from these centres the hydrate formation will subsequently spread to all those sites in the gas source which display characteristic values corresponding to the thermodynamic conditions of hydrate formation under the given circumstances.

If the condition $p_h < p_g$ holds, then the gas source will undergo transition to the gas hydrate structure under hydrate formation conditions similar to those in the case of a free gas–water interface (i.e. a non-porous medium). However, if $p_h > p_g$, and even if crystallization centres are present that originated in the micropores, the transformation of the gas source to the gas hydrate will occur under thermodynamic conditions which correspond to the vapour pressure above the hydrate, p_h. Thus, if $p_h < p_g$, the extents of gas hydrate sources can be determined from the conditions of their formation under thermodynamic parameters which correspond to hydrate formation in the case of the free gas–water interface (i.e. not in a porous medium).

For the prognostication of gas sources involving gas hydrates, and hence for their actual research and exploitation, analytical methods were elaborated by Makogon and his co-workers by means of the statistical thermodynamic procedures detailed in Section 4.2 [158, 472, 495–497]. In their method, from a knowledge of the decrease in the saturation vapour pressure of the water (or of the solution if brine is present) the initial equilibrium conditions of hydrate formation (i.e. the related p and T values) which correspond to the thermodynamic conditions for the existence of the hydrate can be determined for any gas mixture of known composition and for a porous medium with any degree of liquid saturation:

$$\ln \frac{p_{H_2O} - \Delta p_{por}}{p_{H_2O}^{\circ}} = \frac{n}{(1+n)m} \ln \left(1 - \sum_{X,Y,Z} \theta_1\right) + \frac{1}{(1+n)m} \ln \left(1 - \sum_{X,Y,Z} \theta_2\right)$$

$$(6.2)$$

In this equation, p_{H_2O} is the saturation vapour pressure of water, $p_{H_2O}^{\circ}$ is the same above the theoretically vacant crystalline hydrate lattice, Δp_{por} is the decrease in the saturation vapour pressure above the water in the case of a capillary elevation in the porous medium, θ_1 and θ_2 are the degrees of occupancy of the various cavities, n is the ratio of the numbers of large and small cavities, and m is the number of water molecules per cavity.

Equation (6.2) can be solved both analytically and graphically, by taking the temperature-dependence of the left- and right-hand sides of the equation. In this case, the values corresponding to the point of intersection of the curves described by the two functions give the equilibrium conditions of hydrate formation in the porous medium. The calculations may be simplified if the following equation is used:

$$\log p_{H_2O}^{\circ} = a + b \log T - \frac{c}{T}$$

$$(6.3)$$

where the values of a, b and c for the hydrate structure H_I are $-47.35\,031$, $+20.22\,408$ and 299.8385, respectively, and for the hydrate structure H_{II} are $-52.7149, +22.09\,367$ and $+84.09\,853$, respectively.

With the aid of the equilibrium data obtained for the initial conditions of hydrate formation, and from their changes, the possible radius of the hydrate formation zone can therefore be determined in the vicinity of the well, as also can the hydrate quantity collected during the operation of the well for a given time interval, and even the permeability decrease occurring as a consequence of the hydrate formation can be evaluated. The knowledge of all these factors evidently facilitates the planning, and gives a possibility for the practical use of methods suitable for the prevention of gas hydrate formation.

Ivanov and his co-workers [496] found that the relative permeability, i.e. the ratio of the permeabilities in the presence and in the absence of hydrates (K_h

and K_0), can be expressed in terms of the hydrate saturation (S_h) and water saturation (S_w) of the pore volume, according to the following relationship:

$$\frac{K_h}{K_0} = \left(1 - \frac{S_h - S_w}{1 - S_h}\right)^2 = (1 - S^*)^2 \tag{6.4}$$

To summarize, we can now answer the two questions asked at the beginning of this section. In both cases the answer is a positive one: under appropriate conditions, there is a possibility for the formation of gas hydrate in the reservoirs in the depths, and (if $p_h > p_g$ holds) the porous medium does influence the initial conditions of hydrate formation.

Gas hydrate formation in the depths at the same time means a greater gas reserve in a given volume than if the gas were present only in the gaseous state in the pay zone. Makogon and his co-workers [496] established that the gas reserve in the gas hydrate state may be 4–9 times more than that in the pure gas state.

A number of authors have investigated the conditions of gas hydrate formation on the surface of porous rocks, predominantly zeolites [498–503]. Mention must be made in particular of the research by Shcherbakova and Byk [500–503].

6.1.1.2. Conditions of gas hydrate plug formation in the tubing, and the methods employed to prevent this

The problem outlined in the title is important mainly in the gas sources in the northern regions. Here, the relatively thick upper layer of the soil is virtually permanently frozen, and hence the gas migrating upwards from the depths under high pressure, and always containing moisture, can easily attain conditions permitting the formation of gas hydrates. This formation can sometimes occur even in regions with a more moderate climate, particularly at the time of hard frosts. In their monograph, Dyegtyarev and his co-workers [471] present a detailed review of the investigations (their own and those of other authors) connected with the solution of this problem.

The basic activity towards the prevention of gas hydrate formation requires a general knowledge of the conditions of hydrate formation, including hydrate composition, moisture content, pressure and temperature (these and their determination were discussed in Chapters 3–5), from which it is possible to predict the depths at which hydrate formation is likely. For this, of course, the temperature and pressure distributions in the well must also be known.

Let us now examine, in accordance with these data, the diagrams for a given gas, a given gas well and a given yield in the cases of a non-operating well (Fig. 6.4) and an operating well (with a gas volume flow rate of 9×10^5 m^3/day (Fig. 6.5). In these figures, curve 1 shows the temperature distribution and curve 2 the

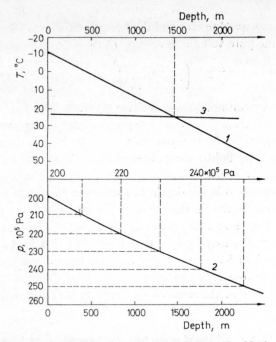

Fig. 6.4. Diagram for determination of the maximum possible depth of hydrate formation in the case of a non-operating gas well [471]

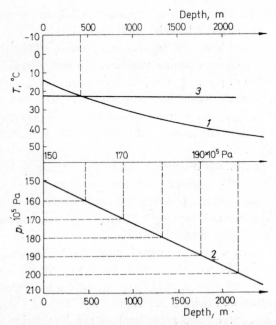

Fig. 6.5. Diagram for determination of the maximum possible depth of hydrate formation in the case of an operating gas well (with a gas volume flow rate of $9 \times 10^5 \, m^3 \, day^{-1}$) [471]

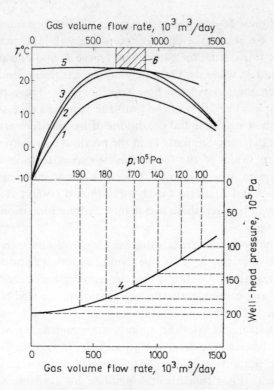

Fig. 6.6. Diagram for determination of the gas volume
flow rate of a gas well that makes it possible to
avoid gas hydrate formation [471]

pressure distribution as functions of the depth of the well (T vs. m and p vs. m
curves), while the p vs. T curve *3* gives the initial conditions of hydrate formation. The T vs. m and p vs. m curves are first recorded separately as in the diagrams, the p values are then transferred to the new abscissa in the middle, and the
p and T values relating to the initial conditions for the given gas (which can be
obtained as described in the previous chapters) are replotted in the upper p vs.
T diagram. This gives curve *3*, and the depth value at which this intersects curve *1*
is the sought maximum depth of hydrate formation. It can be seen from the two
figures that this depth is much greater for a non-operating than for an operating
well.

In the planning concerning the prevention of the development of hydrate plugs,
it must be taken into account that it may be possible to ensure a well gas volume
flow rate at which hydrate formation can be avoided. For the determination of
this, the dependence of the pressure and temperature on the gas volume flow rate

at the well head must be established experimentally or by calculation. The upper diagram in Fig. 6.6 shows the dependence of the well-head temperature on the gas volume flow rate during the experiment (curve *1*) and one and three months later (curves *2* and *3*), while the lower diagram depicts the dependence of the well-head pressure on the gas volume flow rate (curve *4*). These provide the basic data for the graphical solution of the problem. It is next necessary to plot the *p* vs. *T* function showing the initial conditions of hydrate formation for the given gas in the upper diagram. Similarly as in the previous two figures, this is done by transposing the *p* values of the lower diagram so as to form the new abscissa of the upper diagram, where the *p* vs. *T* curve is constructed (curve *5*). It may be seen that curves *3* and *5* intersect each other in two points; at gas volume flow rates in the interval between these two points, hydrate formation can be avoided, or is not probable.

However, the figure also reveals that such a situation does not yet exist when the well has been operating for only one month, as curves *2* and *5* do not intersect each other; the explanation of this is that the temperature of the pay zone around the well-bottom and consequently the temperature of the well-head have not yet risen sufficiently.

In general, therefore, besides the appropriate maintenance of the gas volume flow rate, it is definitely necessary, at least initially, to vary a minimum of one of the three basic conditions of hydrate formation (high pressure, low temperature, free water content). Four fundamental methods are available for this: the decrease of the pressure, the increase of the temperature, the drying of the gas, and the application of inhibitors; however, other special methods too exist.

The use of inhibitors and the principles behind this have been dealt with in detail in Section 3.2, and here, therefore, we shall mention only some questions relating to the technology. As regards the injection of the inhibitor, it is clearly essential that the inhibitor should enter the flowing system before the site of gas hydrate formation. In the event of the danger of gas hydrate formation in the tubing, this means that the inhibitor (salt solution, glycol, methanol) must be pumped to the well-bottom, either through the space around the tubing, or through a separate pipe (e.g. [504]) at the end of which a suitable device ensures the spraying of the inhibitor solution. The use of spraying is generally accompanied by a higher efficiency.

In the choice of the inhibitor, not only the theoretical efficiencies of the individual inhibitors need to be considered. Roles are played by the most varied factors, e.g. the method and possibilities of production, the transportation capacity, the regenerability, the toxic effect, the possibility of individual or centralized conduction of the inhibitor into the tubing, the corrosion phenomena to be expected, the possibility of the use of the inhibitor solution as a heat carrier, the requirements in the various technological phases (production, collection, preparation, transportation, etc.), the difficulties created by the weather in the North, such as the necessity for thermal insulation, the possible obstacles to transportation, etc. Only in the overall knowledge of the given situation and of the above-mentioned factors is it possible to decide on the inhibitor; a generally valid solution to this problem is not to be expected.

Otherwise, the inhibition method is used mainly for the prevention of hydrate formation. If a hydrate plug has already developed, this method can be applied only if the gas hydrate has not blocked the total cross-section of the tubing. Such a hydrate plug can be eliminated by the single injection of an adequate amount (200–400 dm^3) of inhibitor solution (usually methanol or $CaCl_2$), the obstruction subsequently being blown out into the air after a suitable time.

If the hydrate formed has already blocked the total cross-section of the tubing throughout a long section, and particularly if it contains solid particles such as sand, etc., which hinder the decomposition, some other method (in general thawing-out) must be applied.

In this respect, various technical procedures are available: as the most effective and simplest methods, the thawing-out can be achieved by the circulation of warm gas or warmed inhibitor solution, or with pipe thawing-out equipment built in initially.

In the exploitation of oil with auxiliary gases in the Western Siberian oil-fields, production has been hampered by an unusual phenomenon since the middle of the 1970's. It has been reported by Fazlutdinov and his co-workers [505] that hydrate formation can be experienced in the pump–compressor system at depths of 400 m or even more, primarily in the lower-yield wells. The danger of hydrate formation is enhanced when the wells are put into operation following drilling, and when the wells are started up after repairs. In one well, which produced 72.73 % brine, 18.72 % oil and 8.55 % gas, about 42 % of this gas being propane, after repairs the well could be started only with the aid of heating. In the depth interval 190–280 m, the formation temperature drops to −3 °C and the system cooled down still further because of the Joule–Thomson effect; thus, the possibility of hydrate formation arose, as a consequence of which a pressure drop of 20–60 bar was observed in the tubing. Although the gas used to expel the oil was dried, hydrate formation became increasingly more frequent as the site grew wetter. Fazlutdinov [506] describes attempts with several methods to break down the hydrate plugs formed; the most effective proved to be the incorporation of a 10 kW heater. This was a special heater, with a diameter of 38–44 mm, a length of 1100 mm, and a weight of about 10 kg, operated at a potential of 220 V and a current of 30 A. With this, the liquid was heated up to about 150–180 °C. The heater was brought into operation at a pressure of 60 bar.

Besides the two basic methods mentioned above, there are other special methods too. One of these is the coating of the wall of the tubing with a thin hydrophobic layer. This serves two purposes: it decreases the surface roughness of the pipe and hence mechanically reduces the adhesion between the pipe and any gas hydrate formed; in addition, because of the hydrophobic properties of the coating, the water does not undergo adsorption on the pipe wall (or to a very diminished extent), and the crystalline hydrate can therefore be formed only under stricter conditions which do not occur during normal operation [507, 508]. For this purpose, crude oil has proved to be the most suitable, particularly the types which

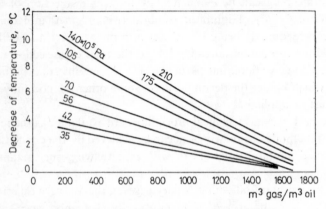

Fig. 6.7. Effect of the addition of crude oil to the gas
on the temperature of gas hydrate formation [471]

contain much asphaltene and tar, but experiments have also been carried out on the use of surface-active materials.

The pipe wall can be coated with a hydrophobic oil film by preliminary treatment with oil. In this case, the treatment must be repeated after a certain time (generally 4–7 days) because the flowing system, the small solid particle in it, and the small hydrate grains formed and moving upwards, gradually wear off the layer in question from the wall.

The hydrophobizing agent can alternatively be applied by direct addition to the gas. Figure 6.7 shows the effect of the addition of crude oil to the gas on the temperature of hydrate formation [471]; it can be seen that the more the oil in the gas and the higher the pressure, the greater the decrease in this temperature. However, the application of this latter method demands a certain degree of caution, for the oil can be the cause of problems in the subsequent processing and preparation.

The other methods serving for the prevention of hydrate formation in the tubing are based not on the effects of the addition of further components to the gas, but on thermal effects, i.e. the warming-up of the whole of the tubing, or of part of it only [509–512]. This can be done by combustion or electric methods, by the circulation of warmed liquid hydrocarbons (petroleum, diesel oil) or naturally hot thermal water, or even without these procedures, by simply preventing cooling by means of adequate insulation of the pipe (with a vacuum mantle possibly also containing porous materials).

Many literature references can be found that give descriptions of the process of hydrate formation in the tubing, and the means of preventing this.

Medovskiy [513] reported the incorporation of a special protecting pipe + lifting pipe pair, which allowed partial substitution of the mud by a solution containing 25–30% $CaCl_2$. Cherskiy and his co-workers [514] similarly used a mud containing $CaCl_2$ in the perforation of the well, but they also injected appropriate quantities of methanol, ethylene glycol or a mixture of these into the pay zone. Kolodezni and Arshinov [515] concluded that 5–10 m^3 inhibitor was necessary in the case of a gas well with a gas flow rate of $1–3 \times 10^6$ m^3/day, when either 30% $CaCl_2$ solution or a 1:9 mixture of methanol $+30\%$ $CaCl_2$ was used, for the well to be operated without the danger of hydrate formation; the temperature of the well-head was then high enough to ensure that no hydrate formation could occur. For the safe operation of a well in a gas field with a temperature of $-60\,^\circ$C, Saifeyev [516] introduced the methanol through a compressor line by means of a counter-flow feeder with an overflow system. A ball-check valve system was built into the line; this began to work if the gas pressure decreased by 10–15 bar. With this feeder, the methanol requirement was 0.5 dm^3/1000 m^3 gas. A continuously operating inhibitor-feeder device was described by Kiyko and his co-workers [517]. Their apparatus operated at a pressure of 320 bar, with 6 and 12 feeder throats. The maximum discharge rate of the pump was 50 dm^3/min, and that of the individual nozzles was 0–1.6 and 1–6.3 dm^3/min. Inhibition with methanol was similarly applied in the recovery of gas hydrates from geological sediments by Vasilyeva and her co-workers [518]. An automatic control method of alcohol injection was reported in the dissertation by Clark [519]. Ciemochowski and Bishop [520] constructed a sensing system which continuously indicated the moisture content and temperature of the gas stream, and also started the injection of hydrate inhibitor when necessary.

6.1.2. Gas hydrates and the transportation of natural gas

As regards natural gas production in the temperate climate zone, the main danger of hydrate formation arises not in the underground tubing, but in the separator or collector and in the preparation or transportation system starting from the well head. The first and most important method for the prevention of gas hydrate formation here is the elimination of the most fundamental conditions for this, i.e. the fixation of the free water content of the gas by adsorption or absorption procedures, possibly through chemical reactions, or by a freezing-out method, where the fixation and separation of the water content of the gas can be performed via gas hydrate formation [81]. If the drying is not complete and the gas does not contain inhibitor, gas hydrate formation can occur because of a local pressure increase, in the cold sections after pipe constrictions, in the case of cooling after the choke valves, or because of poor thermal insulation, etc.

In order that technological troubles connected with the formation of gas hydrates should not arise during the transportation of the gas in the pipelines (and mainly to prevent such formation), in addition to the knowledge of the conditions of hydrate formation (these have already been treated in detail in the previous chapters), it is of the greatest importance for both the designer and the operator to be able to establish the possible or actual sites of hydrate formation, and to estimate the time and rate of hydrate formation, as well as the quantity formed. For this, it is necessary above all that the composition of the gas, and the variations in pressure, temperature and moisture content along the pipeline (corresponding to the technological operation) should be known.

These data are presented in Fig. 6.8 [521] for a pipeline with a length L of 150 km, without any sectional narrowing, transporting an inhibitor-free gas with a relative density of 0.6, with a starting temperature of 40 °C and a starting pressure of 5×10^6 Pa. The curve consisting of alternate points and dashes gives the change in the saturation water vapour content of the gas as a function of the distance (this can be calculated on the basis of Fig. 5.70); the dotted curve illustrates the dependence of the equilibrium temperature of hydrate formation on the distance (this can be calculated on the basis of Fig. 5.79 [65, 361, 521]); while the dashed line presents the dependence of the water vapour content of the gas on the distance, and at the same time clearly shows the first and second possible sites of hydrate formation in the horizontal sections.

It is evident that the condensation of water vapour begins at point 1, without the formation of gas hydrate. The water vapour content of the gas thereby decreases, and the formation of gas hydrate starts at the water vapour content where the T vs. L and the $T_{\text{hydrate form.}}^{\text{equilibrium}}$ vs. L curves intersect each other, i.e. where the two values become equal and the gas is saturated too. In the present case, this will be achieved after a distance of 50 km. In the course of the formation of the gas hydrate, of course, the moisture content will become even lower, as the water is consumed in the formation of the gas hydrate lattice. Gas hydrate formation can similarly be expected at a distance of about 75 km, corresponding to point 4, because the moisture content of the flowing gas here will again be the same as the saturation water vapour content of the gas. At distances greater than this, hydrate forma-

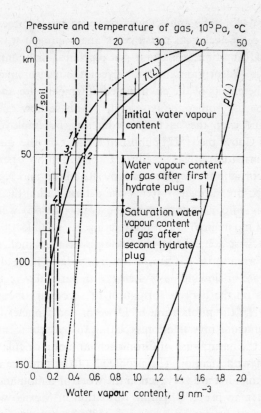

Fig. 6.8. Diagram for determination
of the sites of hydrate formation in a pipeline [521]

tion is no longer possible, for the moisture content of the flowing gas later remains lower throughout than the saturation water vapour content of the gas.

If the quantity ΔV of water condensing from 1 m³ gas at given points of the pipeline, the density of the hydrate ($\varrho = 0.9$ g cm⁻³) and the quantity Q of transported gas (in units of m³ day⁻¹) are known, the rate of hydrate formation can be calculated from the following equation:

$$v = \frac{\Delta V}{\varrho} Q \qquad (6.5)$$

For a gas transported in a quantity of 2×10^6 m³ day⁻¹, this is 0.334 m³ day⁻¹ at the first hydrate plug (where $\Delta V = 0.25$–0.4), while it is 0.177 m³ day⁻¹ at the second hydrate plug (where $\Delta V = 0.17$–0.25).

The possibility of gas hydrate formation in natural gas pipelines was first dealt with from the end of the 1930's. In this the pioneering work was performed by

Deaton and Frost [42]. The subject was reviewed by Hammerschmidt [301] in 1969; he analyzed the technical possibilities and methods of control connected with gas hydrate formation in natural gas pipelines, including the preliminary drying of the gas by adsorption and with glycol, the measurement and regulation of the pressure, the effects of the environmental temperature and of the warming of certain sections for both one-phase and two-phase flow (gaseous and liquid hydrocarbons), and even the safety procedures to be employed.

Kedzierski and Chowaniec [522] published a paper on the quantitative evaluation of hydrate formation in gas pipelines. Information on a new method of predicting hydrate formation was provided by Robinson and Ng Heng-Joo [523]. In the English high-pressure gas systems, the determination of the dew-point and its control (taking into account retrograde condensation) were considered by Cooper [524] a fundamental question relating to reliable operation. Hydrate inhibition was likewise achieved by the addition of methanol. Kinetic investigations on hydrate formation in gas pipelines were made by Babe and his co-workers [525, 526] in the case of flowing gas systems, with particular regard to the evaluation of the effects on the hydrate formation of secondary factors arising in the course of the flow. In the publication of Dewerdt and Roncier [527], an account is given of experimental investigations relating to hydrate formation during the transportation of the natural gas of Groningen, and of a calculation method concerning inhibition with alcohol. The problems of the pipeline transportation of natural gases in Romania are surveyed in the work of Simon and Debrescu [528]. The measures taken to prevent hydrate formation in Mexico were described by Mendez [529], while Khristov and Arizanov [530] have dealt with this topic from a Bulgarian aspect. Of the many patents in the USA, in that of Harber [531] from 1975, hydrate formation from natural gases is prevented by extraction of the moisture content of the gas with 60–85 wt. % methanol as a liquid drying medium. The conditions of the operation are also given.

Detailed information has been provided by Connealy [532] about the operating conditions without hydrate formation in a gas pipeline in Eastern Texas, and about the technological aspects of the inhibition required.

Chernishev and Boiko [533] described a special device connected to the producing equipment; this links a vertical container filled with alcohol or glycol with a valve system at a control point. A tube 30 mm high, with a diameter of 0.08–0.09 mm and with a calibrated opening, is built into this container. As a result of the pressure fluctuation occurring because of hydrate formation, this tube is filled to a corresponding extent with inhibitor, which thus enters the pipeline system. The opening in the feed pipe is determined experimentally in accordance with the inhibitor requirement. Operating experience demonstrated that in one well 1–2 drops alcohol was required in a period of 3–5 minutes, which resulted in an alcohol consumption of 500 cm³ in a month. The inhibitor container was recharged monthly. It proved possible to solve the remote control and automatic checking of the device. Industrial experience is similarly reported in the papers of Malenkov and his co-workers [534], Mamedov [535] and Nikitin and his co-workers [536].

In addition to its inhibitory effect, methanol is also suitable for a certain degree of purification of the gas: thus, primarily the absorption of the inert carbon dioxide and hydrogen sulphide can be ensured under appropriate operating conditions. Such experiments were described by Hinton and Kurata [537] and by Pekhata and his co-workers [538].

Despite the good efficiency of inhibition with alcohol, the problem of regeneration involves difficulties in many cases. Industrial experience relating to this is reported by Herrin and Armstrong [539], Smolyaninov and Khadikin [540], Lisovoder [541] and Tinnikov and his co-workers [542]. Kolesnikova and Legezin [543] drew attention to the problem that methanol exerts a salting-out effect on the salt content of brines originating from gas wells; this results in clogging of the regeneration apparatus, while in addition the danger of corrosion is increased. Various additives were tested in an attempt to eliminate this. The best result was achieved by the addition of 0.5% Na_3PO_4. The presence of phosphate did not influence the regeneration of the alcohol, but it partially prevented the precipitation of the salt and it also formed an effective protective film on the metal surfaces of the regeneration apparatus.

If brine occurs in production wells, this can be mixed with methanol to yield virtually complete additivity for the inhibition of hydrate formation. However, Burmistrov and Krasnov [544] claim that the phenomenon of salting-out is frequent, whereby the inhibitory effect deteriorates and the technological process too is negatively affected. They made a detailed analysis of the conditions of salting-out in the system of the source at Orenburg, and found that 70–80% of the precipitated salt is NaCl. The experience acquired here was utilized in the progressively wetter wells to diminish the salting-out; inhibition was carried out with appropriately diluted aqueous methanol.

Besides inhibition with methanol, the prevention of hydrate formation in pipelines can be solved by other means too. In the case of low-pressure gas pipelines, aqueous solutions of $CaCl_2$ exhibit a good efficiency [545], while at higher pressures the injection of various anti-corrosion additives is frequent. Such solutions are reported in the papers of Khoroshilov and his co-workers [546, 547].

6.1.3. Application of gas hydrates for gas storage purposes

It has already been mentioned in the previous chapters that the volume of a given gas hydrate is much less than the normal volume of the gas from which the hydrate was formed. This fact clearly soon suggested the possibility that natural gas (and other gases too) could be stored at low pressure and temperature in the form of the hydrate. Good surveys of the conceptions and patents referring to this were given by Miller and Strong [50] and particularly by Parent [350] towards the end of the 1940's. According to the calculations of Parent, a natural gas with the composition 80% CH_4, 10% C_2H_6, 5% C_3H_8, 4% n-butane + isobutane and 1% inert gas (in mole %) has a volume of 4.42 m³ at a temperature of 15 °C and a pressure of 1.01×10^5 Pa. In the event of the storage of this gas in the form of its hydrate, only 1/156 of the volume in the free state is needed, i.e. 0.028 m³.

The most important energetic considerations in the production of gas hydrates are the cooling requirements involved in the formation of the gas hydrate and the heat demand in the reconversion to gas. Many efforts have been devoted to the development of economical methods of gas hydrate production, and as a result of these many patents relating to periodical or continuous operation have been granted. In this respect, the much lower storage space is obviously the main advantage. Another significant advantage, however, is that the storage in the form of gas hydrates is much safer than either in the gaseous form or in a state compressed to liquid. The reasons for this are that the hydrates are not inflammable and they do not burn, while in the reconversion to gas, the evolution of gas is relatively slow, it occurs at a not very elevated temperature and it can be controlled too, so that in this process combustion and explosion can be prevented.

This type of gas storage is particularly promising in the permanently frozen regions, where there is no special need for thermal insulation, even in the storage reservoirs at comparatively small depths. The parameters necessary from the aspect of gas storage in hydrate form under such conditions were determined by Dubinin and Zhidenko [548].

One of the applications of the storage of natural gases in the hydrate form is the transportation of these gases in this form; this is a possibility the advantages of which have not been fully exploited in many cases. Means of solving such tasks are reported by Cahn and his co-workers [549], and by Nierman [550] in his patented procedure for the marine transportation of gases.

6.1.4. Methods for the separation of components in the gaseous and liquid states by means of gas hydrates

The principle of the separation of gaseous components by means of gas hydrates is based on the fact that certain components form hydrates, whereas others do not. The most frequent reason why the latter gases do not form hydrates is that they are too large to be accommodated in the cavities in the water lattice, but other of their properties may also play a role, despite their size being satisfactory as regards hydrate formation. This question has been discussed in detail in Chapters 3 and 4. Hammerschmidt [41] reported data on the compositions of natural gases in the hydrate phase and in the gas phase, from which it was evident that the tendencies and extents of the transformations of different components from the gas phase to the hydrate phase are by no means identical; hydrocarbons higher than methane tend to pass over predominantly to the hydrate phase. On the basis of these data, von Stackelberg [69] proposed the separation of N_2 and H_2 in this way.

Utilizing the data compiled by Korotayev and Ponomaryov [551], Musayev [146] carried out calculations relating to three different p and T values, based on the equilibrium constant $K=y/x$ (where y is the mole fraction of the component in the gas, and x is that in the hydrate), and proved that, at a fixed gas composition, the quantities of the natural gas components in a given phase depend on the thermodynamic conditions (p and T): when the pressure and temperature are increased, the methane content of the hydrate phase increases, but the contents of the other components decrease. At the same time, he succeeded in demonstrating that a change in temperature has a greater influence than a change in pressure on the hydrate composition, this situation being more marked at lower temperatures. In accordance with this, as a result of his experiments with natural gases at a relatively high pressure and a relatively low temperature, he succeeded in enriching the initial propane + butane content of 0.04 m³/1 m³ natural gas to a value of 0.5 in the hydrate phase, i.e. a 12-fold enrichment.

In the case discussed above, the separation involved natural gas components all of which were capable of forming hydrates. If components that do not form hydrates are also present, e.g. the large pentane molecule, this cannot be incorporated into the water lattice because of its size, and this is the reason why separation can be achieved. On a similar basis, hydrogen sulphide too can be separated from natural gas [552].

The basic principle of the methods to be followed is that (under suitable conditions) the natural gases are mixed with water, when the C_1–C_4 paraffins form hydrates, as do carbon dioxide and hydrogen sulphide, while the higher paraffins and nitrogen remain in the gas or liquid phase. The solid hydrate is next separated in a filtration apparatus, and the pressure and temperature are subsequently changed appropriately, so that the hydrates are reconverted to the gases. If various organic liquids, such as halogen-substituted low paraffins, are previously added to the water, a better effect can be obtained, because the mixed hydrates are formed at lower pressures. With methods of this kind, satisfactory separations could be attained in mixtures of H_2 and C_2H_6 [553], in mixtures of n-butane and isobutane [121], in mixtures of propane and propylene [89], and in the separation of helium from natural gas by means of a two-stage hydrate-formation process [150].

A number of methods are known for the recovery of the helium content of hydrocarbons; the advantage of the hydrate procedure is that it can be employed economically at low positive temperatures too. In the presence of the hydrate-forming hydrocarbons, helium does not form a hydrate, but its small molecules may be incorporated as inclusions in the crystal lattice. The experiments of Tsarev and Savvin [554] in connection with the separation of such systems led to some interesting conclusions.

If the hydrocarbon–helium mixture is regarded as a two-component system, from thermodynamic considerations relating to the small and large cavities of the solid, crystalline gas hydrate the distribution coefficients of the components are

$$\alpha_1 = \frac{C_A^1}{C_B^1} \quad \text{and} \quad \alpha_2 = \frac{C_A^2}{C_B^2} \tag{6.6}$$

where C_A and C_B are the Langmuir constants of components A and B, which are known from the statistical theory of gas hydrates. In the hydrocarbon–helium system, α has a value of 100–150.

Via the experimental analytical data, the distribution coefficients were determined from the equation:

$$\alpha = \frac{y}{1-y} \frac{1-x}{x} \tag{6.7}$$

where y is the concentration of helium in the equilibrium gas phase, while x is the equilibrium helium concentration of the hydrate phase. When their results were compared with the former relationship, a value of

$$C = \exp\,(26.860 - 0.111\,T) \tag{6.8}$$

was obtained for the temperature-dependence of the Langmuir constant of helium. This research also demonstrated that the distribution coefficient of the helium decreases as the propane content of the hydrocarbon rises or when freon F12 is added to it, for propane and this freon both fit well into the large cavities in the hydrate. As the pressure of the gas is raised, or if components (e.g. hydrogen sulphide) fitting into the small cavities of the hydrate are added to the mixture, the

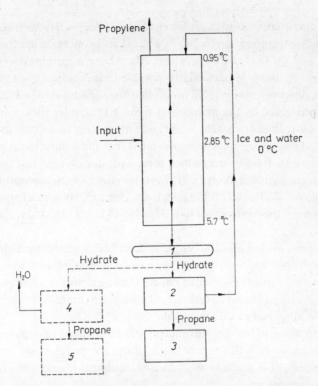

Fig. 6.9. Method for the separation
of liquid components via gas hydrate formation [147]

value of α increases. Their studies indicated that the value of the distribution coefficient was not influenced by an increase in the helium content of the mixture.

In a suitable apparatus, the method involving gas-hydrate formation is suitable for the separation of liquid components too. This possibility was first investigated by Barrer and Ruzicka [100], who achieved separation in the case of binary chloroform–benzene (toluene, carbon tetrachloride, methylene chloride, methylene iodide) mixtures in the presence of various auxiliary gases (air, methane, Kr, CO_2). A method involving gas hydrate formation was used by Glew [555] to separate a propane–propylene mixture. With his method, 98% of the propane was converted to the hydrate, while nearly pure propylene accumulated in the form of a liquid layer in the upper part of the column at the critical temperature of hydrate formation, 1 °C (Fig. 6.9 [147]). In the scheme to be seen in the figure, *1* is the container in which the hydrate mud is collected, the hydrate is decomposed in *2* and *4*, while gaseous propane accumulates in *3*, and liquid propane in *5*.

The thermodynamic conditions governing the possibility of separating hydrocarbons in the form of gas hydrates were investigated by Musayev [556] on the basis of the distribution conditions for the C_2–C_4 systems. He verified experimentally that the distribution coefficients decrease in proportion to the increase in temperature and the decrease in pressure. His calculations indicated that separation via the gas hydrates is particularly suitable in the case of the C_3–C_4 systems. Gebhardt an dhis co-workers [557] found that the separation of n-butane and isobutane was promoted by the addition of freon F11. Under these conditions, the ratio iso-C_4/n-C_4, the enrichment factor, was more than twice that in the absence of freon F11. Their experiments demonstrated that the separation is promoted by all freons (e.g. F21, F142B, etc.) which form hydrates of type H_{II}, but halomethanes which form hydrates of type H_1 have no effect on the separation of the butanes. Bukhgalter [558] studied the selectivity due to $CHCl_3$ (and also other halohydrocarbons) in the systems CH_4–C_2H_6, H_2–C_2H_4, C_2H_6–C_2H_4, and He–N_2–hydrocarbon.

In their patent, Afdah and Barber [559] described a method which permits the separation of higher hydrocarbons in the case of inhibition with alcohol. Further details about this topic are given in papers dealing with the separation of liquid hydrocarbons in the form of hydrates, and with the hydrate-forming properties of hydrocarbons absorbed in oil [560–565].

Among the more important advantages of the method of separating gaseous and liquid components in the form of gas hydrates are the relatively simple apparatus, the possibility for the use of standard equipment, and the comparatively low energy requirements.

Gukman and Kasperovich [566] published a paper on statistical calculations relating to the separation of gas mixtures in the form of gas hydrates. With their method, the maximum concentrations of those components of the mixture that do not form hydrates can be determined, and conclusions can be drawn regarding the sharpness of the separation.

6.2. Desalination of water via gas hydrate formation

In general, the methods discussed so far serve either for the prevention of gas hydrate formation or for the utilization of gas hydrate formation to recover some gas or liquid component that will act as guest molecule. The aim of the desalination of water is the recovery of the water of the gas hydrate lattice in a pure, practically salt-free state.

As long ago as 1961, Karnofsky [567] concluded that, of the many possibilities for the desalination of water, and mainly sea-water (e.g. electrodialysis, electroosmosis, reverse osmosis, ordinary freezing-out, freezing-out by means of hydrocarbons, distillation, etc.) the method involving the use of butane appears to be

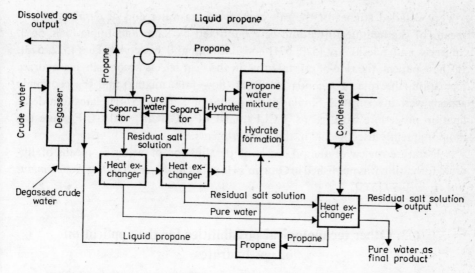

Fig. 6.10. Method for the desalination of sea-water via gas hydrate formation [158]

the most promising from the point of view of the efficiency of practical realization. In this method, butane is passed into the water, and the butane hydrate formed is frozen out, separated, and subsequently melted, whereby the hydrate decomposes to gaseous butane (which can be re-used for hydrate formation) and to pure water.

The mutual solubilities of butane and water are extremely small, and the gas undergoes practically no chemical reaction with water. However, it can form a hydrate at a pressure of about 5 bar. This hydrate can be separated from water on the basis of the differences in their densities. In this case, therefore, filtration does not pose great difficulties.

Figure 6.10 [158] shows the principle of one of the first and simplest practical solutions of this task of desalting water, by means of open circulation with propane as hydrate-forming agent.

In general, not only butane or propane, but also methane and various freons can be used for this purpose. The requirements to be met by the material chosen [158] are as follows: if possible, a high water/gas mole ratio, a low heat of phase transition, a low hydrate formation pressure and a high hydrate formation temperature, a high hydrate formation mass velocity, a low gas solubility, a non-poisonous gas, no danger of explosion, the lowest possible energy and material costs, and still other specific requirements too. If all these factors are considered jointly, a decision can be reached about the hydrate-forming agent to be used and, together with this, the method to be followed (besides the principle presented above, there are a very large number of continuously developing versions).

Very detailed studies of the salt content of sea-water and of desalination by means of gas hydrate formation were published by Saito and Iijima [568, 569], Smirnov and Kleshunov [570, 571] and Barduhn and his co-workers [572, 573]. In these papers, the rate of crystal growth, the heat effects accompanying hydrate formation, the role of secondary factors (flow rate, mixing) and the economic aspects were investigated. Similarly, the patent of Kirkley [574] relates to a desalination procedure with freons. CCl_2F_2, $CHCl_2F_2$, $CHClF_2$ and $CClF_3$ were the most frequently used agents in these experiments.

A literature review of the processes of desalination of water by means of hydrate formation was published in Gmelin's Handbuch, *Water Desalting Supplement*, Vol. 1, in 1979 [575].

6.3. Other technological possibilities for the application of gas hydrates

Besides the possibilities for the application of gas hydrates discussed above, there are a number of other methods which frequently present the engineer with very surprising opportunities. Indeed, by these means certain processes can be interpreted in a manner previously not even considered, and hence appropriate (possibly much more effective) measures can be taken, in the light of different aspects. Of these special possibilities, three will be mentioned here: the increase of pressure without the use of a compressor; the possible role of gas hydrates in the interpretation of coal-mining gas blow-outs; and the possibility of the elimination of fogs and clouds by means of gas hydrates, this being of importance from the point of view of environmental pollution control.

Let us consider *the increase of gas pressure without the use of a compressor*. A hydrate with volume V, molar mass M_h and density ϱ corresponds to

$$n = \frac{V\varrho 10^3}{M_h} \tag{6.9}$$

mol hydrate, and therefore the volume of gas in the hydrate under normal conditions is

$$Q_h = nV_g = \frac{V\varrho V_g 10^3}{M_h} \tag{6.10}$$

where V_g is the molar volume of the gas in m^3.

For the free real gas of the same amount

$$Q_g = \frac{VpT_0}{p_0 TZ} \tag{6.11}$$

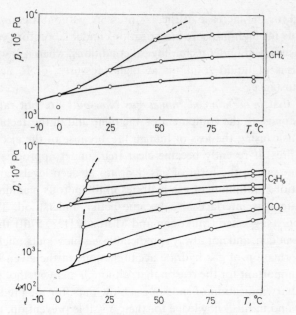

Fig. 6.11. Increase of the gas pressure in a fixed volume
during the decomposition of methane hydrate,
ethane hydrate and CO_2 hydrate [158]

where p is the pressure keeping up the gas volume Q_h and the volume V occupied by the hydrate, p_0 and T_0 are the normal values of the pressure and temperature, and Z is the deviation (compression) factor at pressure p and temperature T. If Equations (6.10) and (6.11) are made equal, the pressure of the originally free gas now in the gas hydrate can be calculated:

$$p = Z \frac{22.4 \times 10^3 \varrho T}{M_h T_0} \approx \frac{2.24 \times 10^4 \varrho}{M_h} Z \qquad (6.12)$$

Since, in general, $\varrho \approx 0.9$ and $M_h \approx 125$:

$$\frac{p}{Z} \approx 160 \qquad (6.13)$$

It is evident, therefore, that in the course of gas hydrate formation the gas pressure is increased to 160 times the original value without any machine compression involving an external work input. This phenomenon provides a possibility for the considerable increase of the gas pressure in a fixed closed volume by means of the decomposition of the gas hydrate, without the use of a compressor [158]. Figure 6.11 illustrates the increase of the gas pressure in a closed volume during the decomposition of methane hydrate, ethane hydrate and carbon dioxide hydrate.

This method for the increase of the gas pressure without a compressor is particularly suitable for high-pressure investigations under laboratory conditions, but it can also be employed under technological conditions, when it is wished to carry out the reactions of liquid solutions at high pressures, or to accelerate these reactions, for instance.

It is known that *in certain coal mines gas blow-outs* are not rare; they often cause severe damage to the equipment of the mine and in the technological processes, as well as causing the loss of human life, because of the ejected rocks and especially the fires. It recently became clear that, under appropriate conditions, gases can be present in the form of gas hydrates at deep levels, and the altered stability conditions and limits of gas hydrates in the porous medium and relating to the free water–gas interface also became better understood. The possibility was then strongly suggested (Makogon and Morozov [158, 576]) that the occurrence of these sudden, and not always predictable gas blow-outs might be connected with the formation of gas hydrate accumulations in the given location. This is particularly important for the reason that, although many methods are available for the preliminary indication of sudden gas blow-outs, for the prediction of their probable sites, and in this knowledge for their possible prevention, these methods are not completely reliable; in the event of a new and better theory, safety methods based on this could be introduced.

Setting out from the observation that sudden gas blow-outs always occur below a definite depth in a given area, for the coal mine he investigated in the Don basin Makogon calculated that at this depth the vapour pressure, the pressure and temperature relations in general and the geological factors are such that there is a possibility for the formation of methane hydrate. Consequently, in the study or the prediction of gas blow-outs, the role of gas hydrates cannot be rejected. Naturally, the complicated petrological conditions in the depths, the existence or non-existence of layers that are permeable or impermeable as regards the gases at the given site, tectonic fissures, etc. can all contribute in a complex way to the possible formation of gas hydrates. The methods of defence against gas blow-outs must include those which take into account the role of gas hydrates. Since the critical temperature for the existence of methane hydrate is 55 °C, if the temperature of layer is above this, then gas blow-outs due to gas hydrates cannot in general be expected, but if the temperature of the layer is below this limit, then blow-outs are possible. Methods that can be recommended, therefore, are the warming to 55–60 °C of limited zones in which the danger of gas blow-outs exists, and the injection of inhibitors (electrolytes, methanol) through suitable bore-holes, or even perhaps the addition of materials in which methane dissolves well.

The linkage of gas blow-outs in coal mines with a possible hydrate decomposition process has given rise to lively debate between the specialists. Opinions have been put forward by both the opponents and the supporters of this theory [577–581].

It has been shown experimentally by Zenin [582] on the basis of gas yield analysis and gas balance calculations that the coal-seam may contain gas hydrates; these undergo decomposition during the drilling, and this may lead to the sudden release of gas. Gas hydrate was also synthesized in coal samples under laboratory conditions. It was found that blow-out zones occur where compressed impermeable layers and gas accumulations form in the coal-seam, the plastic layer therefore containing stresses. It was estimated that one ton of coal (moisture content 2.5–3%) includes 30 kg water, from which about 33 mol hydrate can form. This quantity of hydrate extracts about 2100 kJ heat from the surrounding mass, and hence it is capable of performing 2×10^6 J work on its decomposition. If such gas accumulations are not incorporated into plastic rocks, they decompose practically instantaneously and lead to the elastic deformation of the rock.

The elimination of fogs (in which the dispersed water particles have a radius of about 0.1–100 µm) *and clouds* by means of gas hydrate formation is based on the fact that the water comprising the fog or cloud is removed from the system as the water lattice of the gas hydrate. Despite the method being similar in theory and in essence to the process of the desalination of water by means of gas hydrates, where it was likewise a question of the incorporation of the water into the gas hydrate lattice, the problems appear in a different manner here: in this case a small quantity of water must be removed from the system, for the medium contains only about 500–600 microdrops of water per cm³, which means only 0.05–1 g water in 1 m³ air.

The gas-hydrate formation method is particularly applicable if a high-temperature fog is involved, the temperature of which lies in the interval between -3 and $+3 °C$.

The essence of the method is that the vapour pressure of the liquefied hydrate-forming material dispersed into the fog (mainly butane, various freons, SO_2 or Cl_2) is very high, and it therefore evaporates rapidly. This results in a strong cooling in the microzones around the dispersed particles, the supersaturation of the microzones increases, and condensation and gas-hydrate formation occur. In the process, because of the formation of solid gas hydrate, the vapour pressure of the water decreases still further, so that the moisture migrates more and more from the liquid microdrops to the surface of the hydrate, which thereby grows.

This entire process is promoted if active crystallization nuclei (ready prepared gas hydrate grains) are introduced into the fog zone. It is obvious that these will be most effective if they remain for as long as possible under normal atmospheric conditions, i.e. their stability is high, and the given volume contains as many as possible of them.

Chapter 7

FUNDAMENTAL METHODS
FOR THE EXPERIMENTAL
DETERMINATION OF THE CONDITIONS
OF FORMATION AND DECOMPOSITION
OF GAS HYDRATES

The formation of gas hydrates is a surface phenomenon, and thus the basic task in the construction of the apparatus to be used for the investigations is to ensure adequate contact between the water and the gas. This can be achieved in two ways: by bubbling the gas through the water, or by introducing the gas under pressure into an apparatus partly filled with water. In the latter method, the maintenance of the conditions of hydrate formation requires the continuous renewal of the gas–water interface too. With regard to the state of the gas during the process, these two methods are referred to in the literature as the dynamic and static methods, respectively.

The *static method* is the more widely employed in practice. An appropriate means of achieving the renewal of the gas–water interface is internal mixing in the hydrate cell. Possibilities for this are satisfied by cell arrangements permitting rotation, shaking or oscillation. This leads to a satisfactory solution, though it allows only the comparatively slow mixing of the gas and the water. Its disadvantage lies in the relative difficulty of the practical realization. Primarily from the aspect of the practicability, a better method appears to be magnetic stirring within the cell. Here, the dissipation of the heat originating from the warming-up of the stirrer can cause minor experimental difficulties, but via the solution of the temperature measurement in the interior of the hydrate cell, these errors of measurement can be minimized. The literature contains a reference to the fact that the mixing of the gas and the water was achieved without mechanical means, relying only on diffusion. Such measurements, however, are extremely time-consuming, for the rate of mixing of the gas and the water by means of diffusion only is very low, and accordingly the errors in the reproducibility are appreciable.

Dynamic methods for hydrate investigations (by bubbling the gas through the liquid) are applied where the modelling of the pipeline phenomena too is required in the course of the examinations.

The scheme of the experimental apparatus used by Hammerschmidt [41] in 1934 can be seen in Fig. 7.1. The compressed gas (*1*) was led through a long copper tube (*2*), which was kept at the same temperature as that of the thermostated bath

(6) containing the pyrex glass tube (5) serving for the measurements. In this copper tube, flow rates corresponding to the industrial operating conditions could be modelled. The water was introduced by gravitation from a separate container (3), similarly through a copper tube, into the interior of the pyrex tube (5). The temperature of the gas was measured with an iron–constantan thermocouple (7) and a millivoltmeter (8). The gas left the system at atmospheric pressure, after passing through a water drain receiver, a gas meter (12) and a reducer. The

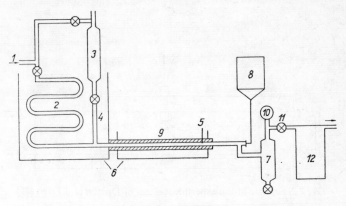

Fig. 7.1. Scheme of the experimental apparatus of Hammerschmidt [41]. Adapted by permission of the *Am. Chem. Soc.*

gas pressure was measured with a Bourdon tube gauge (10) calibrated during the measurement. The hydrates developing in the pyrex tube were snow-like in appearance. For the observation of the melting-point of the hydrate at various temperatures, the gas flow in the tube was stopped, and the gas was allowed to flow out until the desired pressure was reached. The temperature of the bath was then raised gradually until the decomposition of the hydrate began.

In 1937, Deaton and Frost [42] similarly dealt with the investigation of hydrate formation under pipeline conditions. The scheme of their apparatus can be seen in Fig. 7.2. The apparatus consisted of a high-pressure hydrate cell, the main dimensions being $2.5 \times 3.2 \times 15.2$ cm, provided with a window made from flat glass, the thickness of which was 2 cm and which resisted a pressure of 1.8×10^7 Pa. The quantity of gas flowing through the system was regulated and checked by means of a valve system. The cell could be rotated around a horizontal axis, but the observations were always made with the window in the horizontal position. The device was suitable for both static and dynamic measurements. With the cell located in the thermostating bath, the temperatures of the bath and the interior of the cell were each measured with a thermocouple, with an accuracy of about

0.03 °C. Three thermocouples were built into the cell, one in the bottom part, one immersed in the water, and the third in the gas phase. Hydrate was produced by the appropriate choice of pressure and temperature, and the system was then kept at a pressure higher than the decomposition pressure for a prolonged period, regardless of whether the operating mode was static or involved a gas flow, and the decomposition of the hydrate was later started by decreasing the pressure. With some practice, the decomposition pressure of the hydrate could be reproduced within an interval of $0.1–0.2 \times 10^5$ Pa via this method. The apparatus was also

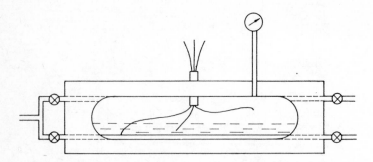

Fig. 7.2. Scheme of the experimental cell of Deaton and Frost [42]

suitable for investigations of the conditions of formation and decomposition of hydrates at constant pressure by changing the temperature. These laboratory experiments showed that the hydrates accumulated on the surface of the water. If the gas flow was passed above the water, a foamy hydrate was formed. This foamy layer covered the whole of the water surface, thereby preventing the possibility of further contact between the water and the gas, and the rate of hydrate formation therefore decreased strongly. When the dynamic method of investigation was employed, a hydrate film developed on the surface of the gas bubbles, and these gas bubbles filled the hydrate cell.

The same cell type was applied by Frost and Deaton [51] in later work. The primary aim of their experiments was the determination of the molar ratio of the gas and water bound in the hydrate; for this, the hydrate cell was about one-quarter filled with water, and the gas was then bubbled through it at the desired temperature and pressure until the cell was practically filled with a hydrate mass riddled with pores. The gas flow was next stopped, but the static pressure was kept at a level above the equilibrium value for the hydrate. The water not bound in the hydrate was then removed from the cell. For several days, the apparatus was kept under conditions suitable for hydrate formation, so that the water trapped between the bubbles should also be able to undergo conversion to the hydrate.

The pressure fall occurring as a consequence of this process was compensated from an external gas reservoir. Subsequently, by means of slow warming-up, the hydrates were decomposed under constant pressure, and the quantities of gas and water liberated were determined in a mercury-seal burette and by gravimetry, respectively. During the experiments, the temperature could be kept constant with an accuracy of 0.05 °C, and the pressure with an accuracy of $0.05-0.5 \times \times 10^5$ Pa.

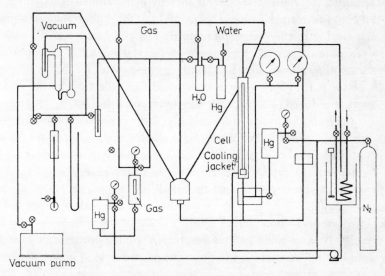

Fig. 7.3. Scheme of the experimental apparatus of Roberts and his co-workers [44]

The flow diagram of the experimental apparatus used by Roberts and his co-workers [44] is shown in Fig. 7.3. The measuring cell of the apparatus had an inner diameter of 3 mm and an outer diameter of 10 mm; it consisted of a glass tube, fitted into a steel block, the tube initially being filled with mercury, the level of which was regulated via the nitrogen gas pressure. The cell was filled and emptied by means of stainless steel capillary tubes. Separate capillaries were used to introduce the water and the gas, and others for the emptying of the cell. The intermixing of the water–gas mixture in the tube was ensured by the up and down movement of a stainless steel platelet. The quantity of water in the capillary was determined with a cathetometer, while the gas quantity was calculated.

In their four-phase equilibrium measurements, Carson and Katz [359] used a rotating cell fitted with a glass window, which was placed in a fluid bath. The development of the four-phase equilibrium was accompanied by the formation of a large quantity of hydrate in the tube. The gas and the liquid phase rich in water were then expelled with mercury, the hydrate was decomposed, and the liberated gas was analyzed, while the composition of the mixture was determined on the basis of density measurement.

Unruh and Katz [64] worked with a similar apparatus, but the gas components were mixed in two separate mixing cells, from which the required amounts of gas were admitted into the measuring cell by displacement with mercury, the quantities of the added individual components thereby being known exactly.

The cell could be turned around a horizontal axis through an angle of about 30°, in this manner the mixing being guaranteed. The cell baths was illuminated in order to facilitate the visual observation. In the starting-up of the operation of the apparatus, the cell was first filled with distilled water by means of a pump, and then about half of this water was expelled from the tube by the admission of gas. Next, the thermostating bath was overcooled by 3–4 °C relative to the expected temperature of hydrate formation, and the cell was kept in motion until hydrate formation began. When the hydrate was formed, the pressure (measured by means of a Bourdon tube gauge) was decreased slowly until the hydrate began to decompose and the pressure therefore began to increase. The temperature of the bath was then slowly raised until almost all of the hydrate had disappeared. With the subsequent decrease of the temperature of the bath by a few tenths of a degree, hydrate formation started again because of the presence of the inoculating crystals. At a constant cell temperature, the components were mixed until the pressure became constant. In the meanwhile, the p vs. T data pairs relating to hydrate formation for the investigated gas composition were recorded. The possibility of serial measurements was given by means of an increase in the pressure, or the addition of further water.

Important experiments were carried out by Unruh and Katz [64] to determine the conditions of hydrate formation in the carbon dioxide–methane system. They were the first who observed that in the presence of a gas mixture the formation of CO_2 hydrate is substantially slower than from CO_2 in the pure state, and that the density of the CO_2 hydrate is apparently lower than that of the water, for in the pure state it accumulates mainly on the surface of the water, while the mixed hydrate is distributed in the form of fine flakes throughout the entire liquid phase.

Reamer and his co-workers [89] employed two different methods for the determination of the hydrate-forming properties of paraffins and olefins. In the first method, a glass capillary was used within which the hydrocarbon and the aqueous phase were situated above a mercury seal. This process is illustrated schematically in Fig. 7.4. The mixing within the capillary was solved by means of a steel spiral, which was moved on an axle. The experimental temperature was measured here too by an iron–constantan thermocouple. The temperature of the external surface

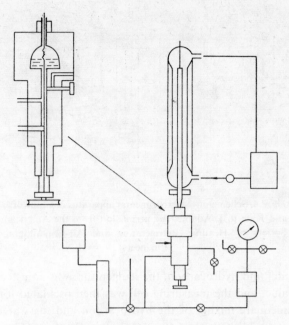

Fig. 7.4. Scheme of the experimental apparatus of Reamer
and his co-workers [89], involving a glass capillary cell.
Adapted by permission of the *Am. Chem. Soc.*

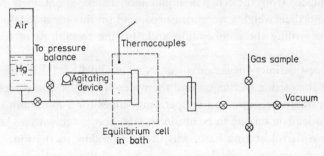

Fig. 7.5. Scheme of the experimental apparatus
of Reamer and his co-workers [89], involving a steel U-tube.
Adapted by permission of the *Am. Chem. Soc.*

of the capillary could be maintained within a value of 0.1 °C by the circulation
of low-viscosity oil. The quantities of the gaseous and liquid components enclosed
in the capillary were determined with a cathetometer.

In the other method, a steel U-tube was used as a measuring cell, and the
"weight-bomb" technique was employed. In experiments with the arrangement to
be seen in Fig. 7.5, the double-walled equilibrium vessel was filled with the necessary

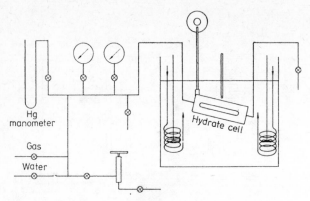

Fig. 7.6. Scheme of the experimental apparatus of Chinworth and Katz [63]. Adapted by permission from the American Society of Heating, Refrigerating and Air-Conditioning Engineers

quantities of water and hydrocarbon, the temperature was adjusted to the desired experimental value, and the measuring cell was then oscillated for several hours to ensure the adequate mixing of the hydrocarbon and the water. The hydrate equilibrium was established when the pressure in the system became constant. At this point, the liquid phase was expelled from the cell by means of the addition of mercury, the gas hydrate was decomposed, and the gas mixture thus formed was also displaced from the cell. The liquid and gas phases obtained from the cell were dried and their weights were measured, and on this basis conclusions could be reached regarding the composition and also the conditions of formation of the hydrate.

In their investigations on gas mixtures, Noaker and Katz [113] employed only diffusion for the mixing of the gas and the water. The filled cell therefore had to be kept at constant pressure and temperature values for a minimum of 24 hours in order for adequate mixing to be attained. Here too, a cell with a glass window, immersed in a thermostating bath, was used, with slow oscillation.

Figure 7.6 presents a simple flow diagram for hydrate investigations with a cell kept in motion by an eccentric swinging arm. As demonstrated by the work of Chinworth and Katz [63], this technique is suitable for studies on the conditions of formation of freon hydrates. The high-pressure cell, containing a glass observation window, was operated in a water-bath. In the case of the freons, the equilibrium pressure developing under atmospheric conditions or in the vicinity of these was determined with a mercury manometer.

A steel cell, similarly with a built-in glass observation window, was applied by Scauzillo [320] to study the effects of hydrate inhibitors. The measuring cell had a volume of about 90 cm³ and was connected with a high-pressure mercury

pump, whereby the pressure in the cell could be varied. Thermostating was ensured and regulated via the quantity of ethylene glycol passed through a copper
cooling coil built around the cell.

The filled cell was kept at the desired temperature for about 15 minutes; then,
after thermal equilibrium had been established, the pressure in the cell was progressively increased, in steps of approximately 1.5×10^5 Pa. After each pressure
increase, the contents of the cell were thoroughly mixed. This process was repeated until the separation of the hydrate could be detected by eye. The conditions of
hydrate formation were also investigated at constant pressure, the temperature
then being progressively lowered step by step until the hydrate began to form.

Here too, the evaluation of the experimental processes proved quite clearly
that under isobaric conditions an overcooling of 2–5 °C was necessary, while
under isothermal conditions an overpressure of $2–10 \times 10^6$ Pa was necessary for
hydrate formation to occur in a pure system not containing inoculating crystals.

The parameters of hydrate formation in the methane–propylene–water system
were similarly determined by Otto and Robinson [115] in a steel cell of variable
volume, containing a glass window. The cell with a volume of about 120 cm³ was
calibrated to a pressure of 1.4×10^7 Pa. Mercury was used as sealing fluid for the
change of volume. For equilibrium to be established, the cell was rotated around
an axle. In their first experiments, the temperature of the thermostating liquid
circulating in the pipe surrounding the cell was accepted as the cell temperature,
but it turned out that this method can be applied only in the case of a very long
thermostating period, and thermocouples were therefore built into the inside of
the cell. The cell was evacuated and then filled with mercury; next, about 4 cm³
distilled water was sucked in, followed by the desired quantity of the hydrocarbon.
The mixing of the gas and the water was ensured by the motion of the cell, during
which the temperature was decreased. As a consequence of the repeated processes
of formation and decomposition of the hydrates, satisfactorily reproducible
results were obtained.

In their later work, the same group [116] mention the modification of the apparatus. A cell was used which had a window in only one side, and the determination
of the decomposition point of the hydrate was therefore more difficult here than
in the cell type in which there were windows on the two opposite walls of the cell.
In the case of the one-window cell, a polished stainless steel mirror was built into
the inside of the cell, by means of which a narrow light beam was projected onto
the hydrate crystal. The possibility of observing the phase changes was ensured
by the good reflectivity of the steel mirror. The better mixing of the gas–water
mixture in the cell was solved too: instead of the rotation of the cell, a revolving
stainless steel tube sealed with a teflon O-ring and built into a fixed steel block
was used.

The mixing was achieved by Clarke and his co-workers [117] with a magnetic mixing device. The cell was placed in a thermostating bath, the temperature of which could be regulated with an accuracy of $\pm 10^{-3}\,°C$. The temperature of the cell was measured in relation to that of a stainless steel block immersed in the same bath. The temperature of the steel block was determined with a platinum resistance thermometer.

The scheme of the high-pressure apparatus made by Marshall [124] for hydrate investigations is shown in Fig. 7.7. His measuring cell was a rotating steel bomb with a volume of 95 cm³, calibrated to a pressure of 7×10^8 Pa. A special probe-type Cr–Al thermocouple served for the measurement of the temperature.

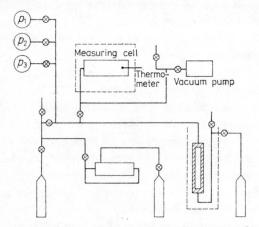

Fig. 7.7. Scheme of the experimental apparatus of Marshall and his co-workers [127]

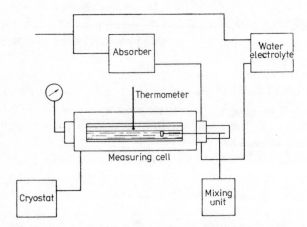

Fig. 7.8. Scheme of the experimental apparatus of Andryushchenko and Vasilchenko [145]

Andryushchenko and Vasilchenko [145] described an apparatus suitable for the study of the changes occurring in the conditions of formation of natural gas hydrates on the action of inhibitors. The hydrate cell here was a steel-covered organic glass cylinder with an inner diameter of about 2 cm and an internal volume of about 100 cm³. Visual observability was ensured by an opening in the steel cover. Special steel heads were connected to the covering cylinder; these contained the electromagnetic mixer, the thermometer connections, and the liquid and gas inlet and outlet valves. The gas sample passed into the measuring cell was conducted through a filter and an oil absorber, and separate feed containers served for the input of the water and inhibitor solutions. The measuring cylinder was cooled with an alcohol solution circulated from a cryostat. The temperature of the cell was measured with a platinum resistance thermometer, connected to a recording unit. Equilibrium measurements were made under isobaric conditions, and the cell temperature was recorded as a function of the cooling time. The scheme of the apparatus is illustrated in Fig. 7.8.

Makogon [158] used special equipment which was also suitable for the determination of the rate of formation and growth of the crystal nuclei over wide pressure and temperature intervals. Visual observation was possible through a plexiglass window with a thickness of 42 mm and a diameter of 100 mm in a bathyscaphe

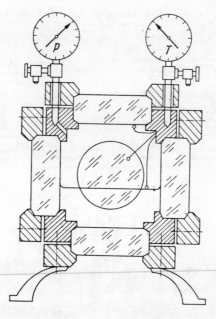

Fig. 7.9. Scheme of the experimental
apparatus of Makogon [158]

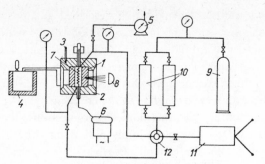

Fig. 7.10. Scheme of the experimental
apparatus of Khoroshilov [471]

chamber with an internal volume of 1000 cm³. The working pressure in the chamber exceeded 2×10^7 Pa. The isothermal measuring conditions were ensured by the use of an air chamber with a temperature regulated to an accuracy of ± 0.1 °C. In measurements under isobaric conditions, the rate of cooling could be maintained at a value of 0.5 °C h⁻¹, so that the temperature fluctuations caused by the heat of crystallization could be eliminated. The investigations were made under static conditions, and the processes taking place were followed by film recording. The experiments proved that the rate of hydrate formation is optimum at an overcooling of about 3 °C, but an overcooling of 6–8 °C slows down the process. The

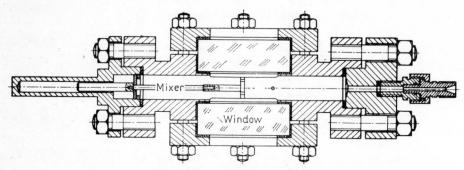

Fig. 7.11. The experimental cell of Berecz and Balla-Achs [296]

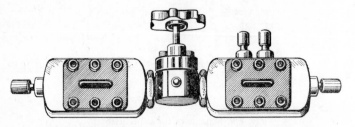

Fig. 7.12. Double hydrate cell of Berecz and Balla-Achs [437]

frontal view of the measuring cell to be placed in the air thermostat is shown in Fig. 7.9.

The flow diagram for the glass-windowed apparatus used by Khoroshilov (cit. [471]) can be seen in Fig. 7.10; this had a volume of 40 or 90 cm³.

Figure 7.11 depicts the structure of the measuring cell (with a window for examination) used for hydrate equilibrium investigations by Berecz and Balla-Achs [296]. In measurements carried out under static conditions, the measuring cell was placed in an alcohol thermostating bath. The good contact of the gas and water was ensured with a mixer operated by a multivibrator. The temperature of the cell was measured with a copper–constantan thermocouple. The apparatus was suitable for the investigation of hydrate equilibria under both isothermal and isobaric conditions.

The apparatus for the investigation of hydrates at higher pressures was transformed by the authors [437] into a double-cell one. This allowed the possibility of sampling separately for analytical purposes from both the gas and the hydrate phases without causing any change in the cell pressure or the equilibrium conditions. The modified apparatus is shown in Fig. 7.12, and a schematic diagram of it in Fig. 7.13.

An outline of the measuring system of van Berkum and Diepen [404], used for the investigation of sulphur dioxide hydrates, is demonstrated in Fig. 7.14.

Figure 7.15 shows the scheme of the experimental apparatus of Aoyagi and his co-workers [346] for the determination of the water content of gaseous methane in equilibrium with the hydrate.

A device permitting measurements in the field on hydrocarbon systems containing more than 45% C_{5+} was reported by Korotayev and his co-workers [584].

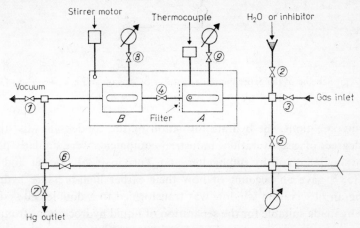

Fig. 7.13. Schematic flow diagram for double cell of Berecz and Balla-Achs [437]

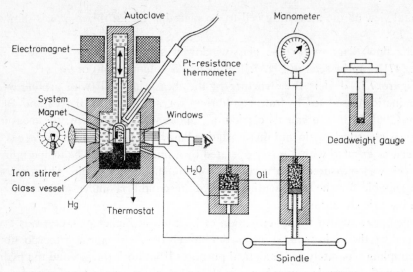

Fig. 7.14. Scheme of the experimental apparatus of van Berkum and Diepen [404]

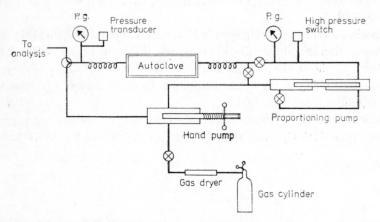

Fig. 7.15. Scheme of the experimental apparatus of Aoyagi and his co-workers [346].
Adapted by the permission of the authors

In their investigations, the hydrate formation by the condensate was studied at various degrees of gas saturation. Industrial equipments were similarly described by Medovskiy [585] and by Bukhgalter and Turikin [586]. Verma and his co-workers [587] gave an account of how their earlier high-pressure hydrate cell containing an observation window was transformed to a double-cell system and was thereby made suitable for the separation of liquid hydrocarbons from natural gas components by means of hydrate formation. The apparatus could also be employed for the determination of the bubble-points of the liquid hydrocarbons.

REFERENCES

[1] Schroeder, W., *Sammlung Chem. und chem. techn. Vorträge.* Vol. XXIX. F. Enke, Stuttgart, 1927.

[2] Davy, H., *Phil. Trans. Roy. Soc. London,* **101**, 1 (1811).

[3] Faraday, M., *Phil. Trans. Roy. Soc. London,* **113**, 160 (1823).

[4] Löwig, G., *Pogg. Ann.* **14**, (D.g.R. 90), 114, 487 (1828).

[5] Alexeyeff, W., *Ber.* **9**, I. 1025 (1876).

[6] de la Rive, A., *Ann. Chem.,* **2**, 40 (D.g.R. 92) 376 (1829).

[7] Pierre, J., *Ann. Chem.* **3**, 23 (D.g.R. 194) 416 (1848).

[8] Schoenfeld, F., *A. Ch.,* Neue Reihe, **19** (D.g.R. 95) 19 (1855).

[9] Ditte, A., *Compt. Rend.,* **95**, 1283 (1882).

[10] Maumené, E., *Chem. N.,* **47**, 154 (1883).

[11] Berthelot, M., *Ann. Chem.,* **3**, 46 (D.g.R. 217) 490 (1856).

[12] Millon, H., *Compt. Rend.,* **51**, 249 (1860).

[13] Duclaux, E., *Compt. Rend.,* **64**, 1099 (1867).

[14] Wartha, V., *Ber.* **3**, 80 (1870); **4**, 180 (1871).

[15] Balló, M., *Ber.* **4**, 118, 160 (1871).

[16] Mijers, J., *Maandbl.,* **3**, 17 (1870).

[17] Decharme, C., *Ann. Chem.,* **4**, 29 (D.g.R. 269) 415 (1873).

[18] Tanret, C., *Compt. Rend.,* **86**, 765 (1878).

[19] Stokris, B. I., *Maandbl.,* **6**, 82 (1870).

[20] Chancel, G., and Parmentier, F., *Compt. Rend.,* **100**, 27 (1885).

[21] Wroblewski, S. V., *Compt. Rend.,* **94**, 212, 1355 (1882).

[22] Cailletet, L., *Compt. Rend.,* **85**, 851 (1877).

[23] Cailletet, L., and Bordet, H., *Compt. Rend.,* **95**, 58 (1882).

[24] Wöhler, F., *Pharm. Zentralbl.,* 1840, 4.

[25] de Forcrand, R., *Compt. Rend.,* **95**, 129 (1882).

[26] Loir, A., *Compt. Rend.,* **34**, 547 (1852).

[27] Roozeboom, H. W. B., *Rec. Trav. Chim.,* **2**, 98 (1883); **3**, 29, 59, 87 (1884).

[28] de Forcrand, R. and Villard, P., *Compt. Rend.,* **106**, 849 (1888).

[29] de Forcrand, R., and Villard, P., *Compt. Rend.,* **106**, 1357 (1888).

[30] Villard, P., *Compt. Rend.,* **106**, 1602 (1888).

[31] Villard, P., *Compt. Rend.,* **111**, 302 (1890).

[32] Villard, P., *Compt. Rend.,* **123**, 377 (1896).

[33] de Forcrand, R. and Thomas, S., *Compt. Rend.,* **125**, 109 (1897).

[34] Scheffer, F. and Meijer, G., *Versl. Kon. Akad. Wetensch. Amsterdam,* **27**, 1104, 1305 (1919).

[35] de Forcrand, R., *Compt. Rend.,* **176**, 355 (1923).

[36] de Forcrand, R., *Compt. Rend.*, **181**, 15 (1925).

[37] Villard, P., *Compt. Rend.*, **176**, 1516 (1923).

[38] Bouzat, A., *Compt. Rend.*, **176**, 253 (1923).

[39] Bouzat, A., and Aziniers, L., *Compt. Rend.*, **177**, 1444 (1923).

[40] Tammann, G. and Kriege, G., *Z. Anorg. Allg. Chem.*, **146**, 179 (1925).

[41] Hammerschmidt, E. G., *Ind. Eng. Chem.* **26**, 851 (1934).

[42] Deaton, W. M. and Frost, E. M., *Am. Gas Journal*, **146**, 17 (1937).

[43] Deaton, W. M. and Frost, E. M., *Gas Age*, **81**, 33 (1938).

[44] Roberts, O. L., Brownscombe, E. R. and Howe, L. S., *Oil and Gas Journ.*, **39**, (30) 37 (1940).

[45] Parent, J. D., *Inst. of Gas Techn. Res. Bull.*, No. 1 (1948).

[46] Carson, D. B., and Katz, D. L., *Petr. Tr. AIME*, **146**, 150 (1942).

[47] Katz, D. L., *Petr. Tr. AIME*, **160**, 140 (1944).

[48] Brown, H., *Proc. Nat. Gas Ass. Am.* May, p. 54 (1940).

[49] Sage, B. H. and Lacey, W. N., *Petr. Tr. AIME*, **186**, 143 (1949).

[50] Miller, B. and Strong, E. R., J., *Am. Gas Ass. Monthly*, **28**, (2) 63 (1946).

[51] Frost, E. M. and Deaton, W. M., *Oil and Gas Journ.*, **45**, 170 (1946).

[52] Bradley, R., *Sci. Progn.* **31**, 282 (1936).

[53] Godchot, M., Caugil, G. and Calas, R., *Compt. Rend.*, **202**, 759 (1936).

[54] Nikitin, B. A., *Z. Anorg. Allg. Chem.*, **227**, 81 (1936).

[55] Nikitin, B. A., *Nature*, **140**, 643 (1937).

[56] Nikitin, B. A., *J. Gen. Chem. (USSR)*, **9**, 1167 (1939).

[57] Nikitin, B. A., *Bull. Acad. Sci. (USSR)*, **1**, 37 (1940).

[58] Nikitin, B. A., *Akad. Nauk USSR Dokl.*, **29**, 571 (1940).

[59] Palin, D. E. and Powell, H. M., *Nature*, **156**, 334 (1945).

[60] Palin, D. E. and Powell, H. M., *J. Chem. Soc.*, 208 (1947).

[61] Palin, D. E. and Powell, H. M., *J. Chem. Soc.*, 61 (1948).

[62] Palin, D. E. and Powell, H. M., *J. Chem. Soc.*, 298, 300, 468 (1947).

[63] Chinworth, H. E. and Katz, D. L., *J. ASRE*, **54**, 359 (1947).

[64] Unruh, C. H. and Katz, D. L., *Petr. Tr. AIME*, **186**, 83 (1949).

[65] Kobayashi, R. and Katz, D. L., *Petr. Tr. AIME*, **186**, 66 (1949).

[66] von Stackelberg, M. and Müller, H. R., *Z. Elektrochem.*, **58**, 25 (1954).

[67] von Stackelberg, M. and Meinhold, W., *Z. Elektrochem.*, **58**, 40 (1954).

[68] von Stackelberg, M. and Frühbuss, H., *Z. Elektrochem.*, **58**, 99 (1954).

[69] von Stackelberg, M., *Z. Elektrochem*, **58**, 104 (1954).

[70] von Stackelberg, M., and Jahns, W., *Z. Elektrochem.*, **58**, 162 (1954).

[71] von Stackelberg, M., *Naturwiss.*, **36**, 327, 359 (1949).

[72] von Stackelberg, M., *Fortschr. Mineral*, **26**, 122 (1950).

[73] von Stackelberg, M. and Müller, H. R., *Naturwiss.*, **38**, 456 (1951).

[74] von Stackelberg, M. and Müller, H. R., *J. Chem. Phys.*, **19**, 1319 (1951).

[75] Müller, H. R. and von Stackelberg, M., *Naturwiss.*, **39**, 20 (1952).

[76] von Stackelberg, M., and Müller, H. R., *Angew. Chem.*, **64**, 423 (1952).

[77] von Stackelberg, M., *Rec. Trav. Chim.*, **75**, 902 (1956).

[78] Claussen, W. F., *J. Chem. Phys.*, **19**, 259, 662 (1951).

[79] Claussen, W. F., *J. Chem. Phys.*, **19**, 1425 (1951).

[80] Pauling, L. and Marsh, R. E., *Proc. Nat. Acad. Sci. U.S.*, **38** 142 (1952).

[81] Records, L. R. and Seely, D. H., *Petr. Tr. AIME*, **192**, 61 (1951).

[82] Bond, D. C. and Russell, N. B., *Petr. Tr. AIME*, **179**, 192 (1949).

[83] Murphy, G. M., *J. Chem. Phys.*, **5**, 637 (1937).

[84] West, J. R., *Chem. Eng. Progr.* **44**, 287 (1948).

[85] Reamer, H. H., Sage, B. H. and Lacey, W. N., *Ind. Eng. Chem.*, **42**, 150 (1950).

[86] Selleck, F. T., Carmichael, L. T. and Sage, B. H., *Ind. Eng. Chem.*, **44**, 2219 (1951).

[87] Lippert, E. L., Palmer, H. A. and Blankenship, F. F., *Proc. Oklahoma Acad. Sci.*, **31**, 115 (1950).

[88] Powell, J. S., *Proc. Pacific Gas Assoc.*, **30**, 52 (1939).

[89] Reamer, H. H., Selleck, F. T. and Sage, B. H., *Petr. Tr. AIME*, **195**, 197 (1952).

[90] Smirnov, A. Sh., *Transport i hranenie gaza*, 112 (1950).

[91] Czaplinski, A., *Nafta (Kraków)*, **12**, 186 (1956).

[92] Hammerschmidt, E. G., *Oil and Gas Journ.*, **37**, 66 (1939).

[93] Lacroix, J. P., *Rev. Inst. Franc. Petrole*, **7**, 34 (1952).

[94] Woolfolk, R. M., *Oil and Gas Journ.*, **50**, 124 (1952).

[95] Pieroen, A. P., *Rec. Trav. Chim.*, **74**, 995 (1955).

[96] von Stackelberg, M., *Rec. Trav. Chim.*, **75**, 902 (1956).

[97] van der Waals, J. H., *Trans. Far. Soc.*, **52**, 184 (1956).

[98] van der Waals, J. H. and Platteeuw, J. C., *Advances in Chemical Physics*. Vol. II (Ed.: Prigogine, I.) 1959.

[99] Barrer, R. M. and Stuart, W. I., *Proc. Roy. Soc. (London)*, A. No. 1233 **243**, 172 (1957).

[100] Barrer, R. M. and Ruzicka, D. J., *Trans. Far. Soc.*, **58**, 2239, 2253, 2262 (1962).

[101] Lennard-Jones, J. E. and Devonshire, A. F., *Proc. Roy. Soc. (London)*, A. **163**, 53 (1937); **165**, 1 (1938).

[102] Evans, D. F. and Richards, R. E., *Proc. Roy. Soc. (London)*, A. **223**, 238 (1954).

[103] McKoy, V. and Sinanoglu, O., *J. Chem. Phys.*, **38**, 2946 (1963).

[104] Kihara, T., *J. Phys. Soc. (Japan)*, **6**, 289 (1951).

[105] McMullan, R. K. and Jeffrey, G. A., *J. Chem. Phys.*, **42**, 2725, 2732 (1965).

[106] Jeffrey, G. A. and McMullan, R. K., "The Clathrate Hydrates". *Progress in Inorganic Chemistry* (Ed.: Cotton, F. A.) Vol. 8, Interscience Publ., New York, 1967.

[107] Child, W., *Quart. Rev. Chem. Soc.*, **18**, 321 (1964).

[108] Wilson, G. J. and Davidson, D. W., *Canad. J. Chem.*, **41**, 264 (1963).

[109] Wilson, G. J. and Davidson, D. W., *Canad. J. Chem.*, **41**, 1424 (1963).

[110] Bertie, J. E. and Othen, D. A., *Canad. J. Chem.*, **50**, 3443 (1972).

[111] Saito, S., Marshall, D. R. and Kobayashi, R., *AICHE J.*, **10**, 734 (1964).

[112] McLeod, H. O. and Campbell, J. M., *J. Petr. Techn.*, June 1962, p. 390.

[113] Noaker, L. J. and Katz, D. L., *Petr. Tr. AIME*, **201**, 237 (1954).

[114] Glew, D. N., *J. Phys. Chem.*, **66**, 605 (1962).

[115] Otto, F. D. and Robinson, D. B., *AICHE J.*, **6**, 602 (1960).

[116] Snell, E., Robinson, D. B. and Otto, F. D., *AICHE J.*, **7**, 482 (1961).

[117] Clarke, E. C., Ford, R. W. and Glew, D. N., *Canad. J. Chem.*, **42**, 2027 (1964).

[118] King, M. B., "Gas Hydrates". *Int. Ser. of Monographs in Chem. Eng.* (Ed.: Danckwerts, P. V.) Vol. 9, p. 163, Pergamon Press, Oxford 1969.

[119] Bhatnagar, W. B., *Clathrate Compounds*. Chem. Publ. Co. Inc., New York, 1970.

[120] Byk, S. Sh. and Fomina, V. I., *Doklady A. N. SSSR*, **204**, 123 (1972).

[121] Fomina, V. I. and Byk, S. Sh., *Gazov. Prom.*, **12**, (3) 50 (1967).

[122] Koshelev, V. S,. Fomina, V. I. and Byk, S. Sh., *Gaz. Delo*, (1) 24 (1971).

[123] Sherwood, P. W., *Australian Gas Bull.*, **26**, 39, 41, 43 (1962).

[124] Marshall, D. R., *NASA Doc.*, N 63-12829, p. 25 (1962).

[125] Trebin, F. A. and Makogon, Yu. F., *Tr. Mosk. Inst. Neftekhim. Gazov. Prom.*, **42**, 196 (1963).

[126] Trebin, F. A. and Makogon, Yu. F., *Mater. Respubl. Konf. po Gazifik. Uzbekistana*, 1961 (published 1963).

[127] Marshall, D. R., Saito, S. and Kobayashi, R., *AICHE J.*, **10**, 202 (1964).

[128] Musayev, R. M. and Chernikhin, V. I., *Gazov. Prom.*, **10**, (5) 36 (1965).

[129] Musayev, R. M., *Gazov. Delo*, (11) 17 (1965).

[130] Trebin, F. A., Khoroshilov, V. A. and Demenko, A. V., *Gazov. Prom.*, **11**, (6) 10 (1966).

[131] Korotayev, I. P., Musayev, R. M. and Khoroshilov, V. A., *Nauchno-Techn. Sb. po Geol. Razr. Transp. Ispol. Prirod. Gazov*, **5**, 114 (1965).

[132] Korotayev, I. P. and Musayev, R. M., *Nauchno-techn. Sb. VNII Prirod. Gazov*, **6–7**, 148 (1967).

[133] Musayev, R. M., and Korotayev, I. P., *Nauchno-techn. Sb. VNII Prirod. Gazov*, **6–7**, 143 (1967).

[134] Fomina, V. I., *Khim. Tekhnol. Masel*, **12**, (10) 1 (1967).

[135] Korotayev, I. P., *Nauchno-techn. Sb. po. Geol. Razr. Transp. Ispol. Prirod. Gazov*, **5**, 125 (1965).

[136] Fomina, V. I., *Gazov. Prom.*, **12**, (3) 50 (1967).

[137] Susummi, S., *Nippon Kaisni Gakkai-Shi.*, **22**, (2) 209 (1968).

[138] Dzhavadov, A. D., *Azerb. Neft. Khoz.*, **46**, (9) 32 (1967).

[139] Yorizane, M. and Nishimoto, Y., *Kagaku Kogaku*, **22**, 158 (1968).

[140] Hildebrand, J. H., *J. Am. Chem. Soc.*, **37**, 970 (1915).

[141] Ichiro, H., *Nippon Kaisni Gakkai-Shi.*, **22**, (3) 129 (1968).

[142] Bukhgalter, E. B., *Gazov. Delo*, (5) 8 (1969).

[143] Kedzierski, S. and Chowaniec, A., *Gaz, Woda, Techn. San.*, **44**, 117 (1970).

[144] Aliyev, A. G., Musayev, R. M. and Ismailov, I. A., *Gazov. Prom.*, (5) 16 (1969).

[145] Andryushchenko, F. K. and Vasilchenko, V. P., *Gazov. Prom.*, (10) 4 (1969).

[146] Musayev, R. M., *Gazov. Delo*, (3) 25 (1970).

[147] Bukhgalter, E. B., *Gazov. Delo*, (2) 23 (1970).

[148] Hutchinson, A. I., USA Patent No. 2363549 (1944).

[149] Hutchinson, A. I., USA Patent No. 2410583 (1946).

[150] Kinney, P. I. and Kahre, L. C., USA Patent No. 30940924 (1963).

[151] Chersky, N. and Makogon, J., *Oil and Gas Int. (London)*, **8**, 82 (1970).

[152] Belov, V., *Oil and Gas Int. (London)*, **8**, 84 (1970).

[153] Katz, D. L., *Soc. of Petr. Eng. J.*, 3061, 7 (1970).

[154] Evrenos, S. I., Heatman, J. and Ralstin, J., *Soc. of Petr. Eng. J.*, 2881, 8 (1970).

[155] Banks, R. B., *World Oil*, March, 91 (1970).

[156] Robinson, D. B. and Mehta, B. R., *J. Canad. Petr. Techn.*, **1**, 32 (1971).

[157] Makogon, Yu. F. and Sarkisyants, G. A., *Preduprezhdenie obrazovaniya gidratov pri dobiche i transporte gaza*, Moscow, 1966.

[158] Makogon, Yu. F., *Gidraty prirodnykh gazov*, Moscow, 1974.

[159] van Cleeff, A. and Diepen, G.A.M., *Rec. Trav. Chim.*, **79**, 582 (1960).

[160] Gilliland, E. R., *Ind. Eng. Chem.*, **47**, 2410 (1955).

[161] Wiegandt, H. E., *Symp. on Saline Water Conv.* (Nov. 1957).

[162] Hendrickson, H. M. and Moulton, R. W., *OSW. Prog. Rep.*, 10 (Aug. 1956).

[163] Barduhn, A. J., Towlson, H. E. and Ye Chien-Hu, *AICHE J.*, **8**, 176 (1962).

[164] Knox, W. G., Hess, G. E. and Smith, H. B., *Chem. Eng. Progr.* **57**, 66 (1961).

[165] Barduhn, A. J. and Towlson, H. E., Prog. Rep., No. 44 (1960). Office of Saline Water, US Dept. of Interior.

[166] Wittstruck, T. A., *J. Chem. Eng. Data,* **6**, 343 (1961).

[167] Miller, S. L., *Proc. NSA,* **47**, 1798 (1961).

[168] Brown, H., *Astrophys. J.,* **111**, 641 (1950).

[169] de Marcus, W. C., *Handbuch der Physik.* Springer Verlag, 1959, Berlin.

[170] Gore, R., *The National Geographic Magazine,* **157**, (1), 2 (1980)

[171] Sagan, C. and Strong, J., *Science,* **133**, 849 (1961).

[172] Scholander, P. F., *J. of Glaciology,* **3**, 813 (1961).

[173] Urey, H. C., *Handbuch der Physik.* Springer Verlag, 1959, Berlin.

[174] Miller, S. L. and Smythe, W. D., *Science,* **170**, 531 (1970).

[175] Whipple, F. L., *Astrophys. J.,* **111**, 375 (1950); **113**, 464 (1951).

[176] Delsemme, A. H. and Swings, P., *Ann. Astrophys.,* **15**, 1 (1952).

[177] Delsemme, A. H. and Miller, D. C., *Planet. Space Sci.,* **18**, 717 (1970).

[178] Delsemme, A. H. and Wenger, A., *Planet. Space Sci.* **18**, 709 (1970).

[179] Mendis, D. A., *Nature (London),* **249**, 536 (1974).

[180] Delsemme, A. H., *Colloq. Inst. C.N.R.S. 1972,* (Publ. 1974) 207, 305.

[181] Pena, J. A. and de Pena, R. G., *J. Geophys. Res.,* **75**, 2831 (1970).

[182] Miller, S. L., *Proc. NSA,* **47**, 1515 (1961).

[183] Frank, H. S. and Evans, M. W., *J. Chem. Phys.,* **13**, 507 (1945).

[184] Eley, D. D., *Trans. Far. Soc.,* **40**, 184 (1944).

[185] Claussen, W. F. and Polglase, M. F., *J. Am. Chem. Soc.,* **74**, 4817 (1952).

[186] Klotz, I. M., *Science,* **128**, 815 (1958).

[187] Klotz, I. M. and Lubowsky, S. W., *J. Am. Chem. Soc.,* **81**, 5119 (1959).

[188] Szent-Györgyi, A., *Bioenergetics.* Academic Press, 1957. London, New York, San Francisco

[189] Pauling, L., *Science,* **134**, 15 (1961).

[190] Pauling, L. and Hayward, R., *The Architecture of Molecules.* W. H. Freeman, San Francisco, 1964. p. 44.

[191] Dorsch, R. R. and de Rocco, A. G., *Physiol. Chem. Phys.,* **5**, 209 (1973).

[192] Dorsch, R. R. and Distefano, V., *Physiol. Chem. Phys.,* **5**, 225 (1973).

[193] Hertl, E. and Römer, G. H., *Ber.,* **63**, B 2446 (1930).

[194] Robertson, J. M., *Organic Crystals and Molecules.* Cornell University Press, 1953.

[195] Clapp, L. B., *The Chemistry of the Coordination Compounds.* (Ed.: Bailar J. C.) Reinhold Publ. Co., New York, 1956.

[196] Ketelaar, J. A. A., *Chemical Constitution.* Elsevier, Amsterdam, 1958.

[197] Schlenk, W., *Ann. Chem.,* **565**, 204 (1949).

[198] Baron, M., *Org. Chem. Bull.,* **29**, 1 (1957).

[199] Powell, H. M., *The Structure of Molecular Compounds.* Part IV. *J. Chem. Soc.,* 61 (1948).

[200] de Forcrand, R., *Ann. Chem.,* (5) 28 (D.g.R. 298) 5 (1883).

[201] Villard, P., *Ann. Chem.,* (7) 11 (D.g.R. 341) 289 (1897).

[202] Glew, D. N., *Canad. J. Chem.,* **38**, 208 (1960).

[203] Lippert, E. L., Palmer, H. A. and Blankenship, F. F., *Proc. Oklah. Acad. Sci.,* **31**, 115 (1950).

[204] Bjerrum, N., *Structure and Properties of Ice. Dan. Mat. Fys. Medd.,* **27**, No. 1 (1951).

[205] Mulliken, R. S., *Phys. Rev.,* **43**, 279 (1933).

[206] Lennard-Jones, J. and Pople, J. A., *Proc. Roy. Soc.,* **A 202**, 166 (1950).

[207] Dennison, D. M., *Phys. Rev.,* **17**, 20 (1921).

[208] Bragg, W. H., *Proc. Roy. Soc.,* **34**, 103 (1922).

[209] Barnes, W. H., *Proc. Roy. Soc.*, A **125**, 670 (1929).

[210] Bernal, J. D. and Fowler, R. H., *J. Chem. Phys.*, **1**, 515 (1933).

[211] Bernal, J. D., *J. Chem. Phys.*, **50**, C 1 (1953).

[212] Pauling, L., *The Chemical Bond.* Cornell University Press, New York, 1967.

[213] Fox, J. J. and Martin, A. E., *Proc. Roy. Soc.*, A **174**, 234 (1940).

[214] Samoilov, O. Ya., *Struktura vodnykh rastvorov elektrolitov i gidratatsiya ionov.* Moscow, 1957.

[215] Tracey, M.V., *Austr. J. Sci.*, **31**, 418 (1969).

[216] Pople, J. A., *Proc. Roy. Soc.*, A **205**, 163 (1951).

[217] Pauling, L., *Hydrogen Bonding.* Pergamon Press, Oxford, 1959.

[218] Pauling, L. and Marsh, R. E., *Proc. Nat. Acad. Sci.*, **38**, 112 (1952).

[219] Marchi, R. P. and Eyring, H., *J. Phys. Chem.*, **68**, 221 (1964).

[220] Forslind, E., *Acta Politechnica*, **115**, 9 (1952).

[221] Eucken, A., *Z. Elektrochem.*, **52**, 255 (1948).

[222] Frank, H. S. and Wen, W. Y., *Disc. Farad. Soc.*, **24**, 133 (1957).

[223] Davis, C. M. Jr. and Litovitz, T. A., *J. Chem. Phys.*, **42**, 2563 (1965).

[224] Hall, L., *Phys. Rev.*, **73**, 775 (1948).

[225] Némethy, G. and Scheraga, H. A., *J. Chem. Phys.*, **36**, 3382 (1962); *J. Phys. Chem.*, **66**, 1773 (1962).

[226] Hajdu, F., *Acta Cryst.*, **A31**, 157 (1975).

[227] Morgan, J. and Warren, B. E., *J. Chem. Phys.*, **6**, 666 (1938).

[228] Danford, M. D. and Levy, H. A., *J. Am. Chem. Soc.*, **84**, 3965 (1962).

[229] Malenkov, E. V., *Zh. Strukt. Khim.*, **3**, 206 (1962).

[230] Eisenberg, D. and Kauzmann, W., *The Structure and Properties of Water.* Clarendon Press, Oxford, 1969.

[231] Kavanau, J. L., *Water and Solute-Water Interactions.* Holden-Day Inc., San Francisco, 1964.

[232] Némethy, G., *The Structure of Water and the Thermodynamic Properties of Aqueous Solutions.* Annali dell'Istituto Superiore di Sanita. Vol. VI. Fasc. spec. **1**, 491 (1970).

[233] Erdey-Grúz, T., *Transzportfolyamatok vizes oldatokban* (Transport processes in aqueous solutions). Akadémiai Kiadó, Budapest, 1971.

[234] Horne, R. A., *Marine Chemistry. The Structure of Water and the Chemistry of the Hydrosphere.* Part 1. Wiley-Interscience Publ., 1969.

[235] Horne, R. A., *Water and Aqueous Solutions.* Wiley-Interscience Publ., 1972.

[236] Falkenhagen, H., *Theorie der Elektrolyte.* Hirzel Verlag, Leipzig, 1971.

[237] Robinson, R. A. and Stokes, R. H., *Electrolyte Solutions.* 2nd ed. Butterworths Sci. Publ., 1959.

[238] Bockris, J. O'M. and Reddy, A. K. N., *Modern Electrochemistry.* Vol. I. Macdonald and Co., London, 1970.

[239] Luck, W. A. P. (Ed.), *Structure of Water and Aqueous Solutions.* Verlag Chemie, Weinheim/Bergstr., 1973.

[240] Covington, A. K. and Jones, P., (Eds.), *Hydrogen-Bonded Solvent Systems.* Taylor and Francis Ltd., London, 1968.

[241] Krestov, G. A., *Termodinamika ionnykh protsessov v rastvorakh.* Izd. Khimiya, Leningrad, 1973.

[242] Ben-Naim, A., *Water and Aqueous Solutions.* Plenum Press, 1974.

[243] Mishchenko, K. P. and Poltoratskiy, G. M., *Voprosi termodinamiki i stroeniya vodnykh i nevodnykh rastvorov elektrolitov.* Izd. Khimiya, Leningrad, 1976.

[244] Erlander, S. R., *J. Macromol. Sci. Chem.*, **A2**, 595 (1968).

[245] Eley, D. D., *Trans. Far. Soc.*, **35**, 1281, 1421 (1930).

[246] Powell, R. E. and Latimer, W. M., *J. Chem. Phys.*, **19**, 1139 (1951).

[247] Franks, F. and Ives, D. J. G., *Quarterly Rev. (London)* **20**, 1 (1966).

[248] Wada, G., *Bull. Chem. Soc. Japan*, **34**, 955 (1972).

[249] Berecz, E. and Báder, I., *Acta Chim. Hung.*, **74**, 213 (1972).

[250] Frank, H. S. and Wen, W. Y., *Disc. Farad. Soc.*, **24**, 133 (1957).

[251] Mishchenko, K. P., *Zh. Fiz. Khim.*, **26**, 1736 (1952).

[252] Berecz, E. and Báder, I., *Acta Chim. Hung.*, **77**, 285 (1973).

[253] Berecz, E., *Zsch. Phys. Chem.*, **229**, 173 (1965); *Acta Chim. Hung.*, **84**, 353 (1975).

[254] Allen, K. W., *J. Chem. Phys.*, **41**, 840 (1964).

[255] King, R. B., *Theor. Chim. Acta*, **25**, 309 (1972).

[256] Glew, D. N., *Nature*, **184**, 545 (1959).

[257] Klotz, J. M., *Frozen Coll. Symp. 1969* (Publ. 1970), p. 5.

[258] Glew, D. N. and Rath, N. S., *J. Chem. Phys.*, **44**, 1710 (1966).

[259] Villard, P., *Compt. Rend.*, **120**, 1262 (1895).

[260] Scheffer, F. E. C., *Versl. Kun. Ak. Wet. Amsterdam, Wi. en Nat. Afd.*, **19**, 1057 (1911).

[261] Braun, B., Dissertation, Bonn, 1938.

[262] Faraday, M., *Quarterly J. of Sci.*, **15**, 71 (1823).

[263] de Forcrand, R., *Compt. Rend.*, **135**, 959 (1902).

[264] Döpping, V., *Bull. Sci. Acad. St. Petersbourg*, **7**, 100, J. pr. 44, 255 (1848).

[265] Roozeboom, H. W. B., *Rec. Trav. Chim.*, **4**, 65 (1885).

[266] Geuther, A., *Ann.*, **224**, 218 (1884).

[267] Löwig, G., *Mag. d. Pharm.*, **23**, 12 (1828).

[268] Giraud, H., *Compt. Rend.*, **159**, 246 (1914).

[269] Mulders, E. M. J., Thesis, Univ. of Delft, 1937.

[270] Zernike, J., *Rec. Trav. Chim.*, **70**, 784 (1951).

[271] Allen, K. W. and Jeffrey, G. A., *J. Chem. Phys.*, **38**, 2304 (1963).

[272] van der Waals, J. H. and Platteeuw, J. C., *Mol. Phys.*, **1**, 91 (1958).

[273] Harris, W. H., *J. Chem. Soc. (London)*, 582 (1933).

[274] Müller, H. R., Dissertation, Bonn, 1952.

[275] Solacolu, C. and Solacolu, I., *Stud. Cerc. Chim.*, **21**, 1307 (1973).

[276] Roozeboom, H. W. B., *Die heterogene Gleichgewichte vom Standpunkte der Phasenlehre.* Vol. II, 2, pp. 191–206. Vieweg u. Sohn, Braunschweig, 1928.

[277] Bruckner, Gy., *Szerves Kémia* (Organic Chemistry). Vol. I-1. 4th ed. Tankönyvkiadó, Budapest, 1973.

[278] Perry, J. J., *Chem. Eng. Handbook*. McGraw-Hill, 1950, p. 165.

[279] Stuhl, D. R., *Ind. Eng. Chem.*, **39**, 517 (1974).

[280] Houghton, J., *J. Chem. Proc. Eng.*, **46**, 639 (1965).

[281] Mucskai, L., *Kristályosítás* (Crystallization). Műszaki Kiadó, Budapest, 1971.

[282] Malenkov, E. V., *Dokl. 3. Resp. Nauchno-Tekhn. Konf. po Neftekhimii AN Kaz.S.S.R.*, **3**, 320 (1974).

[283] Lutoshkin, G. Sh., Bukhgalter, E. B., Dyegtyarev, B. V., Kobzeyev, Yu. V. and Shulyatikov, V. I., SSSR Pat. No. 309.120. 9. Jul. 1971. *Otkr., Izobr., Prom. Obr., Tov. Znaki*, **48**, 120 (1971).

[284] Milev, N., *Khim. Ind. (Sofia)*, **49**, 303 (1977).

[285] Krasnov, A. A., Klimenok, B. V., *Dokl. Neftekhim. Sekts. Bashkir. Resp. Pravl. Vses. Khim. Obsh.* **6**, 241 (1971).

[286] Arshinov, S. A., Kolodezni, P. A. and Semerikov, A. A., *Gaz. Delo,* (12) 1971.

[287] Arshinov, S. A. and Kolodezni, P. A., *Gaz. Delo,* (11) 1971.

[288] Sumets, V. I., *Gaz. Prom.,* (2) 24 (1974).

[289] Tsarev, V. P. and Mordovskaya, N. I., *Gaz. Prom.,* (10) 33 (1976).

[290] Burnykh, V. S., Burnykh, N. M. and Slesarev, V. A., *Otkr., Izobr., Prom. Obr., Tov. Znaki,* **53,** 20 (1976).

[291] Kolesnikova, L. A., *Razrab. Gaz. Mestorozhd. Transp. Gaza* (3) 197 (1974).

[292] Koshelev, V. S., Fomina, V. I. and Byk, S. Sh., *Gaz. Prom.,* (4) 29 (1974).

[293] Gavriya, N. A., Verteletskaya, A. S., Yegorov, S. A., Bondar, A. D. and Moiseyeva, N. F., *Vestn. Kharkov. Politekhn. Inst.,* **125,** 62 (1976).

[294] Khoroshilov, V. A., Legezin, N. E., Kemkhadze, T. V., Bukhgalter, E. B. and Klepikov, V. V., *Korr. Zashch. Neftegaz. Prom.,* (2) 8 (1972).

[295] Khoroshilov, V. A., Dyegtyarev, B. V. and Bukhgalter, E. B., *Gaz. Prom.,* **15,** 18 (1970).

[296] Berecz, E. and Balla-Achs, M., Research report No. 37 (185-XI-1-1974, OGIL), NME, (Techn. Univ. Heavy Ind.), Miskolc.

[297] Kostyuk, V. I., *Gaz. Prom.,* **9,** (3) 41 (1964).

[298] Berecz, E., Unpublished data, NME, (Techn. Univ. Heavy Ind.), Miskolc, 1976.

[299] Atonov, G. A. and Soldatov, V. D., *Nefteprom. Delo,* (2) 29 (1977).

[300] Hammerschmidt, E. G., *Gas,* **15,** (1939).

[301] Hammerschmidt, E. G., *Brennstoff-Chemie,* **50,** 4 (1969).

[302] Poldermann, L. D., *Gas Cond. Conference.* Univ. of Oklahoma, 1958.

[303] Krasnov, A. A. and Klimenok, B. V., *Dokl. Neftekhim. Sekts. Bashkir. Resp. Pravl. Vses. Khim. Obshch.,* **8,** 116 (1972).

[304] Krasnov, A. A. and Klimenok, B. V., *Zh. Fiz. Khim.,* **44,** 1342 (1970).

[305] Krasnov, A. A. and Klimenok, B. V., *Zh. Fiz. Khim.,* **44,** 1333 (1970).

[306] Berecz, E. and Balla-Achs, M., Research report No. 47 (56-XI-5-1976, OGIL), NME, (Techn. Univ. Heavy Ind.), Miskolc.

[307] Rosen, W. F., *The Oil and Gas J.,* 1975, June, p. 128.

[308] Favoritov, Yu. A., Lukyanov, Yu. P. and Belobrova, A. I., *Neft. Gaz. Prom.,* (2) 47 (1975).

[309] Buleiko, M. D. and Starodubtsev, A. M., *Gaz. Prom.,* (7) 40 (1976).

[310] Bondar, A. D. and Guseinov, Ch. S., *Gaz. Prom.,* (11) 52 (1974).

[311] Pick, P. and Simanek, J., *GWF Gas/Erdgas,* **120,** 503 (1979).

[312] Parkhomenko, A. N., *V. Sb. Pererab. Neft. Gaz.,* (1) 25 (1974).

[313] Dashdamirov, F. A. and Rasulov, A. M., *Geol. Dob., Transp., Pererab. Gaza Kond.,* 1975, 185.

[314] Miroshnikov, A. M., Rikhterman, D. L. and Chudov, A. F., *Gazov. Prom.,* (7) 48 (1979).

[315] Yorizane, M., Sadamoto, Sh., Nasuoka, H. and Eto, Y., *Kogyo Kagaku Zasshi,* **72,** 2174 (1969).

[316] Burnykh, V. S., Burnykh, N. M. and Slesarev, V. A., *Otkr. Izobret. Prom. Obr. Tov. Znaki,* **53,** 20 (1976).

[317] Kedzierski, S. and Krzyzanowski, S., *Gaz. Woda Techn. Sanit.,* **45,** 295 (1971).

[318] Kuliyev, A. M., Musayev, R. M. and Iskenderov, S. M., *Gaz. Prom.,* (7) 51 (1973).

[319] Musayev, R. M. and Kuliyev, A. M., *Gaz. Delo,* (8) 13 (1971).

[320] Scauzillo, F. R., *Chem. Eng. Progr.,* **52,** 324 (1956).

[321] Sampson, J. A., Unpublished Thesis, Univ. of Oklahoma.

[322] Lippert, E. L., *Proc. Okl. Acad. Sci.,* **30,** 221 (1949).

[323] Maass, O. and Boomer, E. H., *J. Am. Chem. Soc.,* **44,** 1720 (1922).

[324] Godchot, J. M., Caugil, G. and Calas, R., *Compt. Rend.,* **202,** 759 (1936).

[325] Schöller, J. F., *Kohlenwasserstoffgase, 2*, 171 (1967).

[326] Ivanov, B. D. and Tsarev, V. P., *Fiz. L'da Ledotekhn.*, 1974, 192.

[327] Tezikov, V. N. and Stupin, D. Yu., *Izv. Vyssh. Uchebn. Zaved. Khim. i Khim. Tekhnol.*, **22**, 1039 (1979).

[328] Davidson, D. W., Garg, S. K., Gough, S. R. and Hawkins, R. E., *Canad. J. Chem.*, **5**, 3641 (1977).

[329] Wilson, G. J. and Davidson, D. W., *Canad. J. Chem.*, **41**, 264 (1963).

[330] Davidson, D. W. and Wilson, G. J., *Canad. J. Chem.*, **41**, 1424 (1963).

[331] Davidson, D. W., Garg, S. K., Gough, S. R., Hawkins, R. E. and Ripmeester, J. A., *Canad. J. Chem.*, **55**, 3641 (1977).

[332] Garg, S. K., Majid, Y. A., Ripmeester, J. A. and Davidson, D. W., *Mol. Phys.*, **33**, 729 (1977).

[333] Gough, S. R., Hawkins, R. E., Morris, B. and Davidson, D. W., *J. Phys. Chem.*, **77**, 2969 (1973).

[334] Ripmeester, J. A., *Canad. J. Chem.*, **54**, 3677 (1976).

[335] Ripmeester, J. A., *Canad. J. Chem.*, **55**, 78 (1977).

[336] Ripmeester, J. A. and Davidson, D. W., *Molecular Crystals and Liquid Crystals*, **43**, 189 (1977).

[337] Davidson, D. W. and Ripmeester, J. A., *J. Glaciology*, **21**, 33 (1978).

[338] de Forcrand R., *Compt. Rend.*, **133**, 368, 474 (1901).

[339] de Forcrand, R., *Compt. Rend.*, **133**, 513 (1901).

[340] de Forcrand, R., *Compt. Rend.*, **133**, 681 (1901).

[341] de Forcrand, R., *Compt. Rend.*, **134**, 835 (1902).

[342] Roberts, O. L., Brownscombe, E. R., Howe, L. S. and Reamer, H. H., *Petr. Eng.*, **12**, 56 (1941).

[343] Sloan, E. D. and Kobayashi R., "*Report Covering the Water Content Measurement of a Simulated Prudhoe Bay Gas in Equilibrium with Hydrates*". Submitted to Canadian Arctic Gas Services Ltd., Atlantic Richfield Co., Exxon Production Research Co. and the Northern Engineering Services Co., December, 1975.

[344] Sloan, E. D., Khoury, F. M. and Kobayashi, R., *J. Ind. Eng. Chem. Fundamentals*, **15**, 318 (1976).

[345] Aoyagi, K. and Kobayashi, R., "*Report on the Water Content Measurement of High Carbon Dioxide Content of a Simulated Prudhoe Bay Gas in Equilibrium with Hydrates*". Work supported by Atlantic Richfield Co. and Exxon Production Research Co., December, 1977.

[346] Aoyagi, K., Song, K. Y., Sloan, E. D., Dharmawardhana, P. B. and Kobayashi, R., "*Improved Measurement and Correlation of the Water Content of Methane Gas in Equilibrium with Hydrate*". Presented at 58th Annual GPA Convention", Denver, 1979.

[347] Ng Heng-Joo and Robinson, D. B., *Ind. Eng. Chem. Fundamentals*, **19**, 33 (1980).

[348] Peng, D. Y. and Robinson, D. B., *Ind. Eng. Chem. Fundamentals*, **15**, 59 (1976).

[349] Eucken, A., *Lehrbuch d. Chem. Physik.*, 1st edition, 1930, p. 483.

[350] Parent, J. D., The Storage of Natural Gas as Hydrate. *Inst of Gas Techn. Res. Bull.*, No. 1 (1948).

[351] Koshelev, V. S., Byk, S. Sh. and Fomina, V. I., *Gaz. Delo*, (11) 21 (1971).

[352] Byk, S. Sh. and Fomina, V. I., *Zh. Fiz. Khim.*, **46**, 994 (1972).

[353] Gritsenko, A. I., Nagayev, V. B., Volodina, L. A. and Plyushchev, L. V., *Izv. VUZ Neft i Gaz*, (7) 57 (1978).

[354] Gritsenko, A. I., *Ref. Sbornik VNIIE Gaz. Prom. ser. Razrab. i Ekspl. Gaz. i Gazokond. Mestorozhd.*, **7**, 25 (1976).

[355] Iskenderov, S. M. and Musayev, R. M., *Gaz. Delo*, (12) 6 (1970).

[356] Wilms, D. A. and van Haute, A. A., *Proc. Intern. Symp. Fresh Water, 4th*, **3**, 477 (1973).

[357] Wilcox, W. L., Carson, D. B., and Katz, D. L., *Ind. Eng. Chem.*, **33**, 662 (1941).

[358] Pick, P. and Simanek, J., *GWF, Gas/Erdgas*, **117**, 35 (1976).

[359] Carson, D. B. and Katz, D. L., *Tr. AIME Techn. Publ.*, No. 1371 (1941).

[360] Deaton, W. M. and Frost, E. M., *Proc. Nat. Gas. Dept. AGA*, p. 112 (1938).

[361] Deaton, W. M. and Frost, E. M., U.S. Bureau of Mines, Monograph 8 (1946).

[362] Krasnov, A. A. and Klimenok, B. V., *Dokl. Neftekhim. Sekts. Bashkir. Resp. Pravl. Vses. Khim. Obshchesta*, **6**, 230 (1971).

[363] van der Waals, J. H., *Disc. Farad. Soc.*, **15**, 261 (1953).

[364] van der Waals, J. H., and Platteeuw, J. C., *Rec. Trav. Chim.*, 912 (1956).

[365] Eucken, A., *Naturwiss.*, **36**, 327, 359 (1949).

[366] Byk, S. Sh., Fomina, V. I. and Norozhenko, A. F., *Gaz. Prom.*, **16**, 35 (1971).

[367] Byk, S. Sh., Fomina, V. I. and Norozhenko, A. F., *Neftepererab. Neftekhim.*, (11) 28 (1969).

[368] Koshelev, V. S., Antonova, G. M., Fomina, V. I, and Byk, S. Sh., *Zh. Prikl. Khim.*, **44**, 2573 (1971).

[369] Frost, E. M. and Deaton, W. M., *Proc. Nat. Gas Dept. AGA*, p. 49 (1946).

[370] Kuliyev, A. M., Musayev, R. M., Novruzova, F. M. and Saniyev, Z. A., *Gaz. Delo*, (9) 23 (1972).

[371] Pople, J., *Phil. Mag.*, **42**, 459 (1951).

[372] Wentorf, H., Büchler, A. J., Hirschfelder, J. O. and Curtiss, C. F., *J. Chem. Phys.*, **18**, 1484 (1950).

[373] Rushbrooke, G. S., *Introduction to Statistical Mechanics.* Clarendon Press, Oxford, 1949, p. 277.

[374] Fowler, R. H. and Guggenheim, E. A., *Statistical Thermodynamics.* University Press, Cambridge, 1939.

[375] Blue, R. W., *J. Chem. Phys.*, **22**, 280 (1954).

[376] Ockman, N. *Adv. Phys.*, **7**, 199 (1958).

[377] Kihara, T. K. *J. Phys. Soc. Japan*, **6**, 289 (1951).

[378] Kihara, T. K., *J. Phys. Soc. Japan*, **14**, 247 (1959).

[379] Kihara, T. K., *J. Phys. Soc. Japan*, **16**, 627 (1961).

[380] Hamann, S. D. and Lambert, P. A., *Austral. J. Chem.*, **7**, 1 (1954).

[381] Sinanoglu, O., *J. Chem. Phys.*, **30**, 850 (1959).

[382] Balescu, R., *Physica*, **22**, 223 (1956).

[383] Stupin, D. Yu., *Izv. VUZ. Khim. i Khim. Tekhnol.*, **23**, 416 (1980).

[384] Kihara, T. and Kobe, S., *J. Phys. Soc. Japan*, **9**, 609 (1954).

[385] Dannon, F. and Pitzer, K. S., *J. Chem. Phys.*, **36**, 425 (1962).

[386] Parrish, W. R. and Prausnitz, J. H., *Ind. Eng. Chem. Process Res. Develop.*, II (1) 26 (1972).

[387] Malyshev, A. G. and Tyushnyakova, G. N., *Trudy Gaz. NIPr. Inst. "Giprotyumenneftegaz"*, **35**, 267 (1973).

[388] Ivanov, B. D., *Fiz. L'da, Ledotekhn.*, 1974, 180.

[389] Ivanov, B. D., *Fiz. L'da, Ledotekhn.*, 1974, 174.

[390] Ng Heng-Joo and Robinson, D. B., *Ind. Eng. Chem. Fundamentals*, **15**, 293 (1976).

[391] Dharmawardhana, P. B., Parrish, W. R., and Sloan, E. D., *Ind. Eng. Chem. Fundamentals,* **19,** 410 (1980).

[392] Holder, G. D., Corbin, G. and Papadopulos, K. D., *Ind. Eng. Chem. Fundamentals,* **19,** 282 (1965).

[393] van Cleeff, A. and Diepen, G. A. M., *Rec. Trav. Chim.,* **84,** 1085 (1965).

[394] Bukhgalter, E. B., and Dyegtyarev, B. V., *Gaz. Delo,* (3) 19 (1972).

[395] Miroshnichenko, V. I., *Tr. Tsent. Aerol. Obshch.,* **104,** 85 (1976).

[396] Ewing, G. J. and Ionescu, L. G., *J. Chem. Eng. Data,* **19,** 366 (1974).

[397] Balla-Achs, M. and Berecz, E., Unpublished data, NME, (Techn. Univ. Heavy Ind.), Miskolc.

[398] Garg, S. K., Majid, Y. A., Ripmeester, J. A. and Davidson, D. W., *Mol. Phys.,* **33,** 729 (1977).

[399] Iskenderov, S. M. and Musayev, R. M., *Gazov. Delo,* (12) 6 (1970).

[400] Bozzo, A. T., Chen Hsiao-Sheng, Kass, J. R., and Barduhn A. J., *Proc. Int. Symp. Fresh Water, 4th,* **3,** 437 (1973).

[401] Glew, D. N., and Hames, D. A., *Canad. J. Chem.,* **47,** 4651 (1969).

[402] Groisman, A. G., *Fiz. L'da, Ledotekhn.,* 158 (1974).

[403] Findlay, A., and Campbell, A. N., *The Phase Rule and its Applications.* Longmans, London, 1938. p. 206.

[404] van Berkum, J. G. and Diepen, G. A. M., *J. Chem. Thermodynamics,* **11,** 317 (1979).

[405] de Forcrand, R., *Ann. Chim. Phys.,* **5,** 5 (1883).

[406] de Forcrand, R., *Compt. Rend.,* **94,** 967 (1882).

[407] de Forcrand, R., *Compt. Rend.,* **106,** 849 (1888).

[408] Scheffer, F. E. C. *Proc. Kon. Nederl. Akad. Wetensch.,* **13,** 829 (1911).

[409] Scheffer, F. E. C., *Z. phys. Chem.,* **84,** 734 (1913).

[410] Korvezee, A. E., Scheffer, F. E. C., *Rec. Trav. Chim.,* **50,** 256 (1931).

[411] Schreinemakers, F. A. H., *Z. phys. Chem.,* **82,** 59 (1913).

[412] Wright, R. M. and Maas, O., *Canad. J. Research,* **6,** 94 (1932).

[413] Villard, P., *Compt. Rend.,* **107,** 395 (1888).

[414] Deaton, W. M. and Frost, E. M., *Gas,* **16,** 28 (1940).

[415] Deaton, W. M., and Frost, E. M., *Proc. Nat. Gas Dept., AGA,* p. 122 (1940).

[416] Campbell, J. M., and McLeod, H. D., *Tr. AIME,* **222,** 590 (1961).

[417] Glew, D. N., and Moelwyn-Hughes, E. A., *Proc. Roy. Soc. A.,* **211,** 254 (1952).

[418] Glew, D. N., *Disc. Farad. Soc.,* **15,** 150 (1953).

[419] Winkler, L. W., *Ber.,* **34,** 1408 (1901).

[420] Morrison, T. J. and Billet, F., *J. Chem. Soc.,* 3814 (1952).

[421] Culberson, O. L. and McKetta, J. J., *J. Petr. Techn. Tr. AIME,* **192,** 223 (1951).

[422] Dorsey, N. E., *Properties of Ordinary Water Substance.* Hafner Publ. Co., New York, 1968.

[423] Krichevsky, I. R. and Kasarnovsky, J. S., *J. Am. Chem. Soc.,* **57,** 2168 (1935).

[424] Krichevsky, I. R. and Ilyinskaya, A., *Acta Physicochim. USSR,* **20,** 327 (1945).

[425] Masterson, W. L., *J. Chem. Phys.,* **22,** 1830 (1954).

[426] Sloan, E. D., Khouty, F. M. and Kobayashi, R., *Ind. Eng. Chem. Fundamentals,* **15,** 318 (1976).

[427] Falabella, B. J. and Vanpee, M., *Ind. Eng. Chem. Fundamentals,* **13,** 228 (1974).

[428] Deaton, W. M. and Frost, E. M., *Oil and Gas J.,* **36,** 75 (1937).

[429] Deaton, W. M., and Frost, E. M., *Gas,* **14,** 31 (1938).

[430] Diepen, G. A. M. and Scheffer, F. E. C., *Rec. Trav. Chim.,* **78,** 126 (1959).

[431] van Cleeff, A. and Diepen, G. A. M., *Rec. Trav. Chim.,* **81,** 426 (1962).

[432] Morlat, M., Pernolet, R. and Gerard, N., *Proc. Int. Symp. Fresh. Water 4th*, **3**, 263 (1976).

[433] Sage, B. H., Schaafsma, I. G., and Lacey, W. N., *Ind. Eng. Chem.*, **26**, 1218 (1934).

[434] Reamer, H. H., Selleck, F. F. and Sage, B. H., *J. Petr. Technol.*, **4**, 197 (1952).

[435] Brown, G. G., *Petr. Eng. Continuous Tables*, Vols. 1–6.

[436] Wu Bing-Jing and Robinson, D. B., *J. Chem. Thermod.*, **5**, 461 (1976).

[437] Berecz, E. and Balla-Achs, M., Research report, No. 66 (232–XI–3–1978, OGIL) NME (Techn. Univ. Heavy Ind.), 1980.

[438] Khoroshilov, V. A. and Bukhgalter, E. V., *Zh. Fiz. Khim.*, **47**, 2393 (1973).

[439] Byk, S. Sh. and Fomina, V. I., *Khim. Tekhnol. Topl. Masel*, **17**, 21 (1972).

[440] Malyshev, A. G., Tyushnyakova, G. N. and Kaptelinin, Ya. D., *Trudy Sib. NII. Neft. Prom.*, **3**, 259 (1975).

[441] van der Waals, J. H. and Platteeuw, J. C., *Rec. Trav. Chim.*, **78**, 126 (1959).

[442] Ng Heng-Joo, Petrunia, J. P. and Robinson, D. B., *Fluid Phase Equilibria*, **1**, 283 (1977–78).

[443] Robinson, D. B. and Mehta, B. R., *J. Canad. Petr. Technol*, **10**, 33 (1971).

[444] Craig, R. D., Dissert. Univ. Microfilms, Ann Arbor, Michigan. Order No. 73–12, 496.

[445] Simanek, J. and Pick, P., *Plyn* **53**, 167 (1973).

[446] Török, J., Fürcht, L., and Balla-Achs, M., Lecture at OGIL Conference, Nagykanizsa, Hungary, 1976.

[447] Katz, D. L., *Handbook of Natural Gas Eng.*, McGraw-Hill Book Co., New York, 1959.

[448] Musayev, R. M. and Saniyev, Z. S., *Tr. Azerb. Fil. Vses. NIIPrG* (2) 187 (1973).

[449] McCarthy, E. H., Boyd, W. L. and Reid, L. S., *Petr. Tr. AIME*, **189**, 241 (1950).

[450] Skinner, W. R., *The Water Content of Natural Gas at Low Temperatures*. Univ. of Oklahoma, 1948.

[451] Kobayashi, R. and Withrow, H. J., *Proc. Nat. Gas Ass.*, (1951).

[452] Hammerschmidt, E. G., *Calculation and Determination of Moisture Content of Compressed Natural Gas.*, Western Gas, 1963.

[453] Brickel, W. F., *Petr. Eng.*, **24**, (12) (1952).

[454] Malenkov, E. V., Nauruzov, M. Kh. Gafarova, N. A. and Makogon, Yu. F., *Trudy Inst. Khim. Nefti i Pr. Solei*, **4**, 21 (1972).

[455] Won, K. W., *Adv. Cryogenic Eng.*, **23**, 544 (1978).

[456] Hemmaplardh, B. and King, A. D., *J. Phys. Chem.*, **76**, 2170 (1972).

[457] Byk, S. Sh. and Fomina, V. I., *Neftepererab. Neftekhim.*, (10) 29 (1969).

[458] Khoroshilov, V. A., Bukhgalter, E. L. and Burmistrov, A. G., *Gaz. Prom.*, (4) 30 (1974).

[459] Malenkov, E. V., *Trudy Inst., Khimii Nefti i Prir. Solei AN Kaz. S.S.R.*, **8**, 45 (1975).

[460] Backhurst, J. R. and Harker, J. H., *J. Inst. Fuel*, **43**, 405 (1970).

[461] Falabella, B. J., Dissertation. Univ. Microfilms, Ann Arbor, Michigan. Order No. 76–5849.

[462] Dzhavadov, A. D. and Tribus, N. A., *Azerb. Neft. Khoz.*, 46 (1967).

[463] Katz, D. L. and Bergman, D. F., *Proc. Gas Cond. Conf.*, 1973. 23 H, p. 16.

[464] Ng Heng-Joo and Robinson, D. B., *Ind. Eng. Chem. Fundamentals*, **15**, 293 (1976)

[465] Ng Heng-Joo and Robinson, D. B., *AICHE J.* **23**, 477 (1977).

[466] Robinson, D. B., Peng, D. Y. and Ng Heng-Joo, *Hydrocarbon Processing*, Sept. 1979, p. 262.

[467] Carson, D. B. and Katz, D. L., *Trans. AIME*, **146**, 150 (1941).

[468] Deaton, W. M. and Frost, E. M., *Proc. Nat. Gas Dept. AGA*, 122 (1942).

[469] Katz, D. L., *Proc. Nat. Gas Dept. AGA*, 41 (1942).

[470] Katz, D. L. *Refiner*, **21**, 64 (1942).

[471] Dyegtyarev, B. V., Lutoshkin, G. S., and Bukhgalter, E. B., *Borba s gidratami pri eksploatatsii gazovikh skvazhin v raionakh Severa.* Izd-vo "Nedra", Moscow, 1969.

[472] Makogon, Yu. F. and Skhalyakho, A. S., *Opredelenie uslovii obrazovaniya gidratov i ikh preduprezhdenie.* VNIIEOPr i TE Inf. v Gaz. Prom. Moscow, 1972.

[473] Ginsburg, G. A., *Sbornik statei po gidrogeologii i geotermii.* Vip. 1. Gidrogeologiya Yeniseiskogo Severa. Leningrad, 1969, pp. 109–128.

[474] Katz, D. L., *J. Petr. Technol.* **23** (April) 419 (1971).

[475] Makogon, Yu. F., Tsarev, V. P., and Cherskiy, N. V., *Dokl. AN S.S.S.R.,* **205**, 700 (1972).

[476] Safir, M. Kh. *Geol. Nefti i Gaza,* (6) 26 (1973).

[477] Safir, M. Kh., Ginsburg, G. A., Kislova, V. I. and Bogatirenko, R. S., *Geol. Nefti i Gaza,* (7) 71 (1973).

[478] Khoroshilov, V. A., Bukhgalter, E. B., and Zaitsev, Ya. Ya., *Gazov. Prom.,* (4) 8 (1973).

[479] Sheshukov, N. L., *Geol. Nefti i Gaza* (6) 20 (1973).

[480] Bily, C. and Dick, J. W. L., *Bull. Canad. Petr. Geol.,* **22**, 340 (1974).

[481] Stoll, R. D., Ewing, J. and Bryan, G. M., *J. Geophys. Res.,* **76**, 2090 (1971).

[482] Cherskiy, N. V., Makogon, Yu. F. and Medovskiy, D. M., *Mater. Vses. Sov. Otsenk. Neftegazonos. Territ. Yakutii.* (Publ. 1968) 458. Ed. Trofimuk, A. A., Izd. "Nedra", Moscow, 1966.

[483] Trofimuk, A. A., Cherskiy, N. V. and Tsarev, V. P., *Dokl. AN S.S.S.R.,* **225**, 936 (1975)

[484] Trofimuk, A. A., Cherskiy, N. V. and Tsarev, V. P., *Dokl. AN S.S.S.R.,* **212**, 931 (1973).

[485] Tucholke, B. E., Bryan, G. M., and Ewing, J. I., *Am. Assoc. Petr. Geol. Bull.,* **61**, 698 (1977).

[486] Makogon, Yu. F., Trofimuk, A. A., Tsarev, V. P. and Cherskiy, N. V., *Geol. Geofiz.,* **4,** 3 (1973).

[487] Makogon, Yu. F., *M.G.P. Express Inform.,* **9**, 11 (1972), Vniizgazprom, Moscow.

[488] Yefremova, A. G. and Zhizhchenko, B. P., *AN S.S.S.R.,* **214**, 1179 (1974).

[489] Makogon, Yu. F. and Khalikov, G., *M.G.P. Express Inform.,* **6**, 6 (1971), Vniizgazprom. Moscow.

[490] Verigin, N. N., Khabibullin, I. L., and Khalikov, G. A., *Izv. AN S.S.S.R. Mekhanika Zhidkosti i Gaza,* No. 1, 174 (1980).

[491] Dyegtyarev, B. V. and Bukhgalter, E. B., *Borba c gidratami pri eksploatatsii gazovikh skvazhin v severnikh raionakh.* Izd. Nedra, 1976.

[492] Bukhgalter, E. B., *Razr. Gaz. Mestor. Pr. Gaz.,* **2**, 129 (1974).

[493] Török, J., Összetétel-eloszlások inerttartalmú földgáztelepekben. (Composition distributions in natural gas sources with inert contents.) Dissertation, Hung. Acad. Sci., Budapest, 1978.

[494] Makogon, Yu. F., *Gazov. Prom.* (5) 15 (1965).

[495] Mirzadzhanzade, A Kh., and co-workers: *Razrabotka gazokondensatnykh mestorozhdenii.* Izd-vo "Nedra", Moscow, 1967.

[496] Ivanov, N. S., Makogon, Yu. F., Tsarev, V.P. and Cherskiy, N. V., *Osobennosti rascheta uslovii nachala obrazovaniya gidratov v poristoi srede.* Sbornik Geologiya YaFSO AN S.S.S.R., 1971.

[497] Shirkovskiy, A. I., *Opredelenie i ispolzovanie fizicheskikh parametrov poristoi sredi pri razrabotke gazokondensatnikh zalezhei.* VNIIEGazProm, 1971.

[498] Evrenos, A., Heatman, J. and Ralstin, J., *J. Petr. Techn.,* **23**, 1059 (1971).

[499] Lakeyev, V. P., *Gaz. Prom.,* **10**, 39 (1973).

[500] Shcherbakova, P. R., and Byk, S. Sh., *Nefteper. Neftekhim.,* **10**, 23 (1972).

[501] Byk, S. Sh., Fomina, V. I. and Koshelev, V. S., *Gaz. Prom.,* **17**, 42 (1972).

[502] Shcherbakova, P. R. and Byk, S. Sh., *Gaz. Delo,* 4, 10 (1972).

[503] Shcherbakova, P. R. and Byk, S. Sh., *Gaz. Prom.*, **16**, 41 (1971).

[504] Moreau, B. L., *Oil in Canada*, **11**, 14 (1959).

[505] Fazlutdinov, A. R., Gorodilov, V. A. and Svezhintsev, V. I., *Nefteprom. Delo*, (5) 27 (1979),

[506] Fazlutdinov, A. R., *Nefteprom. Delo*, (12) 28 (1979).

[507] Saifeyev, T. A., *Tr. Kuibishev. NII NP;* Vyp. 2 (1960).

[508] Sidorovskiy, V. A., Mezhlumov, O. A. and Belov, V. I., *Neft. Khoz.*, **2**, 56 (1967).

[509] Lawrence, C. J., *Oil and Gas*, **62**, 72 (1964).

[510] Sheinman, A. B., Sergeyev, N. I. and Malofeyev, G. E., *Elektroteplovaya obrabotka prizaboinoi zoni neftyanikh skvazhin.* Gostoptekhizdat, 1962.

[511] Pastukhov, I. V., Ilyukov, V. A. and Zvonarev, N. G., *Nefteprom. Delo*, (9) 6 (1966).

[512] Pelenichka, L. G., Polivko, I. N., Stepanchikov, E. A. and Paslavskiy, N. G., *Neftepr. Delo*, (1), 19 (1966).

[513] Medovskiy, D. I., *Trudy Mosk. Inst. NKh i GProm.*, **91**, 376 (1969).

[514] Cherskiy, N. V., Tsarev, V. P., Bubnov, A. V., Yefremov, I. D. and Bondarev, E. A., *Fiz. Tekhn. Probl. Severa, 112,* Nauka, Novosibirsk, 1972.

[515] Kolodezni, P. A. and Arshinov, S. A., *Gaz. Delo*, (7) 3 (1972).

[516] Saifeyev, T. A., *Gaz. Delo*, (12) 27 (1969).

[517] Kiyko, E. K., Byrko, V. Ya. and Karachun, F. M., *Gaz. Prom.*, (7) 11 (1973).

[518] Vasilyeva, V. G., Dinkov, V. A., Korotayev, Yu. P., Makogon, Yu. F., Trebin, F. A. and Cherskiy, N. V., *Otkr. Izobr. Prom. Obr. Tov. Znaki*, **50**, 111 (1973).

[519] Clark, M. O., US Pat. 3,644,107 (C1 48/1901 F, 17d) 1972. II.

[520] Ciemochowski, M. F. and Bishop, W. H., *Amer. Gas. Ass., Oper. Sect. Proc.*, 1968, 220.

[521] Török, J., Pápay, J., Haraszti, E., Berecz, E. and Kassai, L., *Fázisegyensúlyok és olajmérnöki alkalmazásuk* (Phase equilibria and their application in oil engineering). II. Bányaipari Szakirodalmi Tájékoztató. NIMDOK Budapest, 1968, pp. 23–41.

[522] Kedzierski, S. and Chowaniec, A., *Gaz, Woda Techn. San.*, **44**, 117 (1970).

[523] Robinson, D. B. and Ng Heng-Joo, *Hydrocarbon Proc.*, **54**, 95 (1975).

[524] Cooper, L. S., *Inst. Chem. Eng. Symp., Ser.*, **44**, (Nat. Gas Pr. Unt.) 1, 1 (1976).

[525] Babe, G. D., Bondarev, E. A., Groisman, A. G., Kanibolotskiy, M.A., *Inzh. Fiz. Zh.*, **25**, 94 (1973).

[526] Bondarev, E. A., Babe, G. D., Groisman, A. G., and Kanibolotskiy, M. A., Nauka, Sib. Otdel. AN S.S.S.R., 1976, 9. 158.

[527] Dewerdt, F. and Roncier, M., *C. R. Congr. Ind. Gaz*, **88**, 295 (1971).

[528] Simon, D., and Debrescu, J., *Previnirea Formatii Depunerilor de Hidrate in Instalatiile Technologice si Conductele de Transport a Gazelor Naturala.* Bucharest, 1973.

[529] Mendez, G. M., *Rev. Inst. Mex. Petrol.*, **1**, 40 (1969).

[530] Khristov, K. and Arizanov, V., *Gor. Vis. Minno-Geol. Inst. Sofia*, **17**, 91 (1973).

[531] Harber, E. A., US Publ. Pat. Appl. B 421.383 (1975 I).

[532] Connealy, L. E., *Gas,* Nov. 1960, pp. 101–108.

[533] Chernishev, A. E. and Boiko, V. A., *Ispol'z. Prir. Gaza v Prom., Mater. Nauchno-Tekhn. Konf. Mold. Issled. Inst. Gaza AN Ukr. S.S.R. 1968* (published 1969), 339.

[534] Malenkov, E. V., Gafarova, N. A., Makogon, Yu. F., Asilkhanov, S. A. and Umrazkov, U., *Dokl. 3. Resp. Nauchn. Tekhn. Konf. po Neftekhimii. AN Kaz. S.S.R.*, **3**, 75 (1974).

[535] Mamedov, Yu. S., *Bezop. Tr. Prom.*, **14**, 32 (1970).

[536] Nikitin, V. I., Krilov, E. V. and Karpov, V. A., *Gaz. Prom.*, (11) 28 (1973).

[537] Hinton, R. A. and Kurata, F., *Neth. Appl.*, 7503,469 C1 B0ID, C07C F 25 J 29. Sept. 1975 US Appl. 454,740. 25 March 1971.

[538] Pekhata, F. N., Bergo, B. G. and Surkov, Yu. V., *Trudy VNII Prir. Gazov,* (1Ch1), 142 (1974).

[539] Herrin, J. P. and Armstrong, R. A., *Proc. Gas Cond. Conf.* 1972, 22G, p. 10.

[540] Smolyaninov, V. G. and Khadikin, V. G., *Gaz. Delo,* (4) 12 (1972).

[541] Lisovoder, G. K., *Gaz. Prom.,* (1) 42 (1974).

[542] Tinnikov, V. V., Pavlov, V. M., Mlyushchenko, N. D. and Repalov, V. I., *Gaz. Prom.,* (3) 31 (1975).

[543] Kolesnikova, L. A. and Legezin, N. A., *Korr. Zashch. Neftegaz. Prom.,* (5) 20 (1976).

[544] Burmistrov, A. G. and Krasnov, A. A., *Nefteprom. Delo,* (10) 28 (1979).

[545] Atanov, G. A. and Soldatov, V. D., *Nefteprom. Delo,* (2) 29 (1977).

[546] Khoroshilov, V. A., Dyegtyarev, B. V. and Bukhgalter, E. B., *Gaz. Prom.* (15) 18 (1970).

[547] Khoroshilov, V. A., Legezin, N. E., Kemkhadze, T. V., Bukhgalter, E. B. and Klepikov, V. V., *Korr. Zashch. Neftegaz. Prom.,* (2) 8 (1972).

[548] Dubinin, V. M., Zhidenko, G. G., *Gazov. Prom-st., Ser. Transp. Khranenie Gaza* (ref. inf.) (6) 20 (1979).

[549] Cahn, R. P., Johnston, R. H. and Plumstead, J. A., US Pat. 3,514,27 (C148—190) F 17C, F 25 J, 26 May 1970.

[550] Nierman, A. J., US Pat. 3,975,167 (C148)190, F 17C13 1976 VIII.

[551] Korotayev, Yu. P. and Ponomaryov, G. V., (Eds.), *Rukovodstvo po dobiche, transportu i pererabotke prirodnogo gaza.* Izd-vo "Nedra", Moscow, 1965.

[552] Makogon, Yu. F. and Novikova, N., *Ekspr. Inf., Dobicha Gaza* (7) 26 (1970).

[553] Crowther, I. F., US Patent No. 2399723 (1946).

[554] Tsarev, V. P. and Savvin, A. Z., *Gazov, Prom,* (2) 46 (1980).

[555] Glew, D. N., US Patent No. 3235630 (1966).

[556] Musayev, R. M., *Gaz. Delo,* (3) 25 (1970).

[557] Gebhardt, H. J., Makin, E. C. and Pierron, E. D., *Chem. Eng. Progr. Symp. Ser.,* **66**, 105 (1970).

[558] Bukhgalter, E. B., *Gaz. Delo,* (2) 23 (1970).

[559] Afdah, R. L. and Barber, H. W., US Pat. 3,676,981 (C150/30; B01d) 18 July 1972. Appl. 118,411, 24 Febr. 1971.

[560] Tabuchi, K., *Ishikawajima-Harima-Giho,* **10**, 481 (1970).

[561] Izmailova, Kh. I., *Razr. Gaz. Mestor, Transp. Gaza,* (3) 40 (1974).

[562] Verma, V. K., Dissertation. Univ. Microfilms. Ann Arbor, Mich. Order No. 75—10324.

[563] Kern, L. R., Can. Pat. 1,017,667 (C1 166—39) 1977 IX.

[564] Ng Heng-Joo and Robinson, D. B., *AICHE J.,* **23**, 477 (1977).

[565] Yakutseni, V. P. and Barkan, E. S., *Geol. Nefti i Gaza,* **2**, 29 (1977).

[566] Gukman, L. M. and Kasperovich, A. G., *Theor. Osn. Khim. Tekhnol.,* **9**, 281 (1975).

[567] Karnofsky, G., *Chem. Eng. Progr.* **57**, 42 (1961).

[568] Saito, S. and Iijima, M., *Nipp. Kaisui Gakhai-Shi,* **23**, 46 (1969).

[569] Saito, S., *Nipp. Kaisui Gakhai-Shi,* **22**, 257 (1968).

[570] Kleshunov, E. I. and Smirnov, L. F., *Izv. Sev.-Kazk. Nauchn. Tsent. Vyssh. Shkol., Ser. Tekhn. Nauk,* **4**, 86 (1976).

[571] Smirnov, L. F., *Kholod, Tekhn.,* **2**, 28 (1973).

[572] Barduhn, A. J., Roux, G. M., Richard, H. A., Giuliano, G. B., and Stern, S. A., *Desalination,* **18**, 59 (1976).

[573] Colten, S. L., Lin, F. S., Tsao, T. C., Stern, S. A. and Barduhn, A. J., *US Off. Saline Water Res. Develop. Progr. Rep.,* No. 753, pp. 72 (1972).

[574] Kirkley, D. W., S. Afr. Pat. 7106,077 (C1.C.02b) 1972 III.

[575] Gmelin Handbuch, *Water Desalting,* Suppl., Vol 1, Springer Verlag, Berlin, Heidelberg, New York, p. 336 (1979).

[576] Makogon, Yu. F. and Morozov, I. F., *Bezop. Truda v Ugol. Prom.* (5) 12 (1973).

[577] Stefanovich, G. Ya., Nikolin, V. I., and Nedosedkin, V. N., *Bezop. Tr. Prom.,* (18) 57 (1974).

[578] Makogon, Yu. F. and Morozov, I. F., *Bezop. Tr. Prom.,* (17) 26 (1973).

[579] Ettinger, L. L. and Shulman, N. V., *Bezop. Tr. Prom.,* (18) 30 (1974).

[580] Zenin, A. G., *Bezop. Tr. Prom.,* (18) 57 (1974).

[581] Ettinger, L. L. and Shulman, N. V., *Vopr. Teor. Uglya Por. Gaza,* Materiala Sem. 1970 (published 1974), p. 271.

[582] Zenin, A. G., *Ugol Ukr.,* 2, 34 (1980).

[583] Koshelev, V. S., Fomina, V. I., Byk, S. Sh., *Zh. Fiz. Khim.,* 45, 2968 (1971).

[584] Korotayev, Yu. R., Semin, V. I. and Khoroshilov, V. A. *Gaz. Delo,* 9, 3 (1969).

[585] Medovskiy, D. I., *Sb. Statei Mater. Nauchn. Konf. Ukhtins. Ind. Inst.,* 2, 115 (1973).

[586] Bukhgalter, E. B. and Turikin, A. F., *Gaz. Prom.,* 7, 32 (1975).

[587] Verma, V. K., Hand, J. H., Katz, D. L. and Holder, G. D., *J. Petr. Technol.,* 2, 223 (1975).

ACKNOWLEDGEMENTS

The authors wish to express their thanks to the following Publishing Houses, Societies, Organizations and Institutes for permission to reproduce literature data employed in connection with the tables and figures:

American Chemical Society (Washington): [41, 86, 114, 166, 185, 279, 357, 396]
American Institute of Chemical Engineers (New York): [115, 116, 127]
American Institute of Physics (New York): [101, 385, 421]
American Society of Heating, Refrigerating and Air-Conditioning Engineers (New York): [63]
Australian and New Zealand Association for the Advancement of Science (Sydney): [215]
Berichte der Bunsen-Gesellschaft für Physikalische Chemie (Darmstadt): [66, 67, 68, 69, 70]
The Canadian Institute of Mining and Metallurgy (Montreal): [156]
The Chemical Society (London): [222, 420]
W. H. Freeman and Company (San Francisco): [190]
Institute of Gas Technology (Chicago): [45]
Koninklijke Nederlandse Chemische Vereniging (The Hague): [159]
McGraw-Hill Book Company (New York): [278, 357, 447]
National Research Council (Ottawa): [117, 202]
Oil and Gas Journal (Tulsa, Oklahoma): [44, 51, 307]
Pearless de Rougemont and Company (Sussex): [403]
Pipeline and Gas Journal — International Pipeline and Gas Utility Design, Construction and Operation (Dallas): [42]
Redaktion Erdöl und Kohle, Erdgas-Petrolchemie (Leinfelden): [316]
Redaktsiya Zhurnala Azerbaidzhanskoe Neftyanoe Khozyaistvo (Baku): [138, 462]
The Royal Society (London): [99, 209, 216]
The Society of Chemical Engineers, Japan (Tokyo): [139, 377, 378, 379, 384]
Society of Petroleum Engineers of AIME (Dallas): [46, 47, 50, 64, 65, 82, 89, 113, 342, 361, 421, 434, 449]
Springer Verlag K. G. (Heidelberg): [71, 73, 75]
Vsesoyuzniy Nauchno—Issledovatelskiy Institut Organizatsii, Upravleniya i Ekonomiki Neftegazovoi Promyshlennosti (Moscow): [146, 147, 399, 472]

Thanks are also due to the Hungarian Research and Development Institute for the Hydrocarbon Industry (earlier the OGIL) for permission to publish results of research activity obtained during collaboration.

AUTHOR INDEX

Afdah, R. L. 292, 325
Alexeyeff, W. 9, 57, 311
Aliyev, A. G. 21, 90–94, 314
Allen, K. W. 57, 317
Andryushchenko, F.K. 29, 53, 91, 306, 307, 314
Antonova, G. M. 150, 209, 320
Aoyagi, K. 117, 309, 310, 319
Arizanov, V. 286, 324
Armstrong, R. A. 287, 325
Arshinov, S. A. 82, 283, 318, 324
Asilkhanov, S. A. 287, 324
Atonov, G. A. 95, 315, 325
Aziniers, L. 13, 312

Babe, G. D. 286, 324
Backhurst, J. R. 259, 322
Báder, I. 43, 45, 317
Balescu, R. 167, 320
Balla-Achs, M. 83–86, 99, 100–102, 139, 179–187, 194–196, 220–227, 251, 252, 254, 260, 308, 309, 318, 321, 322
Balló, M. 9, 311
Banks, R. B. 22, 314
Barber, H. W. 292, 325
Barduhn, A. J. 23, 184, 187, 284, 314, 315, 321, 325
Barkan, E. S. 292, 325
Barnes, W. H. 34, 316
Baron, M. 30, 315
Barrer, R. M. 17, 53, 104, 141–154, 157, 291, 313
Belobrova, A. I. 106, 318
Belov, V. 22, 282, 314, 324
Ben-Naim, A. 40, 316

Berecz, E. 43, 44, 46, 83–86, 94, 99–102, 179–182, 184–187, 194–196, 214, 220–227, 252, 254, 284, 285, 308, 309, 317, 318, 321, 322, 324
Bergman, D. F. 260, 322
Bergo, B. G. 287, 325
Bernal, J. D. 34, 36, 316
Berkum, J. G. van 189, 190, 309, 310, 321
Berthelot, M. 9, 311
Bertie, J. E. 17, 313
Bhatnagar, W.B. 18, 313
Billet, F. 199, 200, 321
Bily, C. 268, 323
Birckel, W. F. 254, 322
Birko, V. Ja. 240
Bishop, W. H. 283, 324
Bjerrum, N. 34, 315
Blankenship, F. F. 16, 33, 111, 112, 313, 315
Blue, R. W. 104, 164, 320
Bockris, J. O'M. 40, 316
Bogatirenko, R. S. 268, 323
Bogayevskiy, I. 23
Boiko, V. A. 286, 324
Bond, D. C. 15, 81, 87, 88, 89, 108, 193, 313
Bondar, A. D. 83, 106, 318
Bondarev, E. A. 286, 323,
Boomer, E. H. 112, 318
Bordet, H. 9, 311
Bouzat, A. 13, 312
Boyd, W. L. 253, 254, 322
Bozzo, A. T. 184, 187, 321
Bradley, R. 14, 312
Bragg, W. H. 34, 315
Braun, B. 56, 317
Brown, G. G. 213, 322
Brown, H. 14, 263, 312, 315

Brownscombe, E. R. 117, 194–196, 201, 204–206, 301, 312, 319
Bruckner, Gy. 66, 67, 68, 317
Bryan, G. M. 268, 269, 323
Bubnov, A. V. 283, 324
Büchler, A. J. 151, 161, 320
Bukhgalter, E. B. 21–23, 81, 83, 104, 136, 174, 228, 259, 267, 268, 272, 273, 277–279, 282, 287, 291, 310, 314, 317, 318, 321, 322, 323, 325, 326
Buleiko, M. D. 106, 318
Burmistrov, A. G. 104, 259, 287, 322, 325
Burnykh, N. M. 82, 109, 318
Burnykh, V. S. 82, 109, 318
Byk, S. Sh. 18, 123, 124, 149, 150, 183, 209, 232, 256, 277, 290, 313, 318, 319, 320, 322, 323, 324, 326
Byrko, V. Ya. 283, 324

Cahn, R. P. 288, 325
Cailletet, L. 10, 311
Calas, R. 14, 112, 312, 318
Campbell, J. M. 18, 134–136, 153, 188, 197, 201, 261, 262, 313, 321
Carmichael, L. T. 16, 191–193, 245, 313
Carson, D. B. 14, 18, 127–133, 150, 207, 209–211, 213–214, 231, 245, 252, 256, 257, 301, 312, 320, 322
Caugil, G. 14, 112, 312, 318
Chancel, G. 9, 311
Chen Hsiao Sheng, 184, 187, 321
Chernikhin, V. I. 19, 109, 259–261, 314
Chernishev, A. E. 286, 324
Cherskiy, N. 22, 23, 268, 269, 276, 283, 314, 323, 324
Child, W. 17, 313
Chinworth, H. E. 14, 215–218, 304, 312
Chowaniec, A. 21, 251, 286, 314, 324
Chudov, A. F. 108, 318
Ciemochowski, M. F. 283, 324
Clapp, L. B. 31, 315
Clark, M. O. 283, 324
Clarke, E. C. 18, 211, 212, 306, 313
Claussen, W. F. 15, 28, 48, 57, 178, 198–200, 206, 209, 312, 315
Cleef, A. van 23, 33, 171–173, 207–209, 314, 321
Colten, S. L. 294, 325

Connealy, L. E. 286, 324
Cooper, L. S. 286, 324
Corbin, G. 170, 321
Covington, A. K. 40, 316
Craig, R. D. 239, 322
Crowther, I. F. 290, 325
Culberson, O. L. 178, 199–201, 321
Curtiss, C. F. 151, 161, 320
Czaplinski, A. 16, 313

Danford, M. D. 38, 316
Dannon, F. 168, 169, 320
Dashdamirov, F. A. 107, 318
Davidson, D. W. 17, 113, 182, 313, 319, 321
Davis, C. M. Jr. 38, 316
Davy, H. 8, 9, 311
Deaton, W. M. 13, 14, 20, 107, 117, 133, 136, 150, 194–196, 201–204, 205, 209–211, 214, 254, 255, 284, 286, 299, 300, 312, 320, 321, 322
Debrescu, J. 286, 324
Decharme, C. 9, 311
Delsemme, A. H. 26, 28, 315
Demenko, A. V. 19, 314
Dennison, D. M. 34, 315
Devonshire, A. F. 17, 158, 159, 161–163, 171, 231, 313
Dewerdt, F. 286, 324
Dharmawardhana, P. B. 117, 170, 309, 310, 319, 321
Dick, J. W. L. 268, 323
Diepen, G. A. M. 23, 33, 171–173, 189, 190, 206–209, 309, 310, 314, 321
Dinkov, V. A. 283, 324
Distefano, V. 29, 315
Ditte, A. 9, 311
Döpping, V. 57, 317
Dorsch, R. R. 29, 315
Dorsey, N. E. 200, 321
Dubinin, V. M. 288, 325
Duclaux, E. 9, 311
Dyegtyarev, B. V. 23, 81, 83, 174, 267, 272, 273, 277–279, 282, 287, 317, 321, 323, 325
Dzhavadov, A. D. 20, 109, 259, 260, 314, 322

Eisenberg, D. 40, 43, 316
Eley, D. D. 28, 40, 189, 199, 200, 315, 317
Erdey-Grúz, T. 40, 41, 43, 316

Erlander, S. R. 40, 317
Eto, Y. 108, 318
Ettinger, L. L. 297, 326
Eucken, A. 37, 118, 146, 316, 319, 320
Euler, L. 45
Evans, D. F. 17, 313
Evans, M. W. 28, 40, 315
Evrenos, S. I. 22, 277, 314, 323
Ewing, G. J. 176–179, 230, 268, 321, 323
Eyring, H. 37, 316

Falabella, B. J. 204, 259, 321, 322
Falkenhagen, H. 40, 316
Faraday, M. 9, 56, 311, 317
Favoritov, Yu. A. 106, 318
Fazlutdinov, A. R. 281, 324
Findlay, A. 188, 321
Fomina, V. I. 18, 20, 109, 123, 124, 149, 150,
    183, 209, 232, 256, 276, 290, 313, 314,
    318, 319, 320, 322, 323, 326
Forcrand, R. de 10–12, 32, 56–58, 114–116,
    174–176, 178, 191, 311, 312, 315, 317,
    319, 321
Ford, R. W. 18, 211, 252, 306, 313
Forslind, E. 37, 316
Fowler, R. H. 34, 36, 316, 320
Fox, J. J. 34, 316
Frank, H. S. 28, 37, 38, 40, 41, 43, 315, 316,
    317
Franks, F. 41, 317
Frost, E. M. 13, 14, 20, 117, 133, 136, 150, 184,
    185, 194–196, 201, 203–205, 209–211, 214,
    254, 255, 284, 286, 299, 300, 312, 320,
    321, 322
Frühbuss, H. 15, 33, 37, 51, 52, 54, 55, 122,
    192, 193, 245, 246, 247, 248, 312
Fürcht, L. 251, 322

Gafarova, N. A. 254, 287, 322, 324
Garg, S. K. 113, 182, 319, 321
Gavriya, N. A. 83, 318
Gebhardt, H. J. 292, 325
Gerard, N. 209, 322
Geuther, A. 57, 317
Gilliland, E. R. 23, 314
Ginsburg, G. A. 23, 267, 268, 323
Giraud, H. 57, 317
Giuliano, G. B. 294, 325

Glew, D. N. 18, 33, 53, 54, 117, 184, 198, 201,
    211, 212, 291, 306, 313, 315, 317, 321, 325
Godchot, J. M. 14, 112, 312, 318
Gore, R. 25, 315
Gorodilov, V. A. 281, 324
Gough, S. R. 113, 319
Gritsenko, A. I. 124, 319, 320
Groisman, A. G. 188, 286, 321, 324
Gukman, L. M. 292, 325
Guseinov, Ch. S. 106, 318

Hajdu, F. 38, 316
Hall, L. 38, 316
Hamann, S. D. 167, 168, 320
Hames, D. A. 184, 321
Hammerschmidt, E. G. 13, 16, 22, 95–97, 110,
    133, 140, 240, 250, 254, 289, 298, 299, 312,
    313, 318, 322
Hand, J. H. 310, 326
Haraszti, E. 214, 284, 285, 324
Harber, E. A. 286, 324
Harker, J. H. 259, 322
Harris, W. H. 57, 317
Haute, A. A. van 127, 320
Hawkins, R. E. 113, 319
Hayward, R. 29, 176, 177, 315,
Heatman, J. 22, 276, 314, 323
Hemmaplardh, B. 255, 322
Hendrickson, H. M. 23, 314
Herrin, J. P. 287, 325
Hertl, E. 30, 31, 315
Hess, G. E. 23, 314
Hildebrand, J. H. 20, 314
Hinton, R. A. 287, 324
Hirschfelder, J. O. 151, 161, 320
Holder, G. D. 170, 310, 321, 326
Horne, R. A. 40, 43, 316
Houghton, J. 77, 317
Howe, L. S. 194–196, 201, 204–206, 301, 312,
    319
Hutchinson, A. I. 22, 314

Ichiro, H. 20, 314
Iijima, M. 294, 325
Ilyinskaya, A. 200, 201, 321
Ilyukov, V. A. 283, 324
Ionescu, L. G. 176–179, 321
Iskenderov, S. M. 100, 125, 183, 318, 320, 321

Ismailov, I. A. 21, 90, 91, 92, 93, 314
Ivanov, B. D. 113, 170, 319, 320, 323
Ivanov, N. S. 276, 323
Ives, D. J. G. 41, 317
Izmailova, K. H. I. 292, 325

Jahns, W. 15, 33, 37, 51, 52, 54, 55, 57, 247, 312
Jeffrey, G. A. 17, 32, 48, 49, 57, 313, 317
Johnston, R. H. 288, 325
Jones, P. 40, 316

Kahre, L. C. 22, 290, 314
Kanibolotskiy, M. A. 286, 324
Kaptelinin, J. D. 232, 322
Karachun, F. M. 283, 324
Karnofsky, G. 292, 325
Karpov, V. A. 287, 324
Kasarnovsky, J. S. 200, 201, 321
Kasperovich, A. G. 292, 325
Kass, J. R. 184, 187, 321
Kassai, L. 214, 284, 285, 324
Katz, D. L. 13, 14, 18, 22, 127–133, 150, 184, 185, 193, 195, 197, 198, 201, 207, 209–211, 213–221, 228, 229, 231, 252, 254, 256, 257, 260–265, 267, 284, 301, 302, 304, 310, 312–314, 320, 322, 323, 326
Kauzmann, W. 40, 43, 316
Kavanau, J. L. 40, 43, 316
Kedzierski, S. 21, 109, 251, 286, 314, 318, 324
Kemkhadze, T. V. 83, 287, 318, 325
Kern, L. R. 292, 325
Ketelaar, J. A. A. 31, 315
Khabibullin, I. L. 271, 323
Khadikin, F. G. 287, 325
Khalikov, G. 23, 271, 323
Khoroshilov, V. A. 19, 23, 83, 104, 228, 259, 268, 287, 308, 309, 314, 318, 322, 323, 325, 326
Khouty, F. M. 117, 118, 203, 319, 321
Khristov, R. 286, 324
Kihara, T. K. 17, 166, 168, 169, 170, 313, 320
King, A. D. 255, 322
King, M. B. 18, 313
King, R. B. 50, 317
Kinney, P. I. 22, 290, 314
Kirkley, D. W. 294, 325
Kislova, V.I. 268, 323

Kiyko, E. K. 283, 324
Klepikov, V. V. 83, 287, 325
Kleshunov, E. N. 294, 325
Klimenok, B. V. 81, 98, 140, 317, 318, 320
Klotz, I. M. 53, 315, 317
Knox, W. G. 23, 314
Kobayashi, R. 18, 117, 118, 171–173, 175, 176, 195, 197, 198, 201, 203, 204, 214, 215, 254, 261, 284, 306, 309, 310, 312–314, 319, 321, 322
Kobe, S. 168, 169, 320
Kobzeyev, Yu. V. 81, 317
Kolesnikova, L. A. 82, 318, 325
Kolodezni, P. A. 82, 283, 318, 324
Korotayev, I. P. 19, 20, 214, 288, 309, 314, 324, 325, 326
Korvezee, A. E. 10, 191–193, 321
Koshelev, V. S. 72, 123, 124, 150, 209, 277, 313, 318, 319, 320, 323, 326
Kostyuk, V. I. 90, 318
Krasnov, A. A. 81, 98, 140, 287, 317 318, 320, 325
Krestov, G. A. 40, 43, 316
Krichevsky, I. R. 200, 201, 321
Kriege, G. 13, 57, 189, 190, 312
Krilov, E. V. 287, 324
Krzyzanowski, S. 109, 318
Kuliyev, A. M. 109, 150, 318, 320
Kurata, F. 287, 324

Lacey, W. N. 14, 191, 209, 312, 313, 322
Lacroix, J. P. 313
Lakeyev, V. P. 277, 323
Lambert, P. A. 167, 168, 320
Latimer, W. M. 40, 198, 317
Lawrence, C. J. 293, 324
Le Chatelier, H. F. L. 12
Lee Kavanau, L. 316
Legezin, N. E. 83, 287, 318, 325
Lennard-Jones, J. E. 17, 33, 147, 151, 158, 159, 161–163, 165–169, 231, 313, 315
Levy, H. A. 38, 316
Lin, F. S. 294, 325
Lippert, E. L. 16, 33, 110, 112, 313, 315, 318
Lisichkin, V. 23
Litovitz, T. A. 28, 316
Loir, A. 10, 58, 311
Löwig, G. 9, 57, 311, 317

Lubowsky, S. W. 315
Luck, W. A. P. 40, 316
Lukyanov, Yu. P. 106, 318
Lutoshkin, G. Sh. 23, 81, 267, 273, 277–279, 282, 317, 323

Maass, O. 119, 181, 191–193, 318
Majid, Y. A. 113, 182, 319, 321
Makin, E. C. 292, 325
Makogon, Yu. F. 19, 20, 23, 44, 60, 61, 75, 77, 81, 89, 90, 94–96, 98, 101–104, 203, 204, 254, 257, 258, 267–277, 283, 287, 289, 293–297, 307, 314, 322, 324, 325, 326
Malenkov, E. V. 39, 79, 254, 259, 287, 316, 317, 322, 323, 324
Malofeyev, G. E. 283, 324
Malyshev, A. G. 170, 232, 256, 320, 322
Mamedov, J. S. 287, 324
Marchi, R. P. 37, 316
Marcus, W. C. de 24, 315
Marsh, R. E. 15, 37, 198, 312, 316
Marshall, D. R. 18, 19, 171–173, 175, 176, 195, 197, 198, 258, 306, 313, 314
Martin, A. E. 34, 316
Masterson, W. L. 200, 201, 238, 321
Maumené, E. 9, 56, 311
McCarthy, E. H. 25, 32, 54, 164, 322
McKetta, J. J. 199, 200, 201, 321
McKoy, V. D. 17, 141, 162–169, 313
McLeod, H. O. 18, 134–136, 153, 197, 201, 261, 262, 313, 321
McMullan, R. K. 17, 31, 32, 48, 49, 313
Medovskiy, D. I. 269, 283, 310, 323, 324, 326
Mehta, B. R. 22, 184, 185, 209–211, 237–239, 314, 322
Meijer, G. 12, 311
Meinhold, W. 15, 33, 37, 51, 52, 54, 55, 83, 122, 194, 195, 209, 219, 312
Mendez, G. M. 286, 324
Mendis, D. A. 27, 315
Mezhlumov, O. A. 282, 324
Mijers, J. 9, 311
Milev, N. 81, 317
Miller, B. 14, 127, 209–211, 272, 286, 312
Miller, D. C. 27, 315
Miller, S. L. 24, 26, 28, 29, 315
Millon, H. 9, 311

Miroshnichenko, V. I. 115, 175, 321
Miroshnikov, A. M. 108, 318
Mirzandzhanzade, A. Kh. 276, 323
Mishchenko, K. P. 40, 45, 316, 317
Mlyuschenko, N. D. 287, 325
Moelwyn-Hughes, E. A. 199, 321
Moiseyeva, N. F. 83, 318
Mordovskaya, N. I. 82, 318
Moreau, B. L. 280, 324
Morgan, J. 38, 316
Morlat, M. 209, 322
Morozov, I. F. 296, 297, 326
Morris, B. 113, 319
Morrison, T. J. 199, 250, 321
Moulton, R. W. 23, 314
Mucskai, L. 77, 317
Mulders, E. M. J. 97, 317
Müller, H. R. 15, 33, 37, 51, 52, 54, 55, 58, 63, 69, 70, 74, 141, 176–178, 183, 184, 198, 206, 209, 245, 246, 312, 317
Mulliken, R. S. 33, 315
Murphy, G. M. 16, 313
Musayev, R. M. 19–23, 90–93, 109, 110, 125, 150, 183, 214, 254, 259–261, 289, 291, 314, 318, 320, 321, 322, 325

Nagayev, V. B. 124, 319
Nasuoka, H. 108, 318
Nauruzov, M. H. 254, 322
Nedosedkin, V. N. 297, 326
Némethy, G. 38, 40, 41, 43, 316
Nguyen, Heng-Joo, 118, 170, 235, 260, 286, 292, 319, 320, 322, 324, 325
Niermann, A. I. 289, 325
Nikitin, B. A. 14, 22, 312
Nikitin, V. I. 287, 324
Nikolin, V. I. 297, 326
Nishimoto, Y. 20, 136, 137, 314
Noaker, L. J. 18, 193, 228, 229, 304, 313
Norozhenko, A. P. 149, 150, 320
Novikova, N. 289, 325
Novruzova, F. M. 150, 320

Ockman, N. 164, 320
Othen, D. A. 17, 313
Otto, F. D. 18, 206–208, 213, 232–234, 239–243, 311, 313, 315, 316

Palin, D. E. 14, 312
Palmer, H. A. 16, 33, 111, 112, 313, 315
Papadopulos, K. D. 170, 321
Pápay, J. 214, 284, 285, 324
Parent, J. D. 13, 122, 188, 189, 196, 213, 215, 272, 288, 312, 319
Parkhomenko, A. N. 106, 318
Parmentier, F. 9, 311
Parrish, W. R. 170, 235–237, 320, 321
Paslavskiy, N. G. 283, 324
Pastuhov, I. V. 283, 324
Pauling, L. 15, 29, 34, 36, 37, 39, 159, 161, 176, 177, 198, 312, 315, 316
Pavlov, V. M. 287, 325
Pekhata, F. I. 287, 325
Pelenichka, L. G. 283, 324
Pena, J. A. 28, 315
Pena, R. G. de 28, 315
Peng, D. Y. 118, 260, 319, 322
Pernolet, R. 209, 322
Perry, J. J. 69, 71, 72, 317
Petrunia, J. P. 235, 322
Pick, P. 107, 132, 250, 318, 320, 322
Pieroen, A. P. 16, 95, 137, 139, 140, 313
Pierre, J. 57, 311
Pierron, E. D. 246–292, 325
Pitzer, K. S. 168, 169, 320
Platteeuw, J. C. 17, 57, 104, 141, 154, 156–158, 161, 162, 163, 165, 168–171, 202, 230, 231, 234, 246–249, 313, 320, 322
Plumstead, J. A. 288, 325
Plyushchev, L. V. 124, 319
Poldermann, L. D. 95, 318
Polglase, M. F. 28, 199, 200, 315
Polivko, I. N. 283, 324
Poltoratskiy, G. M. 40, 316
Ponomaryev, G. V. 289, 325
Pople, J. A. 33, 36, 151, 315, 316, 320
Powell, H. M. 30, 40, 312, 315
Powell, J. S. 14, 16, 313
Powell, R. E. 40, 198, 317
Prausnitz, J. H. 170, 235–237, 320

Ralstin, J. 22, 276, 314, 323
Rasulov, A. M. 107, 318
Rath, N. S. 54, 317
Reamer, H. H. 16, 117, 191, 204–212, 243–245, 302, 303, 313, 319, 322

Records, L. R. 15, 284, 312
Reddy, A. K. N. 40, 316
Reid, L. S. 253, 254, 322
Repalov, V. I. 287, 325
Richard, H. A. 294, 325
Richards, R. E. 17, 313
Rikhterman, D. L. 108, 318
Ripmeester, J. A. 113, 182, 319, 321
Rive, A. de la 9, 57, 311
Roberts, O. L. 13, 117, 184, 194–196, 201, 204–206, 301, 312, 319
Robertson, J. M. 31, 315
Robinson, D. B. 18, 23, 118, 170, 184, 185, 206–213, 226, 232–235, 237, 239–243, 292, 305, 313, 324,
Robinson, R. A. 40, 205, 206, 314, 316, 319, 320, 322, 325
Rocco, A. G. de 29, 315
Römer, G. H. 30, 31, 315
Roncier, M. 286, 324
Roozeboom, H. W. B. 10, 11, 57, 60, 189, 190, 311, 317
Rosen, W. F. 104, 105, 318
Roux, G. M. 294, 325
Rushbrooke, G. S. 155, 320
Russell, N. B. 14, 81, 87–89, 108, 193, 313
Ruzicka, D. J. 17, 53, 291, 313

Sadamoto, Sh. 108, 318
Safir, M. Kh. 268, 323
Sagan, C. 25, 315
Sage, B. H. 14, 16, 191–193, 200, 206–212, 245, 302, 303, 313, 322
Saifeyev, T. A. 282, 283, 324
Saito, S. 18, 171, 172, 173, 175, 176, 195, 197, 198, 294, 306, 313, 314, 325
Samoilov, O. Ya. 36, 37, 39, 43, 45, 316
Sampson, J. A. 111, 318
Saniyev, E. S. 150, 254, 320, 322
Saniyev, Z. A. 150, 320
Sarkisyants, G. A. 23, 81, 89, 90, 96, 98, 257, 258, 273, 314
Savvin, A. Z. 290, 325
Scauzillo, F. R. 110, 304, 318
Schaafsma, I. G. 209, 322
Scheffer, F. E. C. 12, 56, 191–193, 206–208, 311, 317, 321

Scheraga, H. A. 38, 40, 316
Scherbakova, P. R. 277, 323, 324
Schlenk, W. 31, 315
Schoenfeld, F. 9, 57, 311
Scholander, P. F. 26, 315
Schöller, J. F. 112, 319
Schreinemakers, F. A. H. 191, 321
Schroeder, W. 7, 174–176, 184, 188, 311
Schulman, N. V. 297, 326
Seely, D. H. 15, 284, 312
Selleck, F. T. 16, 191–193, 206–211, 245, 302,
    303, 313, 322
Semerikov, A. A. 82, 318
Semin, V. I. 309, 326
Sergeyev, I. N. 283, 324
Sheinman, A. B. 243–245, 283, 324
Sherwood, P. W. 19, 313
Sheshukov, N. L. 268, 323
Shirkovskiy, A. I. 276, 323
Shulyatikov, V. I. 81, 317
Sidorovskiy, V. A. 282, 324
Simanek, J. 107, 132, 250, 318, 320, 322
Simon, D. 286, 324
Sinanoglu, O. 17, 141, 162–169, 313, 320
Skhalyakho, A. S. 267–276, 323
Skinner, W. R. 253, 254, 322
Sloan, E. D. 117, 170, 203, 309, 310, 319, 321
Smirnov, A. Sh. 16, 313
Smirnov, L. F. 294, 325
Smith, H. B. 23, 314
Smolyaninov, V. G. 287, 325
Smythe, W. D. 26, 315
Snell, E. 18, 206–208, 230, 232–234, 305, 313
Solacolu, C. 59, 317
Solacolu, I. 59, 317
Soldatov, V. D. 95, 318, 325
Song, K. Y. 117, 300, 309, 319
Stackelberg, M. von 15, 17, 32, 33, 37, 51, 52,
    55–58, 63, 69, 70, 73, 74, 83, 118–122, 141,
    146, 176, 178, 183, 184, 192–195, 198,
    206, 209, 219, 245–248, 289, 312, 313
Starodubtsev, A. M. 106, 318
Stefanovich, G. J. 297, 326
Stepanchikov, E. A. 283, 324
Stern, S. A. 294, 325
Stokes, R. H. 40, 316
Stokris, B. I. 9, 311
Stoll, R. D. 268, 323

Strong, E. R. 14, 127, 187, 209–211, 272, 288,
    312
Strong, J. 25, 315
Stuart, W. I. 17, 104, 141, 142, 146, 147,
    149–152, 154, 157, 313
Stuhl, D. R. 69, 71, 317
Stupin, D. Yu. 113, 167, 315, 319
Sumets, V. I. 82, 318
Surkov, J. V. 287, 325
Susummi, S. 20, 314
Swings, P. 26, 315
Szent-Györgyi. A. 29, 315

Tabuchi, K. 292, 325
Tammann, G. 13, 57, 189, 190, 312
Tanret, C. 9, 311
Teodorovich, J. 23, 314
Tesner, V. 23, 314
Tezikov, V. N. 113, 319
Thomas, S. 12, 311
Tinnikov, V. V. 287, 325
Török, J. 214, 251, 272, 284, 285, 322, 323, 324
Towlson, H. E. 23, 314, 315
Tracey, M. V. 35, 316
Trebin, F. A. 19, 23, 283, 314, 324
Tribus, N. A. 259, 260, 322
Trofimuk, A. A. 23, 209, 323
Tsao, T. C. 294, 325
Tsarev, V. P. 82, 98, 113, 228, 229, 235, 240,
    268, 276, 283, 290, 319, 323, 324, 325
Tucholke, B. E. 269, 323
Turikin, A. F. 310, 326
Tyushnyakova, G. I. 170, 232, 256, 320, 322

Umrazkov, U. 287, 324
Unruh, C. H. 15, 184, 185, 219–221, 262, 302,
    312
Urey, H. C. 26, 315

van der Waals, J. H. see Waals, J. H. van der
Vanpee, M. 204, 321
Vasilchenko, V. P. 21, 91, 93, 94, 306, 307, 314
Vasilyeva, V. G. 283, 324
Verigin, N. N. 271, 323
Verma, V. K. 292, 310, 325, 326
Verteletskaya, A. S. 83, 318
Villard, P. 11, 12, 15, 32, 56, 57, 59. 115, 134,
    174, 191, 192, 194, 195, 197, 201, 206, 311,
    312, 315, 317, 321

Volodina, L. A. 124, 319

Waals, J. H. van der 17, 57, 104, 141, 150, 154, 156–158, 160–163, 168–171, 202, 230, 231, 234, 246–249, 313, 317, 320, 322
Wada, G. 43, 317
Warren, B. E. 38, 316
Wartha, W. 9, 311
Wen, W. Y. 37, 38, 43, 316
Wenger, A. 27, 315
Wentorf, H. 151, 161, 320
West, J. R. 16, 313
Whipple, F. L. 26, 315
Wiegandt, H. E. 23, 314
Wilcox, W. L. 127–131, 150, 209–211, 213, 252, 320
Wilms, D. A. 127, 320
Wilson, G. J. 17, 113, 313, 319
Winkler, L. W. 199, 200, 321
Withrow, H. J. 254, 322
Wittstruck, T. A. 24, 215–218, 315

Wöhler, F. 10, 311
Won, K. W. 255, 322
Woolfolk, R. M. 16, 313
Wright, R. M. 191, 192, 321
Wroblewski, S. V. 10, 311
Wu Bing-Jing, 215, 322

Yakutseni, V. P. 12, 292, 325
Ye Chien-Hu, 23, 314
Yefremov, I. D. 23, 283, 324
Yefremova, A. G. 269, 323
Yegorov, S. A. 83, 318
Yorizane, M. 20, 108, 136, 137, 314, 318

Zaitsev, A. A. 268, 322
Zenin, A. G. 297, 326
Zernike, J. 57, 317
Zhidenko, G. G. 288, 325
Zhizhchenko, B. P. 269, 323
Zvonarev, N. G. 283, 324

SUBJECT INDEX

acetylene 9, 11, 115
activated carbon 108
activation energy 78, 79
activity of gas phase 126
— of hydrate 126, 178 209,
— of water 126, 138, 139, 141, 144, 155, 178, 209
— — containing inhibitor 126, 127
adsorption, localized 17, 154
—, physical 118
alcohol, calculation of its demand 96
—, effect on the hydrate stability 80, 221–228
—, removal from the gas flow 107
— -water interaction 42
anaesthesia, its hydrate theory 25, 29
anaesthetic effect of xenone 176
— pressure 28, 29
argon 12, 16, 18, 19, 65, 115, 120, 152, 153, 161, 162, 166, 167, 174, 176, 179
association of monomeric water molecules 37
attraction forces 159
auxiliary gas "Hilfsgase" 17, 20, 80, 171, 281, 291
— —, its stabilizing effect 17, 20, 58, 59, 83–87, 149, 176, 214

boiling point of hydrate forming materials 33, 69, 116, 119–122
Boltzmann probability factor 161, 164, 165
bond energy 114, 118
bromine 9, 57, 115, 120, 121, 163, 183 184, 203
butane 29, 124, 130, 136, 170, 213–215, 250, 251, 255, 257, 259, 260, 263

"cage-like" structures 31
carbon dioxide 9, 10, 20, 23, 25, 26, 83–85, 102, 115, 120, 124, 131, 162, 165, 166, 168, 169, 176, 179, 184–188, 219–227, 237–239, 250, 257
— disulphide 9
— monoxide 26
— tetrachloride 12, 121, 247
— tetrafluoride 159–162, 247
cavity potential 163, 165, 167, 168, 169
— types 48, 49, 50, 154
— size in the host lattice 39, 47, 51, 141, 152
cell partition function 158
— potential 165, 166
chemical potential of hydrate-forming materials 18, 134, 142, 143, 144, 154, 157, 163, 170, 171
— — of water 138, 142, 143, 144, 145, 156, 177
clathrate compounds 16, 30, 31, 50, 52, 53, 122
clathration of hydrophobic compounds 31, 32
— of water-soluble, acidogenic gases 31, 32
— —, polar compounds 31
— —, ternary or quaternary alkylammonium salts 32
chlorine 9, 13, 15, 56, 60, 115, 120, 183, 184, 203
chloroform 9, 16, 57, 111, 112, 115, 120, 217, 218, 245–247
Clausius–Clapeyron relationship 114, 116, 117, 123, 124, 125, 134, 139, 171, 174, 262
clouds, elimination by means of gas hydrates 28, 297
cluster model 38, 78
— monomer mixture theory of water 41
combinatorial factor 142, 154
comets and gas hydrates 26, 27
composition of gas hydrates 9, 13, 48, 51–59, 138, 157

compression (compressibility) factor, see deviation factor
condensed apolar phase, its effect on the hydrate 20, 109–113, 259
conformation of hydrate forming component 66, 68
continuum water model 36
corrosion effect, elimination 83
—, inhibition of 90
critical concentration of salt solutions 93, 94
— nucleus size 75
— point of decomposition 65
— — of hydrate 62, 90, 121, 152, 240
— temperature of hydrates 11
crude oil, its effect on hydrate formation 282
crystal nuclei, formation 75–78
cyanogen 11
cyclopropane derivatives 54

Dalton's law 125
Debray rule 10
deformed-bond water model 36
dehydration of gas 15, 280
densitometry 200
density 124, 185, 203, 204
desalination of sea water 23, 80, 209, 292–294
deuterium 167–168
deviation factor 117, 122, 135, 174, 198, 208, 209, 265, 295
dew-point 21, 196, 205, 206
dielectric investigations 17, 113
diffusion 182, 188, 224
dilatometry 200
dipole moment of the solvent 47
dispersion forces 147, 163
dissociation pressure 11, 26, 120, 123, 166, 167, 176, 178, 215
— —, calculation 166
— temperature 73, 82, 84, 123
dissolved materials, their effect on the structure of solvent 40–47
distribution coefficient 126, 290, 291
double hydrates 10, 12, 53, 54, 55, 56, 58, 64, 74, 112, 122, 193, 235, 247
diethyl ether 9
dynamic method for the study of hydrates 299

energy, internal 156
—, activation 78, 79
—, formation 160
—, interaction 151
—, sorption 151
enrichment of C_3 and C_4 hydrocarbons 250, 251
enthalpy 14, 116, 119, 121, 134, 135, 170, 178, 184, 199, 200–203
—, dissociation 17
—, fusion 184, 202
entropy 14, 118, 119, 122, 135, 159, 178
—, dissociation 17
equilibrium, chemical, conditions of 132
— conditions, see at the individual hydrates
— — in porous media 277
— constant 18, 20, 126–133, 145, 148–151, 153, 174, 178, 193, 202, 219–266, 232–234
— data, see at the individual hydrates
— diagrams, see at the individual hydrates
—, five-phase 246
—, four-phase 51, 171, 228, 230–232, 235, 238–241
—, heterogeneous, conditions, 132
—, mechanical 132
—, shift in the presence of inhibitors 170
—, thermodynamic, conditions 132
—, three-phase 16, 145, 171, 189–191, 194, 205, 206, 209, 230, 239, 241, 246
—, two-phase 15, 142, 209, 241
ethane 11, 13, 16, 61, 115, 117, 120, 124, 128, 136, 137, 162–169, 189, 194, 203–206, 255, 257, 259, 260, 263, 295
ethyl bromide 11
— chloride 11, 14, 20, 203, 218
— iodide 11, 121
ethylene 11, 13, 16, 18, 120, 124, 161, 162, 166–169, 206–209, 232–234
— chloride 121
— oxide 16, 18, 54
Euler's theory 48
eutectic composition of the hydrates 13, 225
evaporation ratio 123–126
expansion, effect on the hydrate formation 14, 262–266
experimental apparatus for studying hydrates
— dynamic method 298
— static method 298

flickering clusters 37, 38
fogs, elimination by means of gas hydrates 28, 297
force constants 29
— law 159, 165
formation pressure 271
— temperature 268
fractionation factor 152–154
free energy of host lattice 142, 143, 154
free enthalpy of the hydrate formation 119, 134, 178
freezing point 91, 226, 227
freons 9, 14, 23, 24, 114, 115, 117, 120, 121, 215–219, 245–249, 292
fugacity 18, 73, 118, 126, 127, 150, 156, 171, 209, 211, 213
full hydration, limit of 46, 93

gas–condensate ratio in natural gases 260
gas hydrate, method for the increase of gas pressure 294
— hydrates and the gas blow-outs in coal mines 296
— —, application for gas storage purposes 14, 288
— —, applications in the technology 267–297
— —, basic types of 51
— —, conditions of formation of 60, 80, 103, 104, 121
— —, historical survey of the literature on their studies 9–29
— — in marine deposits 269
— —, natural occurrence of 22, 24, 25, 26, 267
— — of individual gases 171
— — on the depth of the sea 269, 270
— —, on the surface of Mars 26
— —, role in the fresh-water production 23, 292
— —, structure of 15, 30, 48–56
— —, systematization of 48–56
— —, thermodynamics of 114–170
— sources, prognostication 275
Gibbs phase rule 134, 232
"guest" lattice structure 30

haemoglobin 176
— + water + nitrogen + xenon system 176

halo formation 27
halon number 24, 215–219
halogen elements 9, 11, 12, 13, 15, 56, 57, 60, 115, 120, 121, 163, 183, 184
halohydrocarbons 9, 14, 23, 24, 114, 115, 117, 120, 121, 215–219, 245–249, 292
heat capacity 188, 199, 200
— of adsorption 124
— of crystallization 123
— of decomposition of hydrate 24, 117, 213
— of dissociation 116
— of dissolution 117, 174, 200
— — evaporation 114, 118, 122, 124
— of formation 13, 15, 18, 20, 24, 114–117, 121–125, 138–140, 171, 178, 188, 194, 203
— —, uncorrected 139
— of freezing 114, 118
— of fusion 122, 123
— of intercalation 146–148
— of melting 114, 122, 202
— of phase transition 114, 146, 202
— of sorption 148, 202
— of sublimation 123
heavy water fractionation by means of gas hydrates 112
helium 22, 174, 255, 289, 290
Henry absorption coefficient 73, 118
— constant 73, 108
— 's law 176, 200
Hildebrand's rule 136
"host" lattice structure 30, 79, 141, 142, 163
— -guest bonds 30, 79
hydrate decomposition 60–79, 117, 121–125, 138–140, 171, 178, 194, 201, 249, 269, 270
— formation, conditions 9, 11, 21, 22, 28, 60–79, 104, 136, 249, 257, 262, 264, 270, 280
— —, dependence on the depth of ocean 270
— —, experimental study 298–310
— — from gas mixtures 15, 58, 84, 103, 219–266
— — in deep porous rocks 269, 273–277
— — in gas pipelines 107, 283–287
— — — —, determination of the sites of 285
— — in gas wells 22, 83, 277–283
— — — —, quantitative evaluation 286

hydrate formation in the depths 277–278
— — in the tubing 277, 283
— — in zeolites 277
— —, preventing 19, 79–113, 279
— —, process, two stage 290
— —, protection 269
— —, thermal effect 123, 203
— of acetylene 9, 11
— of argon 16, 18, 19, 65, 115, 120, 152, 153, 161, 162, 166, 167, 174–176, 179
— of bromine 9, 57, 115, 120, 121, 163, 183, 184, 203
— of butane 19, 124, 130, 136, 170, 213–215
— of carbon dioxide 9, 10, 23, 24, 25, 26, 102, 115, 120, 124, 131, 162, 165, 168, 169, 171, 179, 184–188, 219–227, 237–239, 295
— — disulphide 9
— — monoxide 161
— — tetrachloride 12, 121, 159–162
— — — +tetrafluoride+hydrogen sulphide 247
— of chlorine 9, 13, 15, 56, 60, 115, 120, 183, 184, 203
— of chloroform 9, 16, 111, 112, 115, 120, 217, 218
— — +hydrogen sulphide 245–247
— of cyanogen 11
— of cyclopropane 54
— of diethyl ether 9
— of ethane 11, 13, 115, 117, 120, 124, 128, 136, 137, 162–169, 189, 194, 203–206, 209
— — +ethylene 16
— of ethyl bromide 11
— — chloride 11, 14, 120, 203, 218
— — iodide 11, 121
— of ethylene 11, 13, 18, 120, 124, 161, 162, 168, 169, 206–209, 232–234
— — chloride 121
— — oxide 16, 112
— of helium 174
— of hydrogen sulphide 10, 11, 15, 18, 56, 58, 61, 81, 87–89, 108, 109, 115, 120, 122, 124, 131, 191–193, 203, 219
— of isobutylene 13, 124
— of iodine 11, 12, 184
— of krypton 56, 65, 120, 148, 152

hydrate of methane 11, 16, 25, 27, 60, 61, 83–86, 115, 117, 120, 124, 128, 136, 150, 159, 162, 166, 168–170, 192, 194–204, 215–234, 274, 295, 296
— — +carbon dioxide 83–85, 219–227
— — +ethane 136
— — +ethylene 18, 232–234
— — +ethylene+propylene 18
— — +hydrogen sulphide 18, 192, 193, 219, 228–232
— — +nitrogen 86
— — +propane 219, 230–232
— — +propylene 18, 239–243
— of methyl bromide 120, 121, 203
— — chloride 14
— — fluoride 11
— — iodide 203
— of neon 174
— of nitrogen 16, 21, 23, 171–174
— — oxide 11
— of oxygen 12, 21, 23, 159, 161, 162, 166, 169, 171–174, 176, 179, 214
— of phenol 17
— of phosphine 9
— of propane 11, 61, 90, 98, 117, 120, 124, 127–129, 136, 150, 189, 203, 205, 209–211
— — +butane 19
— — +carbon dioxide 20, 237–239
— — +hydrogen sulphide 219, 234, 235
— — +nitrogen 235–237
— — +propylene 16, 20, 22, 219, 243–245, 289–292
— of propylene 11, 124, 211–213
— of silicon hexafluoride+argon 167
— of sulphur dioxide 9, 12–74, 56, 188–191
— of trichloromethane+hydrogen sulphide 245–247
— of xenon 120, 152, 153, 162, 165–166, 169, 174–182
— types 51–59
— vapour pressure, calculation 11, 69, 71, 72, 119
hydrates, systematization 48
— of freons 9, 14, 23, 24, 114, 115, 117, 120, 121, 215–219, 245–247, 292
— — +hydrogen sulphide 245–249
— of halogen elements 183, 184
— of hydrocarbons 16, 18, 194–219

hydrates of natural gases 7, 13, 14, 59, 117, 132, 133, 135, 136, 150, 170, 179, 193, 196, 203, 227, 249–266, 272–292
— of nitrous oxides 11
— of noble gases 18, 19, 120, 148, 152, 153, 159, 162, 165, 166, 169, 170, 174–182
hydration, hydrophobic 41
—, negative 43
— number 18, 45, 46, 52, 198, 209
—, positive 42, 43, 95
hydrocarbons 16, 18–20, 109, 194–219, 290
hydrogen bond 30, 31–38, 41
hydrogen sulphide 10, 11, 15, 18, 56, 58, 61, 81, 87–89, 108, 109, 115, 120, 122, 124, 131, 191–193, 203, 219, 228–232, 234, 235, 245–247, 256, 290
hydrophobic compounds 31
— hydration 41
hydrophobization of pipeline 282
hydroquinone clathrate 17
hypsometric level 268

ice structure 35
iceberg theory 41
inclusion compounds 17, 31, 59
inert gases, effect of 12, 13, 14, 59, 80, 83, 86, 136, 176, 250, 253, 254, 272
infrared spectral studies of hydrates 17, 28, 36
inhibition effect of acetylacetone 112
— — of brine 81, 110
— — of electrolytes 19, 80, 81–83, 87–95
— — of flow velocity 106
— — of glycols 21, 80, 81, 95, 108
— — of methanol 13, 16, 20, 21, 95–109, 283, 286, 287
— — of urea 112
— effectivity of electrolytes, calculation 94
— in gas well 269, 273, 277–283
— — pipeline 90, 95–98, 107, 108, 286, 287
— of corrosion 83
— of hydrate formation, mechanism 12, 16, 21, 46, 79–83, 90–93, 104, 126, 137–140, 149, 280
inhibitors, calculation of required amount to be introduced into the pipeline 96
—, choice 87–109, 280
—, condensed apolar compounds 19, 109–113
—, electrolytes 43, 81, 87–95, 184, 187

inhibitors, non-electrolytes 14, 81, 95–109, 111
—, petroleum 20, 83, 283
intermolecular force constants 18, 168, 189
interstellar dusts 28
iodine 11, 12, 59, 184
ion associations 46, 47
— –ion interactions 40, 41, 45, 46, 47
— –solvent interactions 39, 40, 45, 46, 47
ions, effect on the water structure 47, 48
isobutylene 18, 124

Kihara potential 17, 166, 169, 170
Kirkwood –Müller expression 147
krypton 56, 65, 120, 148, 152, 153, 162, 165, 166, 169, 174–176, 179–181

Langmuir adsorption equation 259
— constants 104, 147, 156, 157, 160, 170, 290
— isotherm 16
lattice constants 15, 50, 51, 52, 53, 55, 58
— defects 37
Lennard-Jones 12–6 potential 147, 151, 162, 165, 166, 168, 169
— — 28–7 — 168, 169
— — –Devonshire theory 16, 147, 151, 158–168, 171
liquid hydrates 55, 57, 64, 73, 74, 111, 209
liquid hydrocarbons, hydrate properties 19, 20, 109, 259, 260
local sealing of holed gas reservoirs by means of gas hydrates 22
localized sorption theory 170

magnetic susceptibility 147
melting point of hydrates 63, 64
mental activity and gas hydrates 176
methane 11, 16, 18, 25–27, 60, 61, 83–86, 115, 117, 120, 124, 128, 136, 150, 159, 162, 166, 168–170, 192, 193–204, 215, 219–234, 228–234, 239–243, 255, 260, 263, 264, 274, 295, 296
methane–water mixture, dew point 196
— —, heat capacity 199
— —, standard enthalpy 199
methanol–water system, structure 81
methyl bromide 120, 121, 203
— chloride 14

methyl fluoride 11
— iodide 203
Miers diagram 77
mixed hydrates 11, 12, 19, 25, 54, 55, 56, 59,
    84, 109, 111, 122, 214, 215, 219, 228, 230,
    234, 247, 249, 250, 257
moisture content, role in the formation of gas
    hydrates 106, 196, 203, 251, 254, 255–258
molecular compounds 30
molecularly dissolved material, their effect
    on the structure of water 40–42
molecule size, its effect on the hydrate forma-
    tion 65, 69, 70, 71,
mud 269, 283

natural gases 7, 13, 14, 59, 117, 132, 133, 135,
    136, 150, 170, 174, 193, 196, 203, 227,
    249–266, 272–292
"negatively hydrating" ions 42, 43
nitrogen 16, 21, 23, 86, 171–174, 235–237,
    250–255, 257
— as auxiliary gas 19, 21, 59, 83, 86, 252,
    254
— oxide 11
nitrous oxide 11
noble gases 18, 19, 120, 148, 152, 153, 159,
    162, 165, 166, 169, 170, 174, 182
nuclei, formation 75–78
—, size, critical 75
—, their increase 75–78

occlusion compounds 31
occupancy limit 54
— factor 54
— of the cavities in the host lattice 57, 65,
    120, 122, 142–148, 152, 154, 156, 162, 163,
    167, 174, 198, 202, 209, 231, 276
overburden pressure, its role in the hydrate
    formation 269
oxygen 12, 21, 23, 159, 161, 162, 166, 169,
    171–174, 176, 179, 214

paraffin–olefin mixtures 232, 239, 243
partial pressure of gas component, its calcula-
    tion by means of the molar partition
    function 144
partition functions 142, 144, 155–163, 170
pentane 260

permafrost layer and gas hydrates 83, 267,
    268, 271
permeability 276
permittivity 39, 46, 47
petroleum, its effect on hydrate formation
    20, 83, 283
phenol 17
phosphine 9
planet atmosphere 24
— chemistry 25, 26
Pople's continuum model of water 36
pores, distribution of 275
"positively hydrating" ions 42, 43
potential energy 159–161
— field 159
propane 11, 16, 20, 22, 61, 90, 98, 117, 120,
    124, 127–129, 136, 150, 189, 203, 205,
    209–211, 219, 230–232, 234–239, 243–245,
    255, 257, 260, 263, 289–292
propylene 11, 16, 18, 20, 211–213, 219,
    239–245, 289–292

quadruple point 62, 171, 172, 174, 179, 182,
    184, 190, 191, 206, 213, 239, 240

Raman spectral studies of water 39
Raoult's law 17, 125, 149
repulsion forces 147, 163
retrograde condensation 223, 260
Rosen nomogram 104, 105
rotation of gas molecules in the cavities of gas
    hydrates 141, 159, 164
rotational isomerism 69

salting-out activity 44, 94
— effect 44, 82, 94, 287
Samoilov's theory of water 36
saturation water vapour content of gases
    253–255
seismic reflection 269
separation of components by means of gas
    hydrates 22, 289–292
silicon hexafluoride 167
simple gas hydrates 10, 11, 56
solar system and gas hydrates 24–27
solid solutions 13, 16, 112, 127, 171, 219, 247
solubility data 23, 73, 75, 76, 78, 117, 123, 149,
    176, 179, 185, 198, 199, 200, 202, 203,
    209, 219, 225, 243, 254

solute–solvent interaction 39, 46
solvation 39
solvent–solvent interaction 39, 45
stability of several hydrates 20, 59, 64, 65, 69, 70, 71, 74, 79–113
state-diagrams of gas–water systems, general interpretation 62
statistical thermodynamics, application to the gas hydrates 117, 123, 140–176
storage of gas in hydrate form 14, 288
structural preordering in electrolyte solutions 46
structure breaker ions 42, 43
— maker ions 40, 42
— of electrolyte solutions 42–47
— of gas hydrates 15, 30, 48–56
— of water 40–43
Stuart–Briegleb models 66, 67
sulphur dioxide 9, 12–14, 56, 188–191
supercooling 75, 78, 186, 187
supersaturation 76
surface active materials, effect on the hindering of formation of the hydrate plugs 282
— charge density of ion 95
— energy, specific 75
sweet-water production and gas hydrates 272, 292–294

ternary mixtures 219–249
thermal gradient 267

third component, effect on the hydrates 16, 21, 79–113, 137–141
tridymite lattice structure 34
triple point 241
Trouton's rule 12, 114, 119, 120, 136

unit cell of hydrate 50, 51, 52, 74
urea as inhibitor 112

van der Waals forces 31, 66, 73, 124
— radius 159, 161
Vapour pressure 116, 119, 120, 121, 135, 145, 146, 148, 149, 150, 156, 157, 159, 160, 161, 162, 172, 183, 184, 207, 210, 212, 233, 234
— — of water 148–150, 211
vibration in the lattice structure 164
virial coefficients 166

water hydrate 37
— models 33–39
well-head pressure 278, 279
— temperature 278, 279
working diagrams for hydrate investigations 186, 187

xenon 120, 152, 153, 162, 165, 166, 169, 175, 176–182
X-ray investigations of gas hydrates 50, 54
— — of water 34